Milestones in Drug Therapy
MDT

Series Editors
Prof. Michael J. Parnham, PhD
Senior Scientific Advisor
PLIVA Research Institute Ltd
Prilaz baruna Filipovića 29
HR-10000 Zagreb
Croatia

Prof. Dr. J. Bruinvels
Sweelincklaan 75
NL-3723 JC Bilthoven
The Netherlands

Drugs Affecting Growth of Tumours

Edited by H.M. Pinedo and C.H. Smorenburg

Birkhäuser Verlag
Basel · Boston · Berlin

Editors

Herbert M. Pinedo
Carolien H. Smorenburg
Department of Medical Oncology
VU Medical Center
De Bolelaan 1117
1081 HV Amsterdam
The Netherlands

Library of Congress Cataloging-in-Publication Data

Drugs affecting growth of tumours / edited by H.M. Pinedo and C. Smorenburg.
    p. cm. -- (Milestones in drug therapy)
  Includes bibliographical references.
  ISBN-13: 978-3-7643-2196-3 (alk. paper)
  ISBN-10: 3-7643-2196-2 (alk. paper)
   1. Antineoplastic agents. I. Pinedo, H. M. II. Smorenburg, C. (Carolien), 1965- III. Series.

  RS431.A64D785 2006
  616.99'4061--dc22

               2006045977

Bibliographic information published by Die Deutsche Bibliothek
Die Deutsche Bibliothek lists this publication in the Deutsche Nationalbibliografie; detailed bibliographic data is available in the internet at http://dnb.ddb.de

ISBN 3-7643-2196-2 Birkhäuser Verlag, Basel - Boston - Berlin

Contents

List of contributors

Alessandra Balduzzi, Research Unit of Medical Senology, Department of Medicine, European Institute of Oncology, Via Ripamonti 435, 20141 Milano, Italy

Ferry A.L.M. Eskens, Erasmus University Medical Center Rotterdam, Department of Medical Oncology, PO Box 2040, 3000 CA Rotterdam, The Netherlands; e-mail: f.eskens@erasmusmc.nl

Hans Gelderblom, Leiden University Medical Center, Department of Clinical Oncology, Albinusdreef 2, 2300RC Leiden, The Netherlands; e-mail: a.j.gelderblom@lumc.nl

Aron Goldhirsch, Department of Medicine, European Institute of Oncology, Via Ripamonti 435, 20141 Milano, Italy; and Oncology Institute of Southern Switzerland, Bellinzona & Lugano, Switzerland

Kenneth R. Hande, Vanderbilt/Ingram, Cancer Center, 777 Preston Research Building, Vanderbilt University School of Medicine, Nashville, Tennessee 37232-6307, USA; e-mail: kenneth.hande@vanderbilt.edu

Manon T. Huizing, Antwerp University Hospital, Department of Oncology, Wilrijkstraat 10, 2650 Edegem, Belgium; e-mail: manon.huizing@uza.be

Felix Kratz, Tumor Biology Center at the Albert-Ludwig University Freiburg, Breisacher Strasse 117, 79106 Freiburg i.Br., Germany

Bart C. Kuenen, Dpt. Medical Oncology, De Boelelaan 1117, 1081 HV Amsterdam, The Netherlands; e-mail: b.kuenen@vumc.nl

Ulrich Massing, Tumor Biology Center at the Albert-Ludwig University Freiburg, Breisacher Strasse 117, 79106 Freiburg i.Br., Germany

Klaus Mross, Tumor Biology Center at the Albert-Ludwig University Freiburg, Breisacher Strasse 117, 79106 Freiburg i.Br., Germany; e-mail: mross@tumorbio.uni-freiburg.de

Igor Puzanov, Vanderbilt/Ingram, Cancer Center, 777 Preston Research Building, Vanderbilt University School of Medicine, Nashville, Tennessee 37232-6307, USA

Carolien H. Smorenburg, Division of Immunotherapy, Department of Medical Oncology, Vrije Universiteit Medical Center, PO Box 7057, 1007 MB Amsterdam, The Netherlands; e-mail: c.h.smorenburg@mca.nl

Alex Sparreboom, Clinical Pharmacology Research Core, Medical Oncology Clinical Research Unit, National Cancer Institute, 9000 Rockville Pike, Bldg. 10/Room 5A01, Bethesda, MD 20892, USA; e-mail: sparreba@mail.nih.gov

Rosalba Torrisi, Research Unit of Medical Senology, European Institute of Oncology, Via Ripamonti 435, 20141 Milano, Italy; e-mail: rosalba.torrisi@ieo.it

Alfonsus J.M. van den Eertwegh, Division of Immunotherapy, Department of Medical Oncology, Vrije Universiteit Medical Center, PO Box 7057, 1007 MB Amsterdam, The Netherlands; e-mail: vandeneertwegh@vumc.nl

Kenneth W. Wyman, Vanderbilt/Ingram, Cancer Center, 777 Preston Research Building, Vanderbilt University School of Medicine, Nashville, Tennessee 37232-6307, USA

Preface

This volume of the series 'Milestones' presents pharmacological, preclinical and clinical data of a wide range of anticancer agents varying from traditional cytotoxic agents to novel targeted small molecules. The chapters have been written by experienced pharmacologists and medical oncologists.

This volume emphasizes the multidisciplinary approach and the need for a close collaboration between laboratory and clinic in the development of new anticancer therapies. In recent years, this type of research has resulted in many new anticancer drugs, of which some already are accepted as new standard therapies. The increasing knowledge of molecular biology has resulted in the development of a large number of agents specifically targeting cellular processes of tumor cells. Other strategies have focused on improving traditional chemotherapeutic agents, better tolerability and improved patient compliance. In the field of immunology, advances have been made with novel vaccination techniques, while research on endocrine treatments has been revived due to successful new therapies for breast cancer.

We are grateful to Hans-Detlef Klüber and Karin Neidhart for their support in producing this edition. We would like to thank our colleagues for their critical review and comments.

In the rapidly changing field of oncology, research remains endless. We are only at the beginning of a very exciting period of drug development.

Carolien H. Smorenburg
Herbert M. Pinedo   Amsterdam, April 2006

Drugs Affecting Growth of Tumours
Edited by Herbert M. Pinedo and Carolien H. Smorenburg
© 2006 Birkhäuser Verlag/Switzerland

1

Antimetabolites

Kenneth W. Wyman, Igor Puzanov and Kenneth R. Hande

Vanderbilt/Ingram Cancer Center, 777 Preston Research Building, Vanderbilt University School of Medicine, Nashville, Tennessee 37232-6307, USA

Methotrexate and other folic acid antagonists

Mechanism of action

The synthesis of DNA requires reduced folates. Purine synthesis requires 10-formyltetrahydrofolate (CHO-FH$_4$) as a methyl donor and 5,10-methylenetetrahydrofolate (CH$_2$-FH$_4$) as carbon donor in the synthesis of thymidine (Fig. 1). Methotrexate inhibits dihydrofolate reductase (DHFR) depleting cells of reduced folates, including CHO-FH$_4$ and CH$_2$-FH$_4$ [1]. Reduced folate depletion does not account for all inhibition of DNA synthesis seen with methotrexate. Methotrexate is metabolized to methotrexate polyglutamates that contribute to cytotoxicity by directly inhibiting the folate dependent enzymes of thymidylate and purine biosynthesis (TS, AICAR, GAR; see

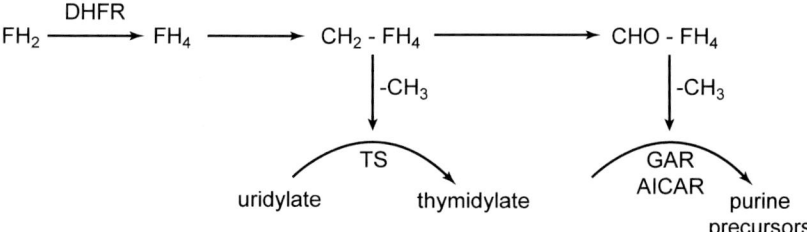

FH$_2$ = dihydrotolate
FH$_4$ = tetrahydrofolate
CH$_2$ - FH4 = 5,10 - methylenetetrahydrofolate
CHO - FH4 = 10 formyltetrahydrofolate

DHFR = dihydrofolate reductase
TS = thymidylate synthetase
GAR = glycineamide ribonucleotide
transformylase
AICAR = amino imidazole
carboxamide ribonucleotide
transformylase

Figure 1. Mechanism of action of methotrexate. Reduced folates (FH$_2$, FH$_4$, CH$_2$-FH$_4$, CH$_0$-FH$_4$) are needed for DNA synthesis. Methotrexate or methotrexate polyglutamates inhibit several enzymes (DHFR, TS, AICAR, GAR) critical in folate metabolism.

Fig. 1) [2, 3]. Pemetrexed (Alimta®), a recently Food and Drug Administration (FDA) approved antifolate analog, is metabolized, like methotrexate, to polyglutamate forms within the cell [4]. Pemetrexed polyglutamate metabolites are also inhibitors of folate-dependent enzyme reactions.

Cellular pharmacology

Folates (and methotrexate) are transported into cells by two carrier systems: 1) a high-capacity, low-affinity reduced folate carrier (RFC) and 2) a low-capacity, high-affinity folate receptor system [5]. The RFC system appears to be the more clinically relevant methotrexate transporter. Cells with defective methotrexate transport are resistant to methotrexate [6]. Pemetrexed and raltitrexed (Tomudex®) can be transported by either system and may be less susceptible to drug resistance. Within the cell, methotrexate is converted to a polyglutamate form. Within 12–24 h, most intracellular methotrexate exists as polyglutamates. Polyglutamates enter and exit cells only sparingly. The selective cytotoxicity of methotrexate may come from increased formation of polyglutamates in neoplastic cells compared to normal tissues. The ability to generate methotrexate polyglutamates correlates with methotrexate response [7].

Methotrexate and methotrexate polyglutamates are both potent tight-binding inhibitors of dihydrofolate reductase. An excess of drug is needed to maintain total inhibition of DHFR [8]. Resistance to methotrexate can occur through increased expression of DHFR, development of a mutant DHFR with reduced affinity for methotrexate and amplification of the DHFR gene [9].

Moreover, decreased activity of folyl polyglutamate synthetase (FPGS), the enzyme which catalyzes polyglutamation, has been described as a mechanism of resistance to methotrexate [10, 11]. Increased activity of folyl polyglutamate hydrolase (FPGH), the enzyme which catalyzes the reduction of number of glutamates, has been suggested as a mechanism of resistance to this drug [12]. For this reason ZD9331 has been developed, a quinazoline TS inhibitor that does not require polyglutamation in order to be active. In Phase I and II studies this drug seems promising.

Another, recently discovered, mechanism of resistance to methotrexate, at least *in vitro*, is overexpression of the multidrug resistance proteins 1 and 2 (mrpl and mrp2) [13].

Leucovorin (a reduced folate) can be given to rescue cells from methotrexate. Leucovorin repletes reduced folate pools and competes with polyglutamate inhibition of TS, GAR, and AICAR.

The concentration of methotrexate within the cell and the duration of cell exposure to methotrexate are critical determinants of cytotoxicity. Cytotoxicity is directly related to time of drug exposure, but doubles only with a log increase in drug concentration. The concentration of reduced folate in the circulation affects cytotoxicity. Higher doses of leucovorin are needed to rescue cells exposed to higher methotrexate concentrations [14].

Clinical pharmacology

Methotrexate can be given orally, intravenously, or by intrathecal injection. Oral bioavailability is dose dependent (87% at doses <12 mg/m^2 *versus* 51% at doses >12 mg/m^2) [15]. Due to variability in oral absorption, methotrexate is usually administered intravenously. Methotrexate distributes into total body water including third-space fluid collections. Third space retention of methotrexate can be associated with a prolonged plasma drug half-life and increased toxicity. Methotrexate is primarily cleared by renal excretion (50–100%) [16]. Dose modifications must be made in patients with reduced creatinine clearance. Urinary methotrexate concentrations may exceed solubility limits after high-dose methotrexate therapy unless hydration and urinary alkalinization are employed. Methotrexate plasma levels must be monitored following high-dose therapy with the dosage of rescue leucovorin adjusted until plasma levels are less than 0.05 µM [17].

Methotrexate can be metabolized to 7-hydroxy methotrexate and 2,4-diamino-N-10 methyl pteroic acid (DAMPA). DAMPA is inactive but can cross-react with methotrexate in plasma assays. The 7-hydroxy metabolite is poorly soluble and can, like methotrexate, precipitate in renal tubule following high-dose therapy.

Toxicity

The primary toxicities of folate antagonists are myelosuppression and gastrointestinal mucositis. Mucositis occurs 3–7 days following therapy and precedes the development of myelosuppression by 1–2 days. High-dose methotrexate (3–12 gm/m^2) can result in reversible renal failure due to precipitation of methotrexate and metabolites in the renal tubule [17]. Hydration, alkalinization of urine, leucovorin rescue and monitoring of methotrexate concentrations are important in preventing toxicity associated with high-dose therapy resulting in renal toxicity.

Other methotrexate toxicities include hepatotoxicity and pneumonitis. Hepatotoxicity with portal fibrosis and occasionally cirrhosis is seen with the chronic use of low-dose methotrexate [18]. Use of 'pulsed' weekly therapy rather than continuous dosing reduces the incidence of hepatotoxicity (25% incidence of fibrosis and 3% cirrhosis). A self-limited pneumonitis with fever, cough, and a pulmonary infiltrate has been associated with methotrexate therapy.

Methotrexate can be given intrathecally to treat or prevent carcinomatous meningitis. A maximum dose of 12 mg is recommended. Three neurotoxic syndromes have been described with intrathecal methotrexate therapy: acute arachnoiditis, a subacute (2–3 weeks) motor paralysis and a demyelinating encephalopathy with dementia and occasionally coma [19]. Treatment with high-dose methotrexate may also result in encethalopathy due to central nervous system (CNS) drug penetration. The etiology of methotrexate-induced neurotoxicity is unknown.

Pyrimidine analogs

Cytosine arabinoside (Cytarabine)

Cytarabine (1-β-D-arabinofuranosylcytosine, Ara-C) is an antimetabolite analog of cytidine, the difference between the two molecules being the inversion of 2'-hydroxyl group from trans position in cytidine to the cis configuration in the cytarabine (Fig. 2). Ara-C is used in the treatment of multiple hematologic malignancies including AML, ALL, lymphoma, and CML [20].

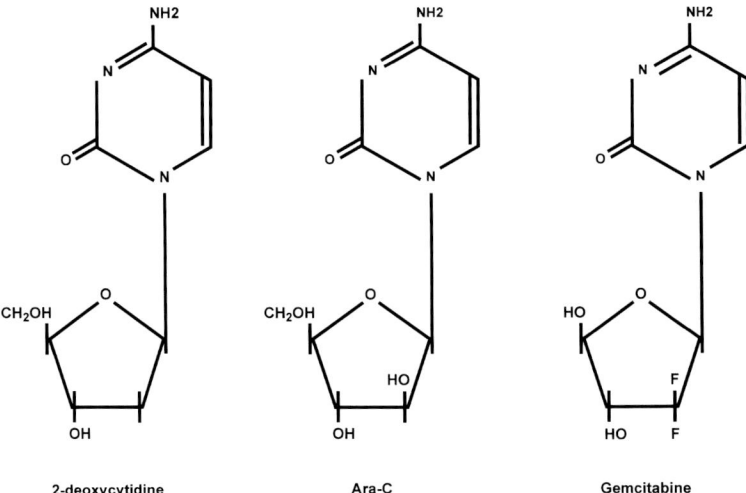

Figure 2. Structure of 2-deoxycytidine, cytarabine (ara-C) and gemcitabine.

Cellular pharmacology and mechanisms of action

The transmembrane transport of ara-C is dependent on nucleoside-specific transmembrane transport proteins [21]. Ara-C is a prodrug and must be activated through serial phosphorylation to its active form, ara-cytidine triphosphate (Ara-CTP) (Fig. 3). The nucleoside triphosphate form of ara-C (Ara-CTP) alters DNA synthesis and DNA strand elongation through several mechanisms. Ara-CTP inhibits DNA polymerases α and δ and interferes with action of DNA polymerase β used to repair DNA damage. However, the inhibition of DNA chain elongation is a basis for the most important cytotoxic effect of cytarabine [22]. Once ara-C is incorporated into DNA, the ara-CTP residue will terminate strand elongation and result in accumulation of DNA fragments, possibly through a repeated synthesis of small duplicated DNA segments [23].

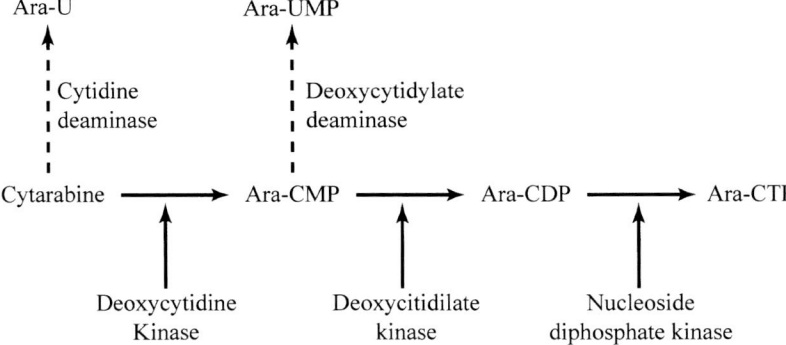

Figure 3. The pathway for intracellular cytarabine activation to ara-CTP. Inactivation occurs through metabolism to ara-U and ara-UMP. Abbreviations: ara-CMP, cytarbine 5'-monophosphate; ara-CDP, cytarabine 5'-diphosphate; ara-CTP, cytarabine 5'-triphosphate; ara-U, uracil arabinoside; ara-UMP, uracil arabinoside 5'-monophosphate.

Ara-C and ara-CMP are inactivated by the action of cytidine and deoxy-cytidylate deaminase, respectively, to form nontoxic metabolites, ara-U and ara-UMP. Ara-U inhibits deamination of ara-C through feedback mechanisms that contribute to increased cytotoxicity of ara-C in high dose regimens.

Clinical pharmacology and pharmacokinetics
As cytarabine is a cell cycle dependent drug, prolonged exposure of cells to cytotoxic concentrations is critical to achieve maximum cytotoxic activity. *In vitro* studies suggest that maximum cytotoxic activity is achieved with administration of cytarabine at concentrations >0.1 mg/L that are maintained for at least 24 h [24]. As noted previously, cytarabine must be phosphorylated to ara-CTP before it can exert its cytotoxic effect. The first enzyme of this pathway, deoxycitidine kinase (dCK), is rate limiting in the process of ara-CTP formation. Low levels of dCK in lymphoblasts have been correlated with ara-C resistance while transfection of hematopoietic cell lines with retroviral vectors containing dCK cDNA substantially increases susceptibility to ara-C. Accumulation of ara-CTP within cells appears to be saturated at plasma concentrations of cytarabine exceeding 8–10 μmol/L [25].

In systemic circulation, ara-C is rapidly catabolized to ara-U, which is subsequently renally excreted. In CNS, due to the lack of cytidine deaminase activity, elimination of cytarabine is similar to CSF bulk flow (0.42 mL/min) with a terminal half-life of 3–4 h that is significantly longer than the plasma half-life. Thus, cytotoxic concentrations of 0.1 mg/L are maintained for 24 h after single intrathecal administration of 30 mg of ara-C [26].

Toxicity
The toxicity of ara-C is primarily determined by the duration of exposure and by drug concentration. With conventional ara-C treatment regimens of 5–7

days, myelosuppression and gastrointestinal epithelial injury are the primary toxicities [27]. Neutropenia and thrombocytopenia start at the end of treatment and last for 2–3 weeks. Nausea, vomiting, and diarrhea may occur during the period of drug administration. Other reported toxicities include mucositis, typhlitis, and cholestasis.

High dose ara-C (3 g/m^2 q 12 h × 6–12 doses) causes cerebellar toxicity in 10% of patients. The risk factors for cerebellar toxicity include age >40 years, renal dysfunction and elevated alkaline phosphatase. Cerebellar toxicity is manifested as slurred speech, unsteady gait, dementia and coma leading to occasional death [28]. Conjunctivitis, responsive to topical steroid eye drops, and neutrophilic eccrine hydradenitis, manifested as skin plaques or nodules, have also been described with high-dose ara-C therapy. Intrathecal administration of ara-C is infrequently associated with arachnoiditis, fever, and seizures occurring within 4–7 days after therapy.

Novel cytarabine formulations
The search for ara-C formulations with improved pharmacokinetic parameters led to the development of several clinically useful compounds. DepoCyt^{TM} is a depot formulation in which ara-C is encapsulated in multivesicular liposomes. This formulation consists of microscopic (3–30 μm) spherical particles (DepoFoam) composed of numerous nonconcentric internal aqueous chambers containing cytarabine [29]. Following intrathecal administration of 25 mg DepoCyt^{TM}, concentrations of free ara-C in the ventricular CSF are maintained above the threshold for cytotoxicity for approximately 2 weeks. A randomized trial to compare safety and efficacy of intrathecal DepoCyt^{TM} 50 mg once every 2 weeks with intrathecal free cytarabine, 5 mg twice weekly revealed improved rates of complete response (71% *versus* 15%), time to neurological progression (median 78 *versus* 42 days) and median survival (99.5 *versus* 63 days) in patients with neoplastic meningitis with DepoCyt^{TM} therapy [30].

Gemcitabine

Gemcitabine (2',2'-difluorodeoxycytidine, dFdC, Gemzar$^®$), an analog of deoxycytidine (Fig. 2), has activity against several solid tumors including pancreatic, lung, breast, and bladder cancer.

Mechanism of action
Although similarities exist between gemcitabine and its analog, ara-C, several important differences in mechanisms of action have been demonstrated. Similar to other nucleoside analogs, gemcitabine requires intracellular phosphorylation to the nucleotide form for biologic activity (Fig. 4). Gemcitabine gains intracellular access through the nucleoside transporter system; resistance to gemcitabine has been demonstrated in transporter system-deficient cells [31]. Intracellularly, gemcitabine is phosphorylated to its active triphos-

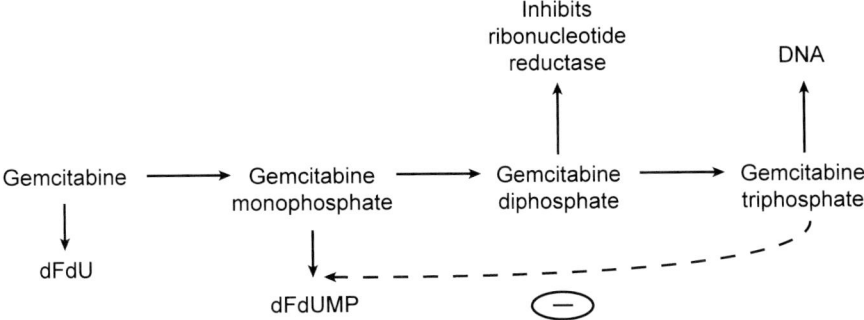

Figure 4. Intracellular gemcitabine activation and catabolism. dFdU and dFdUMP are inactive metabolites. Gemcitabine triphosphate is incorporated into DNA resulting in cytotoxicity and feedback inhibition of drug inactivation. Abbreviations: dFdU, 2 deoxy 2,2 difluorouridine; dFdUMP, 2 deoxy 2,2 difluorouridine monophosphate.

phate form. The first step in the phosphorylation of gemcitabine catalyzed by deoxycytidine kinase (dCK) plays a pivotal role in gemcitabine activation. A clear relation was reported between activity of dCK and sensitivity to gemcitabine [32]. This makes dCK a possible predictive marker of survival in gemcitabine-treated patients and a candidate for gene therapy. Deamination of gemcitabine and its mono- or di-phosphates results in formation of inactive metabolites. Gemcitabine triphosphate is a direct inhibitor of dCMP deaminase, which further increases the accumulation of the active triphosphate form. Gemcitabine diphosphate is an inhibitor of ribonucleoside diphosphate reductase, an enzyme responsible for maintaining the intracellular pools of deoxynucleotide triphosphates [33]. Ribonucleotide reductase (RNR) is emerging as an important determinant of gemcitabine chemoresistance in human cancers [34].

DNA replication and repair is dependent on dCTP and reduced levels of dCTP inhibit these functions. Gemcitabine triphosphate competes with dCTP for incorporation into DNA by the action of DNA polymerase and a decrease in the pools of dCTP favors incorporation of gemcitabine triphosphate [35]. Gemcitabine triphosphate and the reduction in cellular dCTP effectively inhibit dCMP deaminase, resulting in prolonged retention of gemcitabine di- and triphosphate. Lastly, the enzyme responsible for the synthesis of CTP, CTP synthetase, is inhibited by high concentrations of gemcitabine triphosphate [36]. All of these 'self-potentiating' interactions within the cell serve to enhance the cytotoxicity of the drug. Gemcitabine triphosphate is incorporated into DNA. An interesting phenomenon, termed 'masked chain termination' occurs. DNA strand termination does not occur until one additional deoxynucleotide is incorporated into the DNA strand, after addition of gemcitabine triphosphate [35]. Resistance to the 3'–5' exonuclease activity of DNA polymerase by this mechanism inhibits the subsequent DNA repair.

Clinical pharmacology

Gemcitabine is rapidly metabolized by the action of cytidine deaminase to 2',2'-difluorodeoxyuridine (dFdU) that lacks significant antitumor activity. In human ovarian A2780 cell lines, dFdU was 1,000-fold less active than gemcitabine [37]. Gemcitabine's half-life is approximately 8 min [38]. The metabolite, dFdU, is excreted in the urine and demonstrates a biphasic elimination [39]. Accumulation of gemcitabine triphosphate is saturated when gemcitabine levels exceed 15–20 micromoles per liter. These plasma levels are achieved when gemcitabine is infused at a fixed-dose-rate of 8–10 mg/m^2/min [40]. In an effort to maximize the accumulation of gemcitabine triphosphate without saturating deoxycytidine kinase, investigators have evaluated fixed-dose-rate schedules of administering gemcitabine. A recent randomized Phase II trial enrolled 92 patients with locally advanced or metastatic pancreatic cancer to either a fixed-dose-rate of gemcitabine at 10 mg/m^2/min for a total dose of 1,500 mg/m^2 or to a dose-intense 30-min infusion at 2,200 mg/m^2. This study demonstrated higher mononuclear cell concentrations of gemcitabine triphosphate in the fixed-dose-rate group and revealed a longer median survival, more objective responses, and higher one-year survival [41].

Toxicity

At commonly used doses of 0.8–1.2 gm/m^2 weekly, gemcitabine is well tolerated with less than 5% of patients discontinuing therapy due to adverse events [42]. The most common laboratory abnormalities included myelosuppression, elevated transaminases, and proteinuria. Myelosuppression is mild with World Health Organization (WHO) Grade 3 or 4 infections occurring in less than 1% of patients. Elevations in transaminases are transient and do not worsen with additional treatment with gemcitabine. Mild proteinuria has been demonstrated but does not appear to be clinically relevant [43]. Gastrointestinal toxicities (nausea, vomiting, diarrhea, mucositis) are mild and well controlled with supportive measures. Any degree of alopecia occurs in less than 15% of patients. Flu-like symptoms are seen in 20% of patients and peripheral edema is noted in 30%. Although rare, pulmonary toxicity and hemolytic uremic syndrome have been described during treatment with gemcitabine.

Fluoropyrimidines

5-Fluorouracil (5-FU) and other fluoropyrimidines are used for treatment of cancers of the gastrointestinal tract, breast, and head and neck. 5-FU is an analog of the pyrimidine uracil, which is fluorinated at carbon 5 position of the pyrimidine ring [44].

Mechanism of action

After 5-FU transport into the cell, metabolic activation is required for antitumor activity. Transport into the cell is accomplished by a facilitated nucleotide transport system, which is shared by uracil and hypoxanthine. Several pathways responsible for 5-FU activation have been identified (Fig. 5). Formation of metabolites 5-FdUMP and FUTP leads to antitumor activity. 5-FdUMP inhibits thymidylate synthase (TS) and FUTP is incorporated into cellular RNA [45]. Inhibition of TS disrupts DNA synthesis by depleting the pools of thymidine triphosphate (dTTP), an essential compound for DNA synthesis [46]. This inhibition occurs through the tight binding of 5-FdUMP, along with a reduced-folate cofactor, to TS. FdUMP is also incorporated into DNA disregulating DNA synthesis. FUTP is incorporated into RNA. Several mechanisms for 5-FU cytotoxicity resulting from FUTP incorporation into RNA have been proposed [47].

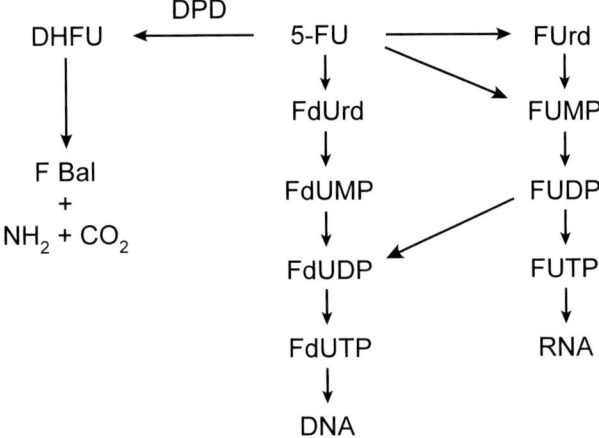

Figure 5. Metabolism of 5-fluorouracil (5-FU). 5-FU is converted within the cell to active metabolites FdUMP, FdUTP and 5FUTP. 5-FU is inactivated by conversion to DHFU which is subsequently broken down to F-Bal, NH_2 and CO_2. Abbreviations: 5-FU (5-fluorouracil); 5-FUDR (5-fluorouridine); 5-FUdR (5-fluorodeoxyuridine); 5-FUMP (5-fluorouridine monophosphate); 5-FdUMP (5-fluorodeoxyuridine monophosphate); FUDP (5-fluorouradine diphosphate); FdUDP (5-fluorodeoxyuridine diphosphate); FUTP (5-fluorouridine triphosphate); FdUTP (5-fluorodeoxyuridine triphosphate); DHFU (5,6- dihydrofluorouracil); F Bal (fluoro B-alanine); DHDP (dihydropyrimidine dehydrogenase).

Clinical pharmacology

5-FU is primarily cleared from plasma by hepatic metabolism. Drug half-life is short (10–15 min) [48]. Continuous infusion of 5-FU is more rapidly cleared from the plasma than bolus administration due to saturation of the primary catabolic enzyme, dihydropyrimidine dehydrogenase or DPD (Fig. 5). The duration of 5-FU infusion is inversely proportional to the tolerated dose, that is, lower doses are necessary for longer duration of infusion. The majority of 5-FU is eliminated by metabolism through the action of DPD, with only 5–10% of drug excreted through the kidney. The liver has the highest level of DPD and is responsible for the majority of 5-FU catabolism [44]. Although prior clinical reports suggested an increase in toxicity with hepatic dysfunction, a more recent Phase I trial demonstrated no alteration in 5-FU in clearance in patients with elevated bilirubin (1.5 mg/dL or greater).

Dihydropyrimidine dehydrogenase (DPD) is the rate-limiting enzyme responsible for metabolizing 5-FU to inactive metabolites [49]. A deficiency in DPD can be life-threatening if a fluoropyrimidine is administered. DPD deficiency is uncommon (<1 in 300 patients). It is inherited through an autosomal recessive pattern. Since a screening test is not established, deficiency is usually suspected when an early, unexpected toxicity is detected.

Hepatic artery infusion (HAI) of fluoropyrimidines has been used in a number of clinical trials in an attempt to improve response rates and survival in patients with hepatic metastases. When delivered by HAI, FUdR has a first-pass extraction by normal liver of 94–99%, whereas 5-FU has a first-pass extraction of only 19–51% [50]. Local toxicities predominate with FUdR HAI with dose-limiting side effects including gastritis, hepatitis, ulceration, or duodenitis. Systemic toxicities are generally mild. Oral fluoropyrmidine formulations have been developed and will be discussed later.

Toxicity
5-FU has a wide array of gastrointestinal side effects, including mucositis, nausea, vomiting, diarrhea, dysphagia, and proctitis. Diarrhea and/or vomiting can lead to severe dehydration requiring vigorous hydration and supportive care. The dose-limiting toxicities of bolus 5-FU are typically mucositis, diarrhea, and myelosuppression. In contrast, severe myelosuppression is an uncommon side effect with the use of protracted venous infusion. The dose-limiting side effects with continuous infusion 5-FU include stomatitis and palmar-plantar erythrodysesthesia [51].

Cerebellar ataxia, somnolence, and other neurologic symptoms attributed to 5-FU toxicity have been described in the literature. Most of these neurologic side effects are reversible with time. The clinical trials demonstrating these toxicities used intensive daily scheduling or 5-FU modulators. Severe neurotoxicity, manifested as encephalopathy, has also been reported in patients with DPD deficiency.

Chest pain, arrhythmia, electrocardiographic changes, and elevated cardiac enzymes have been described in a temporal association with 5-FU infusion [52]. However, coronary angiography performed in some patients after an acute ischemic event demonstrated no evidence for an obstructive lesion, suggesting coronary vasospasm as a possible mechanism. Various dermatologic toxicities from 5-FU have been noted and include hair loss, nail changes, photosensitivity, and dermatitis. An inflammatory reaction can occur in the distribution of actinic keratoses. A number of ocular toxicities have been attributed to 5-FU but the most common is tear duct stenosis.

Drug interactions
Attempts have been made, with varying degrees of success, to augment the cytotoxicity of 5-FU by combining it with other agents or modalities.

Pretreatment with methotrexate augments the cytotoxicity of 5-FU. Reduced folates are important in the formation of the ternary complex, FdUMP-TS-510-CH$_2$FH$_4$. Exogenous leucovorin (5-CHO-FH$_4$) provides expansion of the reduced folate pool and enhances TS inhibition [53].

Oral fluoropyrimidines
The use of oral fluoropyrimidines has been hindered by the poor and erratic bioavailability of 5-FU. Potential advantages for the use of oral agents include ease and flexibility of administration, avoidance of intravenous catheter complications, protracted exposure to 5-FU, and possibly a reduction in healthcare resources. Several fluoropyrimidine produgs (Ftorafur, Capecitabine, S-1) have been developed to improve bioavailabity [54]. Capecitabine (Xeloda) has the most widespread clinical use [55]. Capecitabine is well absorbed from the gastrointestinal tract and is activated through a series of three enzymatic steps to eventually release 5-FU within tumor cells. Dose-limiting toxicities have included diarrhea, nausea, vomiting, and palmar-plantar erythrodysesthesia syndrome [56]. Another approach to circumvent the degradation of oral 5-FU by DPD is the addition of ethynyluracil (Eniluracil), a potent irreversible inactivator of DPD. Ethynyluracil, when given prior to oral 5-FU, significantly increases the oral bioavailability and decreases 5-FU catabolism [57]. Unfortunately, the combination of eniluracil and oral 5-FU is less effective than intravenous 5-FU and leucovorin in the treatment of colorectal cancer [58].

Purine analogs

Guanine analogs

Mechanism of action
Azathioprine, 6-mercaptopurine (6-MP) and 6-thioguanine (6-TG) are guanine analogs used as immunosuppressants and antineoplastic agents. Azathioprine is a prodrug of 6-MP which is converted by non-enzymatic mechanisms to 6-MP and methyl-4-nitro-5-imidazole. The imidazole metabolite of azathioprine may contribute to the immunosuppressive activity of this drug. 6-MP undergoes one of three routes of metabolism. The activation pathway leads to 6-thioguanine triphosphate (6-TGTP) incorporation into DNA (Fig. 6). The cytotoxicity of 6-TG requires: (a) incorporation of 6-TG into DNA (b) miscoding during DNA replication and (c) recognition of the abnormal incorporated base pairs by proteins of the postreplicative mismatch repair system [59]. Similar to 6-MP, 6-TG is incorporated into DNA where fraudulent nucleotides lead to defective DNA replication.

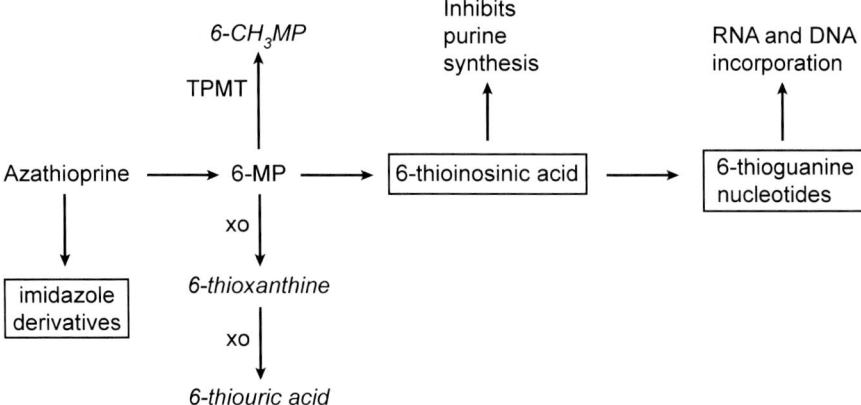

Figure 6. Mechanism of activation and catabolism of azathioprine and 6-mercaptopurine (6-MP). Active metabolites are indicated by surrounding boxes. Inactive (or less active) metabolites are indicated by italic print. (Abbreviations: 6-CH₃MP, 6-methyl mercaptopurine; TPMT, thiopurine methyltransferase; XO, xanthine oxidase.)

Clinical pharmacology

6-MP can be given intravenously or orally. Oral absorption is incomplete and highly variable. Bioavailability averages 16% (range 5–37%). Clearance occurs through two metabolic routes. 6-MP is oxidized to an inactive metabolite, 6-thiouric acid, by xanthine oxidase (Fig. 6). Poor oral bioavailability is due to a large first pass effect as drug is absorbed through the intestinal wall into the portal circulation and metabolized by xanthine oxidase in intestine and liver before entering the systemic circulation [60]. The concomitant use of allopurinol (an inhibitor of xanthine oxidase) significantly increases 6-MP bioavailability and toxicity. 6-MP also undergoes S-methylation by the enzyme thiopurine methyltransferase (TPMT) to yield inactive 6-methylmercaptopurine (Fig. 6). Patient-to-patient variation in TPMT activity results in significant variation in 6-MP metabolism and drug toxicity among patients. One in 300 subjects has very low TPMT activity; 11% of the population has intermediate activity and the rest have high enzyme activity. A single genetic locus with two alleles (one for low and one for high activity) is responsible for the trimodal distribution [61]. Patients with absent TMPT have increased toxicity and require a 10–15-fold reduction in 6-MP dosage.

Thioguanine is not a substrate for xanthine oxidase, but is converted to 6-thioinosine (an inactive metabolite) by the action of the enzyme, guanase. Inhibitors of xanthine oxidase, such as allopurinol, do not interfere with 6-TG metabolism. Methylation of thioguanine, via thiopurine methyltransferase (TPMT), to an inactive metabolite is more extensive than is that of 6-MP.

Toxicity

Myelosuppression is the dose limiting toxicity of 6-MP, azathioprine and thioguanine [62]. Platelets, granulocytes and erythrocytes are all affected. Purine antagonists are immunosuppressants leading to an increased rate of infection. Approximately 25% of treated patients experience nausea, vomiting, and anorexia. Gastrointestinal side effects are more common in adults than in children. Hepatotoxicity is infrequent, usually mild and reversible, with a clinical picture consistent with cholestatic jaundice. Increased transaminase levels are noted in roughly 15% of patients. Frank hepatic necrosis can occur. An increased incidence of myelodysplasia and AML following azathioprine and 6-MP therapy has been reported in children who have low TPMT activity [63].

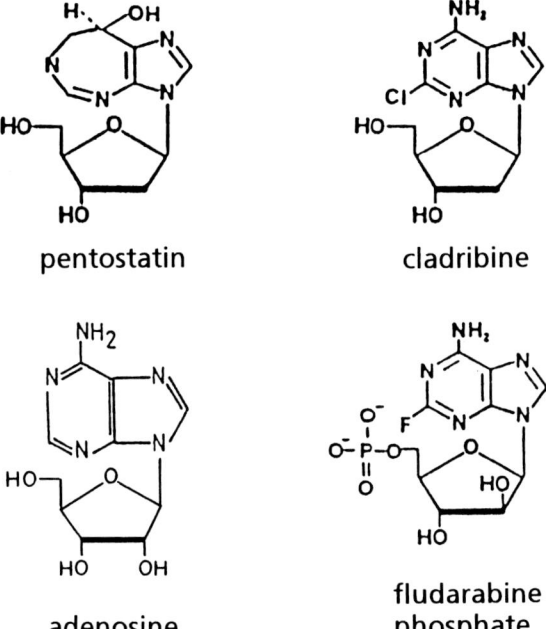

Figure 7. Structure of adenosine and adenosine analogs.

Adenosine analogs

Three adenosine analogs are in current clinical use; fludarabine, cladribine and pentostatin (Fig. 7). All have activity against indolent lymphomas and leukemias.

Mechanism of action

Both fludarabine (9-β-d-arabinofuranosyl-2-fluoroadenine or F-ara-A) and cladribine (2'-chlorodeoxyadenosine or 2CdA) are phosphorylated within the cell to their active triphosphate metabolite [64]. F-ara-ATP inhibits several intracellular enzymes important in DNA replication including DNA polymerase, ribonucleotide reductase, DNA primase and DNA ligase I. In addition, F-ara-ATP is incorporated into DNA. Excision of a 3'-terminal F-ara-AMP in DNA does not easily occur and the presence of this false nucleotide leads to apoptosis. The amount of fludarabine incorporated into DNA is linearly correlated with cytotoxicity. 2-CdATP is incorporated into DNA and produces DNA strand breaks and inhibition of DNA synthesis. High intracellular concentrations of 2-CdATP also inhibit DNA polymerases and ribonucleotide reductase causing an imbalance in deoxyribonucleotide triphosphate pools with subsequent impairment of DNA synthesis. The mechanism of adenosine analog cytotoxicity in non-dividing cells is less well understood. Cytotoxicity has been attributed to inhibition of DNA repair, NAD^+/ATP depletion, p53 mediated apoptosis and inhibition of mitochondrial depletion [65].

Pentostatin cytotoxicity is believed to be due to inhibition of adenosine deaminase with the accumulation of deoxyadenosine and dATP. Abnormally high levels of deoxyadenosine triphosphate (dATP), which accumulate with ADA inhibition, exert a negative feedback on ribonucleotide reductase resulting in an imbalance in deoxynucleotide pools. The imbalance inhibits DNA synthesis and alters DNA replication and repair [66].

Clinical pharmacology

Fludarabine is phosphorylated to increase its solubility. Following IV administration, fludarabine rapidly loses its phosphate group to produce 9-β-D-arabinofuranosyl-2-fluoroadenine (F-araA). Both F-ara A and cladribine are primarily cleared by renal excretion (≈50%) [64, 67]. Dose reductions are needed for patients with renal dysfunction. Oral bioavalability of both fludarabine and cladribine is good (50–75%) and oral formulations are under development. Only a small amount of pentostatin is metabolized. Most pentostatin (40–80%) is excreted unchanged in the urine. Although not carefully studied, pentostatin dose reductions are likely needed for patients with renal insufficiency.

Toxicity

Myelosuppression and immunosuppression are the primary toxicities of the adenosine analogs [68]. Up to 25% of patients treated with adenosine analogs will have a febrile episode. Many will be fevers of unknown origin, but one-third will have a serious infection documented. Platelet nadirs of less than 50–100,000/mm³ are seen in 20% of patients. Fludarabine and cladribine are immunosuppressive. Therapy is associated with an increased risk of opportunistic infections [69]. CD4 and CD8 T-lymphocytic subpopulations decrease to levels of 150–200/mm³ after three courses of therapy. Infections with

Cryptococcus, Listeria monocytogenes, Pneumocystis carinii, CMV, *Herpes simplex* virus, *Varicella zoster* and *Mycobacterium*, organisms associated with T-cell dysfunction, are seen. Other side effects of fludarabine and cladribine include renal failure, hemolytic anemia and neurotoxicity, which are uncommon but documented [70].

Hydroxyurea

Mechanism of action

Hydroxyurea (HU) is primarily used as a myelosuppressive agent for a variety of myeloproliferative disorders. HU inhibits ribonucleotide reductase, the enzyme responsible for converting ribonucleotide diphosphates to their deoxyribonucleotide form. Inhibition of DNA synthesis correlates closely with decreased deoxyribonucleotide pools [71]. The inhibition of ribonucleotide reductase results from the inactivation of the tyrosyl free radical on the M-2 subunit of the enzyme with disruption of the iron-binding center [72]. Cells enter S phase at a normal rate but accumulate there as a result of the inhibition of DNA synthesis, due to reduced deoxyribonucleotide pools.

Clinical pharmacology

HU has excellent bioavailability (80–100%) and is generally administered orally. The elimination half-life is roughly 4 h with renal clearance being the primary route of drug elimination [73]. Precise dosing guidelines for patients with renal insufficiency are not available. HU distributes readily into tissues including the CSF, ascites, or pleural effusions. Drug clearance is not linear with dose.

Toxicity

The dose-limiting toxicity of HU is myelosuppression. In patients with non-hematologic malignancies, the peripheral white blood cell count begins to fall in 2–5 days. Patients with leukemia or a myeloproliferative syndrome experience a more rapid fall in white blood cell counts. The rapidity of the effect on the circulating leukemia cell population and the brief duration of its action have been the basis for the use of HU in patients with acute nonlymphocytic leukemia who present with markedly elevated peripheral blood blast counts or platelet counts [74]. Reversal of HU's effect on myelocytes occurs rapidly, but platelet recovery may be delayed (7–10 days).

At commonly used doses (0.5–2.0 gm/d), nausea, vomiting and anorexia are usually mild. Patients who have taken HU for an extended period may develop one of several dermatologic changes. These include hyperpigmentation, erythema of the face and hands, a diffuse maculopapular rash, dry skin with atrophy, multiple pigmented nail bands, an ulcerative dermatitis, and skin ulcerations usually in the legs [75]. Liver function abnormalities are seen, usually mild and transient, but may progress to jaundice. Acute lung injury has been reported.

References

1 Baram J, Allegra CJ, Fine RL et al. (1987) Effects of methotrexate on intracellular folate pools in purified myeloid precursor cells from normal human bone marrow. *J Clin Invest* 79: 692–697
2 Allegra CJ, Drake JC, Jolivet J et al. (1985) Inhibition of phosphoribosyl aminoimidazole carboxamide transformylase by methotrexate and dihydrofolic acid polyglutamates. *Proc Natl Acad Sci USA* 82: 4881–4885
3 Chu E, Drake JC, Boarman D et al. (1990) Mechanism of thymidylate synthase inhibition by methotrexate in human neoplastic cell lines and normal human myeloid progenitor cells. *J Biol Chem* 265: 8470–8478
4 Goldman ID, Ahao R (2002) Molecular, biochemical and cellular pharmacology of pemetrexed. *Sem Oncol* 29: 3–17
5 Fan J, Vitols KS, Huennekens FM (1991) Biotin derivatives of methotrexate and folate. Synthesis and utilization for affinity purification of two membrane-associated folate transporters from L1210 cells. *J Biol Chem* 266: 14862–14865
6 Moscow JA (1998) Methotrexate transport and resistance. *Leuk Lymphoma* 30: 215–224
7 Galvin AJ, Schuetz JD, Masson E et al. (1997) Differences in folylpolyglutamate synthetase and dihydrofolate reductase expression in human B-lineage *versus* T-lineage leukemic lymphoblasts: mechanisms for lineage differences in methotrexate polyglutamylation and cytotoxicity. *Mol Pharmacol* 52: 155–163
8 White JC, Loftfield S, Goldman ID (1975) The mechanism of action of methotrexate. Requirement of free intracellular methotrexate for maximal suppression of ^{14}C formate incorporation into nucleic acids and protein. *Mol Pharmacol* 11: 287–297
9 Curt GA, Carney DN, Cowan KH et al. (1983) Unstable methotrexate resistance in human small-cell carcinoma associated with double minute chromosomes. *N Engl J Med* 308: 199–202
10 Roy K, Mitsugi K, Sirlin S et al. (1995) Different antifolate-resistance L1210 cell variants with either increased or decreased folylpolyglutamate synthetase gene expression at the level of mRNA transcription. *J Biol Chem* 270: 26918–26922
11 McCloskey DE, McGuire JJ, Russel CA et al. (1991) Decreased folylpolyglutamate synthetase activity as a mechanism of methotrexate resistance in CCRF-CEM human leukemia sublines. *J Biol Chem* 266: 6181–6187
12 Rots MG, Pieters R, Peters GJ et al. (1999) Role of folylpolyglutamate synthetase and folylpolyglutamate hydrolase in methotrexate accumulation and polyglutamylation in childhood leukemia. *Blood* 93: 1677–1683
13 Hooijberg JH, Broxterman HJ, Kool M et al. (1999) Antifolate resistance mediated by the multidrug resistance proteins MRP1 and MRP2. *Cancer Res* 59: 2532–2535
14 Pinedo HM, Zaharko DS, Bull JM et al. (1976) The reversal of methotrexate cytotoxicity to mouse bone marrow cells by leucovorin and nucleosides. *Cancer Res* 36: 4418–4424
15 Balis FM, Savitch JL, Bleyer WA (1983) Pharmacokinetics of oral methotrexate in children. *Cancer Res* 43: 2342–2345
16 Shen DD, Azarnoff DL (1978) Clinical pharmacokinetics of methotrexate. *Clin Pharmacokinet* 3: 1–13
17 Ackland SP, Schilsky RL (1987) High-dose methotrexate: a critical reappraisal. *J Clin Oncol* 5: 2017–2031
18 Mackenzie AH (1998) Hepatotoxicity of prolonged methotrexate therapy for rheumatoid arthritis. *Cleve Clin Q* 52: 129–135
19 Blaney SM, Balis FM, Poplack DG (1991) Current pharmacological treatment approaches to central nervous system leukaemia. *Drugs* 41: 702–716
20 Garcia-Carbonero R, Ryan DP, Chabner BA (2001) Cytidine analogues. In: BA Chabner, DL Longo (eds): In: *Cancer Chemotherapy and Biotherapy: Principles and Practice*, 3rd ed. Lippincott, Williams and Wilkins, Philadelphia, 265–294
21 Wiley JS, Jones SP, Sawyer WH et al. (1982) Cytosine arabinoside influx and nucleoside transport sides in acute leukemia. *J Clin Invest* 69: 479–484
22 Kufe DW, Major PP, Egan MM et al. (1980) Correlation of cytotoxicity with incorporation of araC into DNA. *J Biol Chem* 255: 8997–9000
23 Grant S (1998) Ara-C: cellular and molecular pharmacology. *Advances Cancer Res* 72: 197–233
24 Graham FL, Whitmore GF (1970) The effect of 1-β-D-arabino furanosylcytosine on growth, via-

bility, and DNA synthesis of mouse L-cells. *Cancer Res* 30: 2627–2635

25 Plunkett W, Liliemark JO, Estey E et al. (1987) Saturation of ara-CTP accumulation during high-dose ara-C therapy: pharmacologic rationale for intermediate-dose ara-C. *Semin Oncol* 14 Suppl 1: 159–166

26 Zimm S, Collins JM, Miser J et al. (1984) Cytosine arabinoside cerebrospinal fluid kinetics. *Clin Pharmacol Ther* 35: 826–830

27 Slavin RE, Dias MA, Saral R (1978) Cytosine arabinoside-induced gastrointestinal toxic alterations in sequential chemotherapeutic protocols. *Cancer* 42: 1747

28 Rubin EH, Anderson JW, Berg DT et al. (1992) Risk factors for high-dose cytarabine neurotoxicity: an analysis of a cancer and leukemia group B trial in patients with acute myeloid leukemia. *J Clin Oncol* 10: 948–953

29 Murry DJ, Blaney SM (2000) Clinical pharmacology of encapsulated sustained-release cytarabine. *Ann Pharmacother* 34: 1173–1178

30 Glantz MJ, Jaeckle KA, Chamberlain MC et al. (1999) A randomized controlled trial comparing intrathecal sustained-release cytarabine (DepoCyt) to intrathecal methotrexate in patients with neoplastic meningitis from solid tumors. *Clin Cancer Res* 5: 3394–3402

31 Mackey JR, Mani RS, Selner M et al. (1998) Functional nucleoside transporters are required for gemcitabine influx and manifestation of toxicity in cancer cell lines. *Cancer Res* 58: 349–357

32 Bergman AM, Pinedo HM, Peters GJ (2002) Determinants of resistance to 2',2'-difluorodeoxycytidine (gemcitabine). *Drug Resist Updat* 19–33

33 Plunkett W, Hung P, Xu YZ et al. (1995) Gemcitabine: metabolism, mechanisms of action, and self-potentiation. *Semin Oncol* 22: 3–10

34 Duxbury MS, Ito H, Zinner MJ et al. (2004) RNA Interference targeting the M2 subunit of ribonucleotide reductase enhances pancreatic adenocarcinoma chemosensitivity to gemcitabine. *Oncogene* 23: 1539–1548

35 Huang P, Chubb S, Hertel LW et al. (1991) Action of 2',2'-difluorodeoxycytidine on DNA synthesis. *Cancer Res* 51: 6110–6117

36 Heinemann V, Schulz L, Issels RD et al. (1991) Gemcitabine: a modulator of intracellular nucleotide and deoxynucleotide metabolism. *Semin Oncol* 22(4 Suppl 11): 11–18

37 Ruiz van Haperen VW, Veerman G, Eriksson S et al. (1994) Development and molecular characterization of a 2',2'-difluorodeoxycytidine-resistant variant of the human ovaria carcinoma cell line A2780. *Cancer Res* 54(15): 4138–4143

38 Abbruzzese JL, Grunewald R, Weeks EA et al. (1991) A phase I clinical, plasma, and cellular pharmacology study of gemcitabine. *J Clin Oncol* 9: 491–498

39 Storniolo AM, Allerheiligen SR, Pearce HL (1997) Preclinical, pharmacologic, and Phase I studies of gemcitabine. *Semin Oncol* 24(2 Suppl 7): 2–7

40 Grunewald R, Kantarjian H, Du M et al. (1992) Gemcitabine in leukemia: a phase I clinical, plasma, and cellular pharmacology study. *J Clin Oncol* 10: 406–413

41 Tempero M, Plunkett W, Ruiz Van Haperen V et al. (2003) Randomized phase II comparison of dose-intense gemcitabine: 30-minute infusion and fixed dose rate infusion in patients with pancreatic adenocarcinoma. *J Clin Oncol* 21: 3402–3408

42 Storniolo AM, Enas NH, Brown CA et al. (1999) An investigational new drug treatment program for patients with gemcitabine: results for over 3000 patients with pancreatic carcinoma. *Cancer* 85: 1261–1268

43 Green MR (1996) Gemcitabine safety overview. *Semin Oncol* 23 (Suppl 10): 32–35

44 Grem J (2000) 5-Fluorouracil: forty-plus and still ticking. A review of its preclinical and clinical development. *Invest New Drugs* 18: 299–313

45 Longley DB, Harkin DP, Johnston PG (2003) 5-Fluorouracil: mechanisms of action and clinical strategies. *Nat Rev Cancer* 3: 330–338

46 Santi DV, McHenry CS, Sommer H (1974) Mechanism of interaction of thymidylate synthetase with 5-fluorodeoxyuridylate. *Biochemistry* 13: 471–481

47 Santi DV, Hardy LW (1987) Catalytic mechanism and inhibition of tRNA (uracil-5-) methyltransferase: evidence for covalent catalysis. *Biochemistry* 26: 8599–8606

48 Heggie GD, Sommadossi JP, Cross DS et al. (1987) Clinical pharmacokinetics of 5-Fluorouracil and its metabolites in plasma, urine, and bile. *Cancer Res* 47: 2203–2206

49 Diasio RB, Lu Z (1994) Dihydropyrimidine dehydrogenase activity and fluorouracil chemotherapy. *J Clin Oncol* 12: 2239–2242

50 Ensminger WD, Rosowsky A, Rosa V et al. (1978) A clinical-pharmacological evaluation of hepat-

ic arterial infusions of 5-fluoro-2'-deoxyuridine and 5-fluorouracil. *Cancer Res* 38: 3784–3792

51 Lokich JJ, Ahlgren JD, Gullo JJ et al. (1989) A prospective randomized comparison of continuous infusion fluorouracil with a conventional bolus schedule in metastatic colorectal carcinoma: a Mid-Atlantic Oncology Program Study. *J Clin Oncol* 7: 425–432

52 Freeman NJ, Costanza ME (1988) 5-Fluorouracil-associated cardiotoxicity. *Cancer* 61: 36–45

53 Wright JE, Dreyfuss A, El-Magharbell I et al. (1989) Selective expansion of 5,10-methylenete-trahydrofolate pools and modulation of 5-fluorouracil antitumor activity by leucovorin *in vivo*. *Cancer Res* 49: 2592–2596

54 Kuhn JG (2000) Fluorouracil and the new oral flouorouinated pyrimidines. *Ann Pharmacother* 35: 217–227

55 Ishitsuka H (2000) Capecitabine: preclinical pharmacology studies. *Invest New Drug* 18: 343–354

56 Mackean M, Planting A, Twelves C et al. (1998) Phase I and pharmacologic study of intermittent twice-daily oral therapy with capecitabine in patients with advanced and/or metastatic cancer. *J Clin Oncol* 16: 2977–2985

57 Baker SD, Diasio RB, O'Reilly S et al. (2000) Phase I and pharmacologic study of oral fluo-rouracil on a chronic daily schedule in combination with the dihydropyrimidine dehydrogenase inactivator eniluracil. *J Clin Oncol* 18: 915–926

58 Schilsky RL, Levin J, West WH et al. (2002) Randomized, open-label, phase III study of a 28-day oral regimen of eniluracil plus fluorouracil *versus* intravenous fluorouracil plus leucovorin as first-line therapy in patients with metastatic/advanced colorectal cancer. *J Clin Oncol* 20: 1519–1526

59 Swann PF, Waters TR, Moulton DC et al. (1996) Role of postreplicative DNA mismatch repair in the cytotoxic action of thioguanine. *Science* 273: 1109–1111

60 Zimm S, Collins JM, Riccardi R et al. (1983) Variable bioavailability of oral mercaptopurine. Is maintenance chemotherapy in acute lymphoblastic leukemia being optimally delivered? *N Engl J Med* 308: 1005–1009

61 Yates CR, Krynetski EY, Loennechen J et al. (1997) Molecular diagnosis of thiopurine 5-methyl-transferase deficiency: genetic basis for azathioprine and mercaptopurine intolerance. *Ann Intern Med* 126: 608–614

62 El-Azhary (2003) Azathioprine: current status and future considerations. *Int J Derm* 42: 335–341

63 Black AJ, McLeod HL, Capell HA et al. (1998) Thiopurine methyltransferase genotype predicts therapy-limiting severe toxicity from azathioprine. *Ann Intern Med* 129: 716–718

64 Gandi V, Plunkett W (2002) Cellular and clinical pharmacology of fludarabine. *Clin Pharmacokinet* 41: 93–103

65 Pettit AR (2003) Mechanism of action of purine analogs in chronic lymphocytic leukemia. *Brit J Haemato* 121: 692–702

66 O'Dwyer PJ, Wagner B, Leyland-Jones B et al. (1988) 2'-Deoxycoformycin (pentostatin) for lym-phoid malignancies. *Ann Inter Med* 108: 733–743

67 Lilliemark J (1997) The clinical pharmacokinetics of cladribine. *Clin Pharmacokinet* 32: 120–131

68 Adkins PC, Peters DH, Markham A (1997) Fludarabine: An update of its pharmacology and use in the treatment of hematologic malignancies. *Drugs* 53: 1005–1037

69 Cheson BD (1995) Infectious and immunosuppressive complications of purine analog therapy. *J Clin Oncol* 13: 2431–2448

70 Cheson BD, Vena DA, Foss FM et al. (1994) Neurotoxicity of purine analogs: a review. *J Clin Oncol* 12: 2216–2228

71 Nicander B, Reichard P (1985) Relations between synthesis of deoxyribonucleotides and DNA replication in 3 T6 fibroblasts. *J Biol Chem* 260: 5376–5381

72 Nyholm S, Thelander L, Graeslund A (1993) Reduction and loss of the iron center in the reaction of the small subunit of mouse ribonucleotide reductase with hydroxyurea. *Biochemistry* 32: 11569–11574

73 Tracewell WG, Trump DL, Vaughan WP et al. (1995) Population pharmacokinetics of hydrox-yurea in cancer patients. *Cancer Chemother Pharmacol* 35: 417–422

74 Kennedy BJ, Yarbro JW (1966) Metabolic and therapeutic effects of hydroxyurea in chronic myel-ogenous leukemia. *JAMA* 195: 1038–1093

75 Kennedy BJ, Smith LR, Goltz RW (1975) Skin changes secondary to hydroxyurea. *Arch Dermatol* 111: 183–187

Drugs Affecting Growth of Tumours
Edited by Herbert M. Pinedo and Carolien H. Smorenburg
© 2006 Birkhäuser Verlag/Switzerland

DNA-intercalators – the anthracyclines

Klaus Mross, Ulrich Massing and Felix Kratz

Tumor Biology Center at the Albert-Ludwig University Freiburg, Breisacher Strasse 117, 79106 Freiburg i.Br., Germany

History

The anthracyclines are derivatives of rhodomycin B, a red-pigmented polyketide antibiotic, isolated in the 1950s from Gram-positive *Streptomyces* present in an Indian soil sample. Many microorganisms produce and secrete complex antibacterial and antifungal compounds into their surroundings to protect their life-sphere against potential invaders. After the discovery of the antitumor activity and chemistry of rhodomycin B, Farmitalia initiated a program to find new anticancer compounds produced by novel strains of microbes isolated from soil. In 1957, a colony of *Streptomyces* producing a red pigment was grown from a soil sample taken at Castel del Monte near the city of Andria in southeastern Italy. This microbe produced a substance named daunorubicin after a pre-Roman tribe in southeastern Italy; Di Marco demonstrated antitumor activity in 1963. At nearly the same time this compound was isolated by French researchers at Rhône Poulenc, who named it rubidomycin. Later on, it became clear that rubidomycin and daunomycin were identical and daunorubicin became the only name for this compound. In 1969, Arcamone and his co-workers succeeded in isolating and purifying doxorubicin (14-hydroxydaunomycin) from *Streptomyces peucetius* variety *caesius*, a mutant of the original *Streptomyces* strain found near the Adriatic Sea. This is the reason why doxorubicin was named Adriamycin. The story of the two anthracyclines is now nearly half a century old. The clinical development of daunorubicin started in 1964 for the treatment of acute leukemias, and doxorubicin in 1968, and this drug was broadly evaluated in patients with leukemia, lymphoma and most solid tumors. The first clinical experiments with doxorubicin were performed in Milano by Bonadonna which showed remarkable antitumor activity that were later confirmed by studies in the USA. Only 6 years later, in 1974, doxorubicin was approved by the US Food and Drug Administration (FDA). At the end of the 1970s the two anthracyclines dauno- and doxorubicin were the most efficacious anticancer drugs with an enormous impact on the development of anticancer therapy with cytotoxic drugs and medical oncology which grew up to an independent medical discipline within internal medicine.

As a consequence, the search for new anthracyclines was persued that considered:
a) the limited time protection of the patent
b) the remarkable side effect profile which is in some aspects very unpleasant for the patient
c) the separation of cardiotoxicity and antitumor activity
d) the search for anthracyclines active also in resistant tumor cells
e) differences in tissue specificity and the modulation of pharmacokinetic properties of the drug in order to alter either dose-effect or time-concentration relationships, and
f) the minor molecular difference between dauno- and doxorubicin that had shown great influence on the spectrum of antitumor effects

The clinical success of doxorubicin has been the impetus for a diligent search for more effective and less toxic anthracycline analogs. In the mid 1980s the planned successor of doxorubicin was introduced, i.e., epirubicin, and in 1991 idarubicin, the successor of daunorubicin, entered clinical trials. These four compounds that exhibit only minor differences in terms of chemical structure dominated the class of anthracyclines and were exclusively developed by Farmitalia Carlo Erba in Milano, Italy, a company bought by Pharmacia in the mid 1990s which itself was bought from Pfizer in 2002. Other anthracyclines like pirarubicin, zororubicin, aclarubicin and carminomycin have reached the status of registered drugs in a few countries but play no significant role in global terms. Structurally related to the anthracyclines are the anthracediones which were developed in the laboratories at American Cyanamid Laboratories in the late 1970s [6] and the anthrapyrazoles [7] which were synthetized at the Warner-Lambert/Parke-Davis Company in the mid 1980s. The clinical development of mitoxantrone started in 1980, and this drug became registered in the mid 1980s. No anthrapyrazole has been registered up to now, but several clinical studies are still ongoing. The tremendous efforts of developing better anthracyclines have been reviewed for the interested reader [8, 9].

Research groups have developed liposomal formulations of dauno- and doxorubicin hereby changing the pharmacokinetic behavior drastically. Three liposomal formulations have been marketed with limited indications: one daunorubicin and two doxorubicin liposomal formulations, that vary significantly in the composition of the liposomes. This important research field will be extensively reviewed and discussed later on. Another concept for improving the efficacy of anthracyclines is that of 'magic bullets', pioneered by Paul Ehrlich that aim at delivering anticancer drugs selectively to the tumor. Also the specific technology of tumor–drug targeting systems by anthracycline conjugates will be described in detail later on.

Chemistry

Structurally, all anthracyclines share a common four-ringed 7,8,9,10-tetrahy-drotetracene-5,12-quinone structure and usually require glycosylation at specific sites for biological activity. The anthracyclines are a subgroup of the aromatic polyketides that form one of the largest families of naturally occurring bioactive compounds comprising 5,000 members, of which 2,000 belong to the anthracycline-type family. Mathematical approaches that consider the detailed basis of structural diversity of these compounds suggest that more than 10,000 theoretical anthracycline-analogs structures could be possible. The general structure of anthracyclines is depicted in Figure 1 that illustrates the partial planar structure of the tetracyclic ring system (ring B,C,D) which represents the chromophore (anthracyclines are red compounds) and includes the quinone structure.

The 7 and 9 position in ring A are important because the daunosamine sugar moiety is linked glycosidically at the 7-position and at the 9-position a sidechain with a ketone group is tethered. The name anthracycline was created in the late 1950s based on the presence of an anthraquinone chromophore and the polycyclic ring system in the chemical structure, which is similar to that of tetracyclines.

The four major anthracyclines in clinical use differ in the residuals R1 to R4. The smallest difference is found between doxo- (DOX) and epirubicin (EPI) which differ only in the C-4 position of the OH-group: in the case of DOX the hydroxy group has an axial orientation in case of EPI an equatorial orientation. This orientation renders EPI a good substrate for human D-glucuranyl transferases, and EPI is therefore conjugated *in vivo* at the daunosamine sugar moity with glucuronic acid, which is not a metabolite known for DOX.

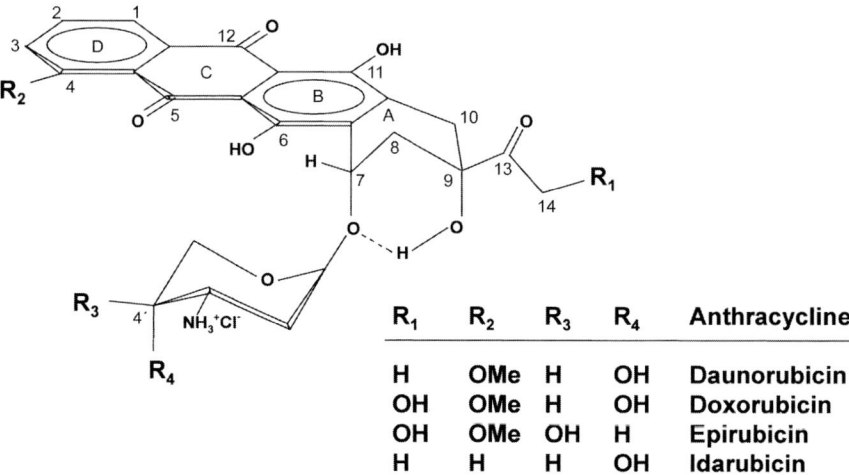

R₁	R₂	R₃	R₄	Anthracycline
H	OMe	H	OH	Daunorubicin
OH	OMe	H	OH	Doxorubicin
OH	OMe	OH	H	Epirubicin
H	H	H	OH	Idarubicin

Figure 1. Chemical structure of the anthracyclines daunorubicin (DNR), doxorubicin (DOX), epirubicin (EPI) and idarubicin (IDA).

The different configurations of the 4'-OH groups in the daunosamine sugar have two important consequences for the pharmacology of the drugs. Differences between DOX and EPI were found in ionization, which affects cell penetration and metabolism. The effect of different configurations of the 4'-OH groups on pKa, lipophilicity and cell penetration have been studied in detail. For DOX the axial 4'-OH group is in close vicinity to the NH_2-group of the daunosamine sugar that allows hydrogen-bond formation between the two groups. The consequence is a higher pKa value for DOX because this interaction facilitates the ionization of the NH_2-group and a lower lipophilicity. In EPI the equatorial configuration of the 4'-OH group prevents its interaction with the amino group and its lipophilicity is therefore not reduced. EPI's 4'-OH group does undergo internal hydrogen bonding with the groups at C5 and D6 of the aglycone system. At physiological conditions anthracyclines with relatively lower pK-values have higher lipophilicity than those with higher pK values. The pKa value of DOX is 8.22 and of EPI 7.7, which is the reason why the partitioning coeffecient (butanol/water at pH 7.4) is different because EPI is more lipophilic. This very small difference in chemical structure has a great influence on the physicochemical properties, the metabolism, and the toxicity of EPI which makes DOX and EPI different in some aspects. In idarubicin, a derivation of its Italian name 4-demetoss**ida**unorubicin, the C-4 methoxy-group in the D ring of the aglycone of DNR is replaced with a hydrogen atom. The missing methoxy group is the only structural difference to daunorubicin. The consequence of this chemical modification is a much higher lipophilicity of IDA compared to DNR and to DOX. This property improves absorption across the gastrointestinal mucosa and enhanced uptake into tumor cells *in vitro*. The same antitumor effect at much lower doses may be related to this fact, IDA is a more potent drug than DNR. Furthermore, IDA is the only anthracycline available for oral administration. Idarubicin is a much better substrate for the ubiquitous located tissue aldoketoreductase than daunorubicin. This is the reason why the biotransformation of IDA to IDAol differs quantitatively from that of DNR to DNRol (see pharmacokinetics).

Pharmacodynamic (PD) properties

In vitro antitumor activity

All anthracyclines discussed so far have demonstrated cytotoxicity against a wide range of animal and human tumor cell lines. Cytotoxicity increases exponentially with both drug concentration and duration of exposure, and maximal lethal effects were demonstrated in the S- and G2-phases of the cell cycle and less or no cell kill in the G1 and M phases. However, at high concentrations cytotoxic effects can be observed in G1 and M phases as well [10].

The cytotoxicity of DOX and EPI in tumor cell cultures (e.g., liver, lung, colon, breast) were nearly identical at equimolar concentrations. No advantage was found with respect to a broader spectrum of activity for EPI except for gastric cancer cells, which were found to be more sensitive to EPI than to DOX [11]. A number of *in vitro* studies with DNR and IDA in animal and human tumor cell lines have demonstrated a higher potency of IDA when cytotoxicity was measured and compared with DNR. IDA was always more potent than DNR at equimolar dose [12]. Interestingly, IDA was tested *in vitro* in several solid tumor cell lines with remarkable cytotoxic efficacy. It was found that idarubicinol, the major metabolite of IDA, had similar activity as the parent drug in these experiments. This phenomenon was not observed for doxorubicinol, epirubicinol or daunorubicinol. It is well known that *in vitro* studies with antitumor agents, and in particular with anthracyclines, do not always predict the antitumor activity *in vivo*. The relevance of the numerous *in vitro* studies with anthracyclines for *in vivo* studies is therefore debatable.

In vivo antitumor activity

In general, the antitumor activity of doxorubicin and epirubicin appears to be similar in various orthotopic tumor models as well as in human tumor xenografts in nude mice. Differences in the spectra of antitumor activity have been noted but it appears that the predictive value for clinical use remains uncertain. Both drugs, DOX and EPI, showed activity against breast carcinoma, small cell lung cancer, and sarcoma and were not active in colon tumors [13]. In non-small cell lung cancer the *in vivo* results showed activity in three quarters of tumors transplanted into nude mice with both anthracyclines, a result which does not correlate with clinical results. The same holds true for melanoma. For this reason *in vivo* evaluations in a large panel of human tumors in nude mice can only give a first indication for future clinical development. There is clearly a limitation of tumor *in vivo* models which do not reflect correctly the tumor biology in humans, e.g., host–tumor interactions in man are not addressed sufficiently in the available models.

For IDA it was shown that this drug has a 4-to-8-fold greater potency than DOX and DNR in leukemias and lymphomas [14]. The evaluation of the antitumor activity of IDA in solid tumors is limited to only a few orthotopic murine tumor models including mammary carcinoma and sarcoma and to human tumor xenografts in nude mice: i.e., breast, lung, melanoma, ovarian and sarcoma. In these *in vivo* models, IDA and DNR showed similar activity. Idarubicinol demonstrated antitumor activity equivalent to that of IDA [15].

Mode of action and molecular biology

The precise mechanism of antitumor action for the anthracyclines is not fully understood. The following chapter summarizes the proposed modes of action of anthracyclines.

Drug–cell membrane interactions

Each drug which is administered iv or po is present with a certain concentration in the central compartment where the amount can be determined (see section pharmacokinetics). To enter the tumor cell, the anthracycline must leave the blood vessels, enter the interstitial tissue und penetrate and cross the cell membrane in order to reach the inner compartment of the cell. The transmembrane movement of the anthracyclines occurs by free diffusion of the non-ionized drug [16]. No active drug carrier is known for the anthracyclines. The daunosamine sugar is partly protonated within the physiologic pH range and therefore both extracellular and intracellular pH has a significant impact on tumor cell uptake of anthracyclines [17]. The uptake of anthracyclines from the extracellular space into the tumor cell is hampered by a pH of 6–6.5 which is often found in tumor masses as small as 1 cm because a protonated anthracycline cannot rapidly diffuse through the cell membrane. If the pH is in the physiologically range in the extracellular space, the anthracycline can cross the cell membrane very easily as non-ionized drug and is then trapped in the cytoplasma/nucleus of the tumor cell by intracellular acidosis as well as rapid binding to intracellular components such as DNA. Interestingly, two other phenomenons with respect to drug–cell membrane interactions are noteworthy. Several tumor cells as well as normal cells feature an efflux pump system, with which several natural products are efficiently pumped out of the cell. This protein, called P170-glycoprotein, is integrated into the cell membrane and has an ATP-binding site in the cell and is an important drug carrier system (from inside to outside) and has been widely discussed as one of the reasons for anthracycline resistance [18]. The second phenomenon is the fact that even anthracyclines which cannot cross the tumor cell membrane show cytotoxic activity. Doxorubicin was covalently coupled to large agarose beads which were unable to enter cells but still exerted strong antitumor effects in cell culture systems. Within this model the antitumor effects are produced at the cell membrane level and could be explained by the generation of reactive oxygen species (ROS) at the cell membrane, which in turn damage the membrane by lipid peroxidation thereby activating important signalling pathways [20]. A semiquinone free radical that is produced by daunorubicin incorporated into the cellular membrane of intact cells has been described [21].

Drug–DNA intercalation

Cytotoxicity mediated by anthracyclines is generally thought to be the result of drug-induced damage to the DNA. Because the drug concentrates in the cell nucleus and is a good intercalator of DNA [22], the drug was thought to exert its activity by DNA intercalation, but this simple explanation is not sufficient to explain the whole spectra of different actions of the anthracyclines. The planar aglycon (without the daunosamine sugar) intercalates with DNA as well, but no antitumor activity was found [23]. The intercalation of anthracyclines with DNA is reversible, no covalent binding is necessary. Hydrophobic interactions, hydrogen bonds to the phosphate groups of the DNA and the insertion of the daunosamine sugar into the small groove of the DNA with an affinity to the CpG-complex and transcriptional active sites of the DNA lead to a fixed drug-DNA-complex with a long half-life [24].

Drug–topoisomerase-II interaction

It has been shown that anthracyclines cause protein-associated breaks and these breaks correlate with cytotoxicity [25]. The reason for these protein-associated breaks are due to fine interactions of the anthracyclines with the topoisomerase-II (TOPO-II), an enzyme that promotes DNA strand breaks and is involved in resealing the breaks [26]. It is possible that the intercalation of anthracyclines induce an alteration in the three-dimensional conformation of DNA that arrests the cycle of TOPO-II action at the point of DNA cleavage, but it may well be that anthracyclines also stimulate TOPO-II-mediated DNA cleavage by nonintercalative mechanisms. A number of studies have shown that anthracyclines induce topoisomerase-II-mediated DNA damage at drug concentrations that are clinically relevant. Furthermore, a good correlation between cytotoxicity and DNA damage was observed. Cell lines which have altered TOPO-II activity exhibit resistance to anthracyclines [27]. Other TOPO-II inhibitors such as VP-16 showed a relative constant relationship between cytotoxicity and protein-associated DNA break frequency, and the anthracyclines exhibits more cytotoxicity per break. Therefore, the interaction of anthracyclines with TOPO-II is an important factor for the cytotoxicity but other mechanisms of action might be important as well. With respect to DNA intercalation and inhibition of topoisomerase-II, the anthracyclines act as chemically inert compounds by their ability to distort the three-dimensional geometry of the targets DNA and TOPO-II. Despite these important modes of actions induced by the unchanged drug, the anthracyclines are chemically very reactive compounds with an extraordinary and fantastic chemistry, not understood in all details yet [8, 28].

One- and two-electron reduction

Free radical formation after anthracycline administration is a major issue for understanding some of the side effects of this class of drugs. The one-electron reduction is crucial for cardiac toxicity. All anthracyclines in clinical use are anthraquinones that can undergo a one- and two-electron reduction to reactive compounds that are able to damage DNA and cell membranes (under certain conditions) [29]. In complex biological systems these reactions are catalyzed by enzymes. Several enzyme systems accept anthracyclines as substrates for a one-electron reduction: NADPH-cytochrome-P-450-reductase in the endoplasmatic reticulum, NADH-dehydrogenase in the mitochondria, xanthinoxidase in the cytoplasma and not identified enzymes in the nucleus. Figure 2 depicts the reaction cascade of this electron transfer.

The one-electron reduction leads to the formation of the semi-quinone free radical which in the presence of oxygen donates its electron to oxygen thus generating a superoxide anion. At neutral pH the main reaction of the super-

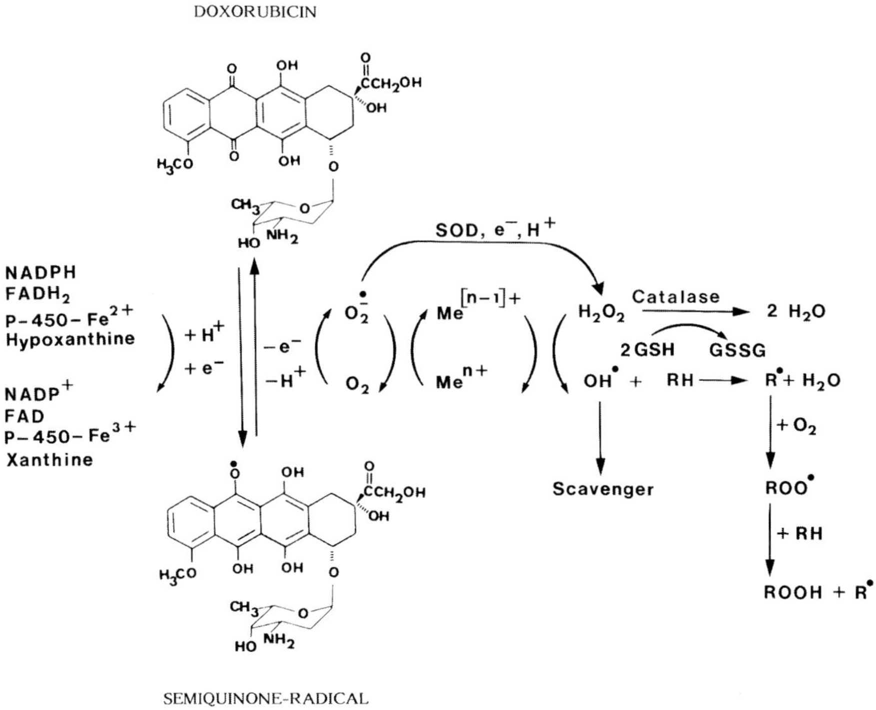

Figure 2. Free radical formation pathway for doxorubicin.

oxide anion is a relatively spontaneous dismutation to yield hydrogen peroxide and oxygen. This reaction can be accelerated by superoxide dismutase. Hydrogen peroxide can undergo reductive cleavage to the hydroxy radical, a very reactive and destructive chemical with an extremely short half-life. The presence of iron seems to be essential for this reaction cascade. Superoxide dismutase, catalase, glutathione peroxide act in concert to reduce superoxide and hydrogen peroxide to water without the formation of the hydroxyl radical. These enzymes are present in many mammalian cells because oxygen radical formation occurs as a result of normal metabolic processes and is a common mechanism of action for many naturally occurring toxins. These enzymes are part of the mammalian defense system against the attack of free radicals. These defense systems are not equally distributed in the various tissues of the body [30]. The activity of these enzymes differs remarkably in human tissues. The unique cardiac toxicity, which is mainly due to free radical attack [31] of anthracyclines, can be explained by lower levels of catalase, high levels of flavin-centered reductases that activate the drug and by low levels of glutathione peroxidase. Taken together, cardiac tissue does not have sufficient defense systems to repel a free radical attack induced by anthracyclines. The hydroxyquinone structure of the anthracyclines represents a site for chelation of many metal irons, especially ferric iron. The overall binding constant of doxorubicin for ferric iron is 10^{33} which is similar to desferroxamine [8]. Iron-anthracyclines complexes can bind to DNA by a mechanism distinct from intercalation. This binding is much stronger compared to mere intercalation. An iron–anthracycline complex is able to react rapidly with hydrogen peroxide to generate hydroxyl radicals that damage DNA. In contrast, the DNA-anthracycline intercalation complex quenches all redox activity of the anthracyclines. Because free radical formations, especially the formation of hydroxy radicals, are strongly dependent on iron, attempts have been made to interfere with the iron metabolism in order to reduce the free radical formation especially in cardiac tissue. The role of radical oxygen species (ROS) in tumor cell kill is not fully understood but there is growing evidence that ROS modulates protein kinase c (PKC), tyrosine kinase activities, contributes to cell cycle block, stimulates Raf-1/ERK mitogen-activated protein (MAP) kinases, and triggers the activation of critical transcription factors, including nuclear factor-κB (NF-κB), a negative regulator of DNA-induced apoptosis [32].

The two-electron reduction of the anthracyclines results in the formation of an unstable quinone methide, which rapidly undergoes further changes to the aglycones as depicted in Figure 3.

These aglycones are formed *in vivo* and do not exhibit anticancer activity. Thus, this pathway leads to inactivation of the drug [28]. The role of the quinone methide as a potential monofunctional alkylating agent and its implications for the antitumor activity of anthracyclines is unknown.

However, in case that cytotoxicity mediated by anthracyclines is exclusively thought to be the result of drug-induced damage to the DNA, mediated by

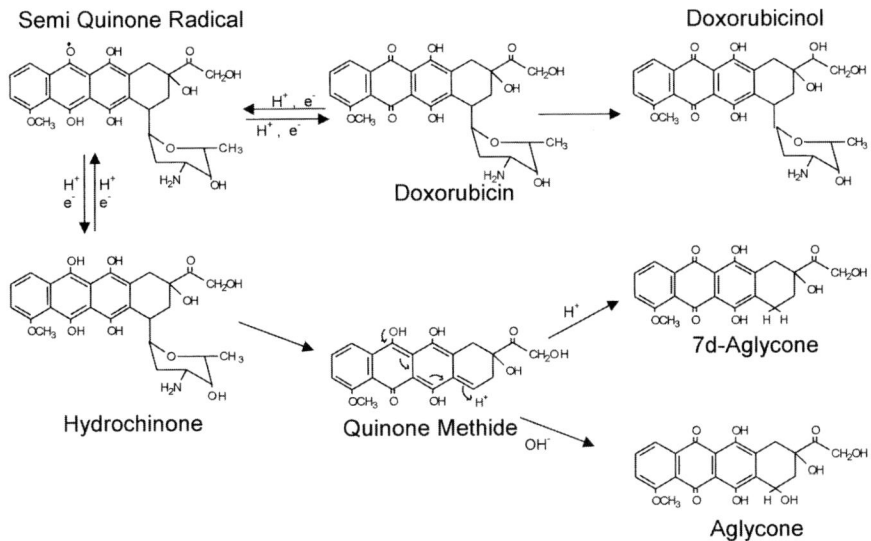

Figure 3. Formation of aglycones via the semiquinone radical-hydrochinon pathway.

quinone-generated redox activity as well as intercalation-induced distortion of the double helix and stabilization of the cleavable complex formed between DNA and topoisomerase-II, it remains unclear how and why such events should induce cell death especially when one considers that DNA interaction may not always be a prerequisite for anthracycline cytotoxicity [33]. Another point of view and explanation for the anticancer activity is that anthracyclines form radical oxygen species (ROS) and trigger apoptotic signals in drug sensitive tumor cells. In a series of research papers [34–36] it was demonstrated that tumor cell response is highly regulated by multiple signalling events and transcription factors including a sphingomyelinase-initiated sphingomyelin-ceramide pathway, mitogen-activated kinases and stress-activated protein/c-Jun N-terminal kinase activation, transcription factors such as nuclear factor-κB (NF-κB) and the Fas/Fas-ligand system. The characterization of pathways involved in the mechanism of action of anthracyclines remains incomplete at present. An overview of the mode of action of doxo- and daunorubicin illustrates some aspects from the intricate field of cell and molecular biology and is depicted in Figure 4.

In this figure the central role of ROS induced by anthracyclines is highlighted. The molecular cellular pharmacology will allow new deep insights

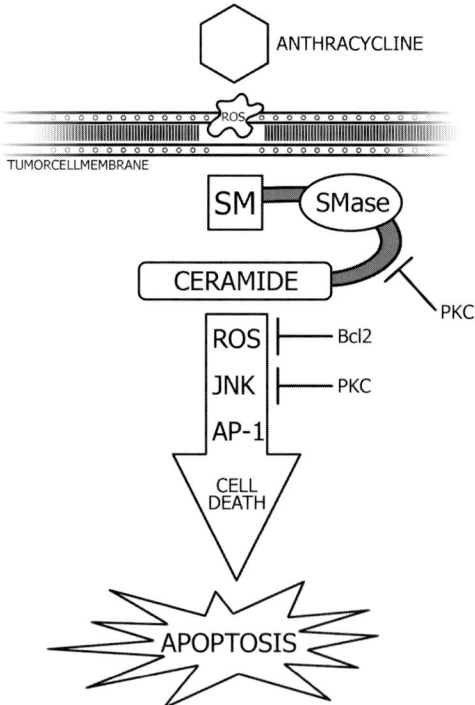

Figure 4. Anthracycline-induced apoptosis pathway.

into the very complex reactions explained in the context of signalling pathways which trigger life and death [20].

Pharmacokinetic (PK) properties of the anthracyclines

Plasma concentrations and distribution

One of the special features of the anthracyclines is the pattern of visible light absorption and fluorescence of each compound. The UV/VIS-spectra of anthracyclines reflect the number and positions of hydroxyl groups in the ring system. The fluorescence spectra are characteristic for each type of ring system and permit a specific detection of anthracyclines in extracts of biological fluids or extracts from tissues following a special extraction procedure. Anthracycline interactions with DNA, RNA and proteins result in fluorescence quenching, which is reversible when the appropriate solvents that release the anthracycline from binding sites on macromolecules are present. The assay procedures in use exploit this feature of the anthracyclines for their very sensitive and specific detection. During the last 25 years, the extraction procedure has been improved by changing from liquid–liquid to liquid–solid procedures, the material of the high performance liquid chromatography (HPLC) columns has been refined and the fluorescence detector systems exhibit higher sensitivity thus lowering the limit of detection. The parent drugs as well as the metabolites (up to seven in case of EPI) can be analyzed with one single run [37, 38].

The pharmacokinetic properties of anthracyclines have been evaluated following intravenous administration to cancer patients with advanced diseases. Although DNR is the oldest drug, PK results for this drug have been primarily generated in comparisons with IDA. The first description of PK of DNR was published in 1971 using tritiated DNR as well as fluoresence detection [39]. The PK of DNR were characterized by a large volume of distribution (about 1,000 L), a long plasma half-life and a urinary excretion of fluorescent substances of about 13% within 7 days. Cellular accumulation of DNR occured quickly, with cell:plasma concentration ratios of about 400 at the end of a DNR infusion and 900–1,600 at the end of sampling 22 h later. The PK results of DNR and DOX of these early days were described and reviewed in 1983 [40].

The pharmacokinetics of DOX was first described during the first clinical studies in 1970s [41] but more sophisticated studies were performed in the 1980s especially in comparison with EPI [42–46]. Thus the most exciting stimulus for detailed PK studies of anthracyclines was the development of anthracycline analogues with systematic studies being performed in order to evaluate differences [47, 48]. The pharmacokinetics of all these drugs are dominated by a fast tissue and plasma protein-binding and different metabolism. During the early distribution phase, drug levels rapidly fall as the anthracycline gains ready access to all tissues except the brain. The blood–brain barrier prevents a distribution into the brain, the cerebral fluid and the meningeal tissue except in the case of IDA [49]. During the short distribution phase most of the drug binds to DNA throughout the body within minutes. The triexponential disposition of all anthracyclines are qualitatively similar, but IDA plasma dis-apperance is fastest for IDA. In addition, plasma protein-binding is high, approximately about 80%. Thus, most of the drug is bound after a short time and the pool of free anthracycline represents a very small fraction which has not been evaluated in depth within pharmacokinetic studies. Every concentra-tion *versus* time curve of anthracyclines can be described with a two- or a three compartment model. In most of the numerous publications on pharmacokinet-ics of anthracyclines a triexponential equation was used although the reasons for this selection were not stated. The number of compartments depends on the used analytical assay, the detection limit and the number of samples used for the calculations. If only one or two samples were taken within the very fast dis-tribution phase, the fast α-half-life will be lost; furthermore, if insensitive assays are used, the elimination phase is inaccurately decribed. Thus, many samples collected over at least 48 h [43, 45] (even better would be a sample period of up to 168 h) [44, 47] and a very sensitive assay are necessary for a valid description of the PK of anthracyclines. The analytical methods have been improved over the last 20 years, and the major analytical procedure is a solid-phase extraction procedure followed by a separation of the anthracy-clines (parent drug and metabolites) on a reversed phase column in a HPLC system with highly sensitive and specific fluorescence detection. The limit of detection is 1 ng/ml, in some laboratories 0.1 ng/ml were reached allowing a longer tracking of the drugs in plasma [33, 38, 50].

Numerous publications on pharmacokinetic parameters exist for all anthra-cyclines. The major route of all drugs is intravenous application as a short infu-sion within 5–10 min. This schedule is the most prominent route and infusion time, but protocols also exist where DOX is given as a 96 h infusion. Table 1 summarizes the major pharmacokinetic parameters after an iv administration of DOX, DNR, EPI and IDA.

The values shown in this table summarize the numerous data published in the last 15 years. The three half-lives reflect the distribution, an intermediate and the terminal/elimination phase, the Cltb means total body clearance, and Vdss means volume of distribution at steady state. The second elimination

Table 1. Key PK data of the four mostly used anthracyclines

Anthracycline	t½ α (min)	t½ β (h)	t½ γ (h)	Cl p (ml/min/m²)	Vdss (L/m²)	t½ γ (h) (-ol)	AUC ratio
Daunorubicin	6	0.9–2.5	30–45	800	2000	27	4.0
Idarubicin	10	1.0–3.0	15–23	1200	1000	58	2.6
Doxorubicin	4	0.5–1.5	24–36	550	1300	29	0.6
Epirubicin	3	0.9–1.6	18–29	1800	1800	30	0.3

half-life shown in the table is that of the 13-dihydrometabolites (DNRol, IDAol, DOXol and EPIol), the area under the curve (AUC) ratio is the AUC of the -ol metabolite devided by the AUC of the parent drug (e.g., DOXol/DOX).

The PK of DNR and IDA have only been studied intensively in leukemia patients, which has become the major field of application of these two drugs. Because IDA was given at a 4–5 times lower dose than DNR, the peak plasma concentration after IDA iv is approximately 5 times lower [51]. The enormous cellular uptake and distribution in deep tissue compartments is reflected by the very large apparent volume of distribution at steady state of approximately 1,725 L/m^2 for DNR and 1,756 L/m^2 for IDA. The terminal half lives were similar: 47,4 and 42,7 h in case of DNR and IDA using a three-compartment model. The AUC was 3–4 times higher in the case of DNR, reflecting the different dosages whereas the clearance of both drugs was similar [48]. The intracellular concentrations of DNR and IDA are similar at equimolar exposure. The amount of DNA single strand breaks at equimolar exposure is highest for IDA when compared to DNR, DOX and EPI. Thus, other factors than just concentration must play a role for explaining the much higher potency of IDA over DNR.

After bolus administration, plasma DOX and EPI levels undergo a decay which can generally be best fitted by a three-compartment model. At equivalent doses, the c(t)-curve of EPI is always below the curve of DOX. The peak plasma concentration after a bolus injection is extraordinary high (time sensitive parameter; an exact time for a bolus has been never stated in the publications) and will fall within minutes by several orders of magnitude, thus distribution into deeper tissue compartments occurs rapidly. The volume of distribution as well as the clearance are high which reflects the rapid fade of the drug from the plasma compartment into deeper tissue compartments. The clearance from the plasma compartment is faster for EPI than for DOX. The AUC of the drug distribution phase is about 40% of the total AUC. The elimination of both drugs is mainly by the bile, and excretion via the kidney is less than 10%. The elimination phase is remarkably long and correlated to plasma levels. EPI has a shorter elimination half-life than DOX due to a higher plasma clearance which is explained by its metabolism. Since most of both drugs are bound to tissue, the total half-life of both drugs (time needed to excrete half of the drug out of the body) is an interesting quantity. In the case of external bile shunting it is possible to calculate such a value. A few cases are described and 50% of DOX is approximately eliminated from the body after about 7 days whereas 50% of EPI is lost after about 4 days [52, 53]. These figures are consistent with results of the PK of both drugs in white blood cells. The tissue (WBC) half-life was about 5 days for DOX and 2 days for EPI [54, 55]. Differences in tissue half-lives were also described in a mice study [56]. The problem with animal PK results is their inability to glucuronidate EPI [57]. That is the reason why animal PK and metabolism studies are not predictive for human PK and metabolism of EPI. A prolongation of the anthracycline administration to 4 h will reduce the peak plasma concentration by a factor of

25 but basic PK parameters such as volume of distribution, clearance, AUC and terminal half-life are not significantly altered by such a change in the time schedule. The same holds true when comparing the PK parameters of DOX after bolus injection and 6½ day continuous infusion [42]. Dose and AUC are correlated up to 150 mg/m^2 in dose escalation studies suggesting linear pharmacokinetics [58].

Metabolism and elimination

Important metabolic pathways have been identified for the anthracyclines. The stereospecific reduction of the ketone at carbon-13 yields 13S-dihydro derivatives which are named after the parent drug with the suffix -ol (dauno-, ida-, doxo- and epirubicinol). This metabolism is catalyzed by ubiquitous cytoplasmatic aldoketo reductases [59]. The aldoketo reductases have different substrate specificities and optimum pH. Dauno- and idarubicin are converted to a higher degree than doxo- and epirubicin (see Tab. 1). Plasma levels of DNRol and IDAol exceed the plasma levels of the parent drugs within a short time (less than 3 h) with longer elimination half-lives compared to DNR and IDA whereas DOXol and EPIol concentrations remain below the c(t)-curve of DOX and EPI with similar or shorter terminal half-lives. The ratio of the AUC of the metabolite and the parent drug is about 2–5 for DNR and IDA and 0.3–0.5 for DOX and EPI. Nucleated blood cells accumulate anthracyclines at 200–500 higher levels than those present in plasma, but the 13-dihydro derivatives are only found at low concentrations in these cells when compared to their parent drugs. These metabolites are obviously not taken up to the same extent as the parent drugs.

Four different aglycones can be detected after injection of DOX and EPI. The deglycosylation of these two anthracyclines can result from a reaction sequence depicted in Figure 3. For DOX and EPI, the aglycone as well as the 7d-aglycone are generated and the same can occur for DOXol and EPIol. All four aglycones can be detected using very sensitive HPLC systems [44, 45]. These metabolites, generated by a complex biotransformation including free-radical formation, has been decribed in mice as well as in man [60, 61]. The importance of the detection of these metabolites in plasma samples are not well understood to date. It is known that these substances can be produced as artefacts during the sample processing. Nevertheless, with modern analytical equipment it is possible to detect these metabolites in nearly all plasma and tissue samples from patients. It is known that the aglycones and 7d-aglycones are not cytotoxic [23]. For IDA and DNR no aglycones have been decribed in publications on pharmacokinetic and metabolism. In one of the first publications on the metabolism of doxorubicin [62], conjugates due to sulfation and glucurondation at the 4-position by demethylation and a Phase-II conjugation at that site has been described in experiments from urine samples. This result has never been reproduced although experiments were performed with sulfatase and glucuronididase to detect these metabolites (Mross and Maessen, 1987; unpublished results). The difference in the PK of EPI compared to DOX (lower AUC and higher clearance of EPI [factor of 2] compared to DOX at equimolar doses) was large and needs to be explained. The first description of an additional metabolism pathway in man was published in 1983 [63]. It took several years to isolate sufficient amounts of the two glucuronides epirubicin-glucuronide (EPI-Glu) and epirubicinol-glucuronide (EPIol-Glu) which are necessary for the calibration of the HPLC methods. Because of the hydrophilicity of these two metabolites, the assay methods had to be adapted (different column material, other buffer systems, and extraction procedures). The AUC of EPI-Glu exceeds that of EPI and is the reason for the much higher clearance of EPI in comparison to DOX. EPIol-Glu and EPIol are relatively minor metabolites. Both glucuronides are excreted by the urine [33, 45]. The metabolism pattern of the four anthracyclines is shown in Table 2. All metabolites leave the body via biliary excretion which is the major excretion pathway and to a much less extent via the urine. The inability to visualize fluorescence anthracyclines in fecal specimens has been presumed to be due to significant alterations of the chromophore of the drugs by intestinal microbial metabolism and the aggressive environment in the gut. Because no clinical relevant toxicity in the gut has been observed after anthracycline administration, it can be assumed that the degradation products in this special compartment are non-toxic.

In summary, the metabolism of clinically established anthracyclines is similar and differs only quantitatively with respect to the reduction at C-13 position by aldoketo reductases. The bioreductive cleavage of the daunosamine sugar moiety leading to the 7-deoxy aglycones has been confirmed in several laboratories and can be linked to the free radical formation chemistry of the chromophore. Finally, the smallest but fundamental difference between EPI and DOX, the epimerization of the 4'-OH group, has remarkable consequences for the pharmacology of EPI which is more susceptible to metabolic conjugation at this site and significantly modifies the pharmacokinetic behavior.

Table 2. Metabolism of the four anthracyclines

Type of Metabolism	DNR	IDA	DOX	EPI
Reduction at C-13 (-ol)	++	+++	++	+
Reduction at C-7 (7-deoxy-aglycon)	?	?	+	+
Hydrolysis at C-7	?	?	+	+
Glucuronidation at 4'-Daunosamine sugar	–	–	–	+++

(– not detectable, + small amounts, ++ significant amounts, +++ large amounts representing the dominating pathway,? not really known)

Effects on healthy tissue

The toxicity of all anthracyclines can be devided in acute, subacute and chronic toxicity. The acute and subacute side effects are haematopoetic (neutrophils > platelets < erythrocytes), gastrointestinal toxicity (mucositis, stomatitis; DOX > DNR), skin necrosis in case of paravasation (DOX > EPI > DNR > IDA) and fatigue. Chronic cumulative toxicities are hair loss, cardiac failure and secondary leukemia. Bone marrow suppression after therapy with DOX and EPI are very similar at equal doses. The maximum toxicity is observed after 7–12 days in the neutrophil counts, less affected are the platelets and the erythrocytes with full recovery after 14–21 days. DNR and IDA are more potent with respect to myelosuppression but the major indication is treatment of leukemias where complete aplasia is still the goal and can be achieved with both drugs. IDA is much more potent, thus less of the drug is necessary to reach this goal. The GI-tract toxicity is most pronounced for DOX and somewhat reduced for EPI. After DNR treatment less gastrointestinal toxicity was seen than after therapy with DOX in a comparative trial. This is one of the reasons why DNR was preferred in the treatment of acute leukemias in order to reduce clinical problems of aplasia and GI-tract toxicity which is a difficult combination to handle because of problems with infections due to the disturbed gut–blood barrier. The mortality rate due to such problems was higher for DOX although efficacy was the same.

Effects of disease and age on anthracycline PK

Renal impairment seems to have have no influence in the clinical use of anthracyclines despite the fact that in early publications it was described for DNR that 10–20% of total fluorescent material was found in the urine [39]. For IDA the analysis of variance indicated a significant correlation between IDA plasma clearance and creatinine clearance. The terminal half-life of IDA and IDAol was somewhat longer [47], but the total amount excreted via the urine is still low, approximately 5% [64, 65]. The excretion of DOX and metabolites is within the same range whereas for EPI the additional glucuronidation pathway with the formation of hydrophilic glucuronidated metabolites leads to a higher excretion of EPI including metabolites into the urine of around 10–15% [44, 45].

An important finding was the recognition of exaggerated toxicities (mucositis and myelotoxicity) of DOX patients with impaired liver function. The first clinical-pharmacological correlation of DOX PK and hepatic function was elucidated in 1974 [66]. In this publication, patients with normal bilirubin as well as patients with bilirubin levels >3 mg/dl received 60 mg/m^2 DOX. The AUC was 3-times higher in case of hepatic function impairment and the terminal half-life was more than 10-times longer (>300 h). In a study of patients with hepatic dysfunction and stepwise reduced DOX dosage, based on the degree of liver abnormalities, the toxicities were indistinguishable from those of patients without hepatic impairment. The terminal half-life of DOX was identical in all patients [67], i.e., the total body clearance of DOX, expressed as the ratio of dose to AUC, progressively declined with increasing bilirubin. Bilirubin is not the only parameter that can influence the pharmacokinetics of DOX. Liver metastasis *per se* together with aspartat-amino-transferase (ASAT) elevation but with normal bilirubin levels have a significant influence on the clearance and the elimination half-life. The clearance was reduced by 40% and the elimination half-life was prolonged by 35% [33, 68]. Up to now it has not been possible to derive universally applicable schemes for DOX dosage reduction in patients with liver impairment and it is a matter of clinical experience for correctly scheduling DOX in patients with liver impairment. For EPI, similar results were observed. In patients with moderate-to-severe hepatic impairment reduced plasma clearance with elevated systemic drug concentration have been described. The clearance was reduced by 60%, but the elimination half-lives were not different [69]. One research group has shown that in patients with elevated AST (SGOT) levels, EPI clearance was significantly impaired and correlated with AST but not with bilirubin. The authors suggested in 1992 that serum AST rather than bilirubin may be the best indicator for dosage adjustment of EPI [70]. A survey of prescription methods for anthracyclines in patients with hepatic impairment used by oncologists in the UK showed a wide variation in the dose that oncologists prescribed [71]. These results from a questionnaire showed the need for a new, widely accepted anthracycline dose modification scheme for patients with liver dysfunction. In a recent paper this group has published results from a population pharmacokinetic project and developed a formula for the EPI clearance, including AST levels which leads to a dosage guideline which is practical and effective [72, 73]. The proposed dosing guideline, which includes AST level as a guiding tool, should reduce variability in systemic exposure to EPI more efficiently than approaches used in the past. In addition, they do not require adjustment according to body surface area which reduces dosage preparation time as well as prescribing and dispensing errors.

In rats, peak plasma levels and AUC in serum and several tissues were 1.5–2 times higher in old rats when compared to young rats. Young rats died with the same rate but at twice the dose of old rats [74]. The effect of age on the PK of anthracyclines in man has not been thoroughly investigated in prospective trials. Altered regional blood flows in different organs in the elderly are known, and it has been shown that initial concentrations of DOX in the distribution phase after intravenous administration are higher in elderly patients. This was explained by a decrease in the clearance in the distribution phase. The volume of distribution remained constant [75]. Possible factors responsible for the variability of PK parameters of anthracyclines can be deduced from population analyses. A significant proportion of the variability in clearance could be attributed to sex and also to age in women. The clearance of a 70-year old women is 35% less on average than the clearance of a 25-year

old man [76]. No recommendations can be made for dose reduction in the elderly. A healthy old person can receive the full dose of anthracyclines if no
severe co-morbidities exist and no other intensive medication is prescribed.
There is a linear correlation with co-morbidity, number of pills and organ dys-
functions. Organ dysfunction as well as a multiplicity of co-drugs have to be a
matter of concern in using full dosages. It is a matter of experience and an
expert decision to treat patients with full dosages.

Insights into the interaction of anthracyclines and other drugs are rare. No
systematic pharmacokinetic studies have been performed. Verapamil has an
influence on PK parameters of EPI as well on the metabolism. The AUC of
EPI is lower under the influence of verapamil whereas the metabolism of the
glucuronides are enhanced [77]. Phenytoin, a drug with a high potential of
interference due to liver enzyme induction, increased the elimination of doxorubicinol in animal experiments and as a consequence the AUC of DOXol
declined. These data indicate that phenytoin induces DOXol metabolism [78].
It can be assumed, that drugs with known liver enzyme induction can influence
not only the metabolism by aldoketo reductases but also glucuronidation.
Thus, alteration of the metabolism of EPI is likely to take place. Another
drug–DOX interaction was described with histamine-2 blockers. In rabbits the
conversion of DOX to DOXol was blocked [79]. The very high variability of
PK parameters of the anthracyclines can partly be explained by the high
amount of co-medication that is used in the complex combination chemotherapy plan for leukaemia, lymphoma and solid tumors.

Clinical toxicity

Hematopoetic toxicity

Bone marrow suppression is a common feature of all anthracyclines and is the
dose limiting toxicity after bolus dose administration. Myelo- and thrombocytopenia are most prominent after each treatment course with maximal toxicity
after 7–10 days (sometimes delayed) with rapid recovery thereafter. The time
to nadir and the recovery are dose dependent. The antiproliferating effect of
the anthracyclines depends on the proliferation status of the bone marrow cells
as well as of the tumor cells [80, 81]. Quiescent, but potentially proliferating
cells are relatively insensitive and can explain the recovery of hematopoesis
after anthracycline-induced bone marrow hypoplasia. DNR and IDA are the
backbone in the treatment of acute leukemia. The dose used for antileukemic
treatment is always a dose which induces full aplasia with a much slower
recovery of all hematopoietic cells. The antiproliferating effect on human bone
marrow clonogenic cells is independent of the infusion rate. For exerting cytotoxic effects on these cells, tightly-bound cellular anthracycline levels are necessary. These levels can be reached after rapid bolus injection as well as after
long(er)-time infusion [82].

Cardiac toxicity

The cardiac toxicity observed after administrations of anthracyclines is unique in terms of pathology and mechanism. Both acute and chronic cardiac toxicity can be observed. The acute toxicity represents a range of arrhythmias which can include a pericarditis–myocarditis combined with congestive heart failure [83]. This kind of toxicity is rare and not dose dependent. Most of all arrythmias will never be seen because most anthracycline administrations in the in- and out-patient setting are performed without any cardiac monitoring. The arrhythmias are only seldom noticed by the patient and occur within a short period of time after administration without any symptoms. This is not the cardiac toxicity which is generally problematic. Quite different is the cumulative cardiac toxicity which is best documented after repeated bolus doses of DOX 60 mg/m^2 every 3 weeks. With this schedule, cardiac toxicity develops as a result of cumulative injury to the myocardium. The pathology of this type of toxicity has been described in detail [84]. With each dose there is a progressive injury to the myocardial tissue that is characterized from grade 0 to grade 3. Grade 0 means no change from normal, grade 1 scanty cells with early myofibrillar loss and/or distended sarcoplasmatic reticulum, grade 2 groups of cells with marked myfibrillar loss and/or cytoplasmatic vacuolization and grade 3 diffuse cell damage with total loss of contractile elements, organelles and mitochondria, and nuclear degeneration. Figure 5 shows the histomorphological changes after doxorubicin therapy representing grade 3 toxicity.

This pathology is unique to the anthracyclines and allows the pathologist to accurately distinguish this cardiac toxicity from other processes. The clinical risk of congestive heart failure (CHF) is small at total doses below certain thresholds. A 5% risk of developing a symptomatic CHF can be deduced from Figure 6. The cumulative doses are 550 mg/m^2 for DOX, 800 mg/m^2 for DNR and 900 mg/m^2 for EPI [33, 85–88]. For IDA no such data are available. The 5% risk for CHF was estimated within the range 120–240 mg/m^2. The above mentioned data were published in the 1970s and 1980s. Figure 6 shows the incidence of CHF related to cumulative anthracycline doses.

The lin-log plot shows in principal similar curves for DOX and EPI but EPI's curve is shifted parallel to higher dose levels which reflects the higher dose necessary to damage the heart to the same degree than after DOX. The slope of the DNR curve is not as steep as those for EPI and DOX.

Results from the 1990s and during the last years have corrected these relative high cumulative doses to lower levels. For EPI a CHF incidence of 14% was described recently at 1,000 mg/m^2 [89] and those patients who had received 850 to 1,000 mg/m^2 EPI had a risk of CHF that further increased from 11% after 1 year to 20% over a 5 year period [90]. For DOX the CHF incidence levels in adults were also recently corrected. Cardiac events were defined as one of three changes in LVEF values compared with baseline as well as clinical CHF. A retrospective analysis of three trials found a risk of

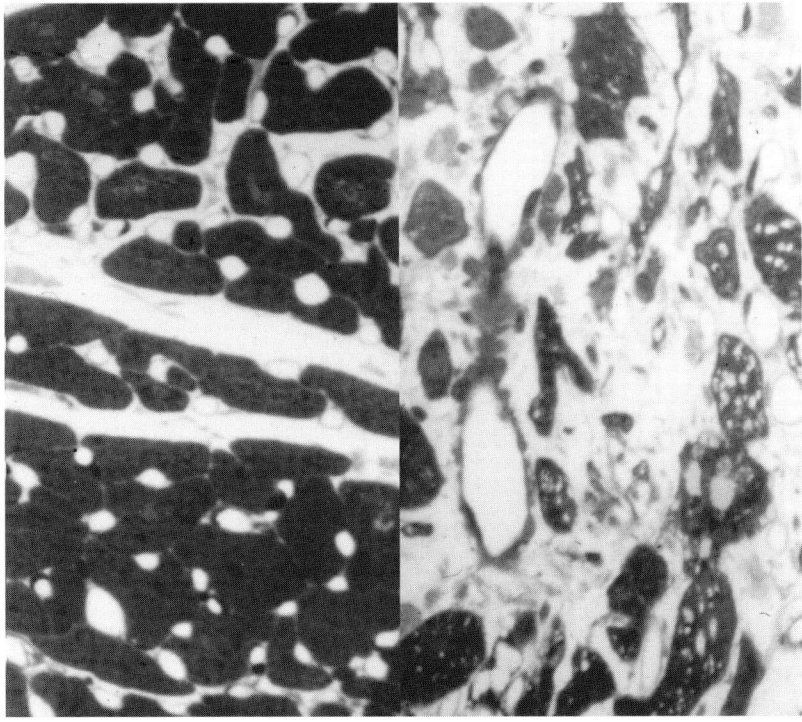

Figure 5. Left side normal cardiac tissue; right side damaged cardiac tissue after anthracycline treatment.

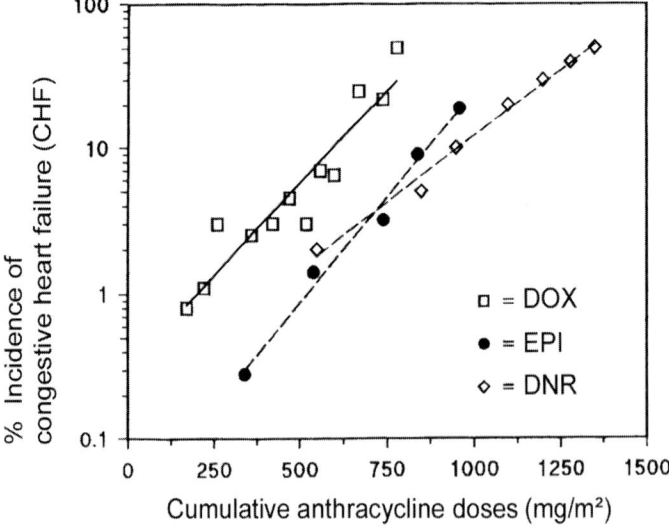

Figure 6. Incidence of congestive heart failure (CHF) in relation to the cumulative anthracycline dose.

26% at 550 mg/m^2 for DOX-related CHF, a 5% incidence has been observed at 400 mg/m^2 [91]. This analysis showed that LVEF measured by multiple gated acquisition scan, a noninvasive tool for assessing heart function (MUGA), which was reported to be a good predictor of CHF [92], may not be a very reliable factor. A reduction of 30% in LVEF was considered to be the cut-off level for increased risk of CHF but two-thirds of all patients who developed a CHF actually had a reduction <30% in LVEF. In pediatric oncology the problem of anthracycline cardiotoxicity is even greater than in adults because nearly all childhood cancers are treated with DOX or DNR-including regimens and two-thirds of children with cancer achieve long-term survival, which means they will experience such toxicity during their lifetime. In a recently published paper reporting results from a prospective longitudinal study, first significant changes of the end-systolic wall stress (ESWS) from cumulative DOX/DNR doses of >250 mg/m^2 were described. A younger age at treatment with these anthracyclines was associated with increased deterioration of ESWS [93]. These results were observed in asymptomatic children. It remains unclear how to interpret such results precisely, but all studies confirm that the dominant predictor of late cardiac dysfunction is cumulative anthracycline dose.

In the USA, many efforts have been undertaken to evaluate cumulative carditoxicity including the morphological monitoring of cardiac tissue by serial endomyocardial biopsies [94] in order to prevent heart failure earlier than by regular monitoring of heart function. The idea was that morphological changes can be seen earlier, especially before CHF symptoms occur, because of the reported structure–function relationship [95]. The best method for studying cardiac performance remains controversial. In the past, cardiac monitoring was performed using resting left ventricular ejection fraction by gated blood-pool imaging (MUGA scans) and left ventricular fractional shortening with echocardiography. Especially, echocardiography has been considerably improved in the last two decades, but longitudinal studies still need experienced investigators. The latest advances in monitoring cardiac function were achieved through use of contrast-enhanced CT-scans as well as dynamic-contrast-enhanced magnetic resonance imaging (MRI) tomography which allows observer-independent evaluation of the cardiac function. Results with these new technologies in the context of anthracycline-induced cardiac monitoring have not been published up to now.

There are further risk factors besides the cumulative dose known for the development of CHF. Age is the only accepted risk factor, younger persons are more vulnerable than adults [88, 93, 94]. Pre-existing heart diseases and cardiac irradiation are other risk factors, but have not been extensively validated.

There are different potential approaches of preventing anthracycline cardiac toxicity: a) alteration of dosing, b) administration of protective agents, and c) the development of less cardiotoxic anthracycline analogs.

Alteration of dosing is obviously the simplest method. Instead of the bolus injection administration mode which results in extremely high peak plasma

levels and thus high concentrations of anthracyclines in cardiomyocytes causing a ROS burst in these sensitive cells which cannot be detoxified with their
own free radical scavengers SOD, glutathione, a prolonged infusion time
would minimize this ROS burst. It has been shown that weekly schedules of
DOX or continuous infusion significantly reduce cardiotoxicity [97, 98]. The
key pharmacokinetic parameters AUC, clearance and terminal half-life are
essentially the same after bolus injection and continuous infusion [42], the
only drastic change is the reduction of peak plasma concentrations in plasma.
As highlighted in the molecular biology of the mode of action and the chemistry of anthracyclines, ROS play a crucial role and free radical formation represents one of the explanatory pathways leading to DOX-mediated apoptosis
of myocytes [99]. During the complex biotransformation of anthracyclines,
superoxide is produced which in turn upregulates nitric oxide synthethase
(eNOS) transcription in endothelial cells and myocytes. Redox-metal chela-
tors inhibit DOX-induced apoptosis, suggesting a central role for ROS for the
deleterious effect of activated-oxygen species resulting from anthracycline-
derived free radicals [100]. This observation led to studies with antioxidants
and other protective agents, including alpha-tocopherol [100], N-acetylcysteine [101] and a bispiperazinedione ICRF-187 (razoxane). The data from a
prospective randomized clinical study support the hypothesis that ICRF-187
protects against the development of chronic doxorubicin-induced cardiac toxicity. ICRF-187 does not add toxicity to the chemotherapy and it does not alter
the antitumor activity [102]. This drug has been registered in the USA and in
France with a restricted indication. In all curative cancers, the drug is not
implemented because it cannot be ruled out that a loss of anticancer activity
(even when it is only a small reduction) can result. In Europe it has never been
used outside clinical trials. The third possibility to reduce the potential of cardiac damage is to develop analogs with less cardiac toxicity. It has been shown
that modifications at the chromophore at position 5 (introduction of an imino
group) leads to analogs with nearly no cardiac toxicity. After such a modification cardiac toxicity is considerably reduced showing clearly that modification of the anthracycline molecule can drastically modify toxic effects [103].
This modification was not successful in clinical practice but epirubicin with
its unique metabolic pathway, described in the section pharmacokinetics and
metabolism, has shown reduced cardiac toxicity [104]. At equimolar dosage
cardiac damage was reduced by 50% compared to doxorubicin. Thus, as long
as clinicians use the same dose and switch from DOX to EPI, a marked reduction in cardiac toxicity is the result. The normal dose of DOX in lymphoma
and breast cancer treatments is 50 mg/m^2 in cyclophosphamide, doxorubicin,
vincristin and prednisone (CHOP) as well as in FAC or 60 mg/m^2 in AC,
which are both combination chemotherapy treatment plans often used with
curative intension.

Dermatologic toxicity

Skin toxicity after systemic anthracycline administration is frequently seen [105]. Skin eruptions as part of a hypersensitivity reaction can occur but are much less common than with other well known notorious sensitizers such as penicillin or phenytoin. The most common cutaneous side effect is diffuse hair loss which is almost complete after 2–3 treatment cycles. Alopecia is a toxic effect of the anthracyclines on the rapidly dividing cells of the hair shaft. Scalp hypothermia can protect from complete hair loss, which often has a severe emotional impact for some patients, especially women. The toxic effects of anthracyclines on the hair bulb are almost reversible but regrowth occurs after a delay of several weeks after completion of anticancer therapy. Scalp hair grows only about 1 cm/month, even if the patient receives no further chemotherapy and maintains good nutritional balance. Stomatitis is another distressing side effect. The normal oral mucosa in maintained by rapidly divid- ing cells, and consequently is very susceptible to the cytotoxic effects of anthracycline therapy. The amount of damage is drug dose related and sched- ule dependent. A damaged mucosa features a high risk for infection because normal mucosa sufficiently protects against entrance of microbes. Thus a clear increase of infection problems correlates with the severity of stomatitis. The only specific treatment for anthracycline related stomatitis is reduction in dose. If this is not possible (in case the cancer can be cured), non-specific measures similar to symptomatic treatment of aphthous ulcers should be used. Extravasation of anthracyclines leads to a burning sensation during infusion lasting for hours or days with different intensity. Anthracycline extravasation can cause a full thickness loss of skin above the affected area. In areas of little subcutaneous fat such as the dorsum of the hand and around joints, severe damage to nerves, tendons and muscle can occur as well as severe local tissue necrosis. The optimal management of anthracycline extravasation remains unclear. Topical applications of DMSO has been reported to prevent doxoru- bicin-induced skin ulceration in the skin of rats and pigs [106]. This procedure was introduced into the clinic and remains the standard care in case of extrava- sation of anthracyclines [107–109], but in case of severe damage with painful necrosis surgical debridement is necessary to interrupt the progressive ulcera- tion process. In case of mediastinal extravasation of daunorubicin due to a malplaced central venous catheter a conservative therapeutic approach is fea- sible [110]. Despite numerous mediastinal complications such as chest pain, cough, pleural and pericardial effusions, dysphagia, thyrotoxicosis and recall phenomenons during additional anthracycline administrations, the only long- term sequelae were moderate costophrenic adhesions. Changes in the finger nails during anthracycline therapy may include pigmentation. The pigment is deposited at the base of the nail and advances outwards as the nail grows. With intermittent therapy, transverse dark bands alternating with bands of normal colour appear and correlate with the time points when drugs were administered as nail material was synthesized [111, 112]. Anthracyclines are capable of severly damaging tissues that have received radiation exposure. This reaction, which differ from a drug's usual toxicity, is termed 'radiation recall' phenomenon. A possible results of radiation-anthracycline interaction is erythema followed by dry desquamation. This can be a painful period if it occurs and needs a sophisticated drug management for pain control.

Secondary cancer induction

As long as anthracyclines are only used in palliative treatment situations, treatment-related AML (t-AML) and their incidence cannot be exactly determined as all patients will die within a period of time which is in general too short to develop an acute leukemia. Because nearly all cytotoxic agents interact with DNA, a mutagenic risk is inherently present. Dauno- and idarubicin are still in use as therapeutic agents for the treatment of acute leukemias and it remains impossible to calculate any incidence of t-AML. Doxorubicin-containing regimens have been used in the adjuvant setting in high-risk breast cancer patients for the last two decades. Epirubicin-containing regimens were used in the last decade, in Europe earlier than in Canada and the USA, because epirubicin became a FDA registered drug only some years ago. Leukemia is a major complication of cancer therapy that has been closely related to chemotherapy with alkylating agents [113], but an increased risk of leukemia was also described for topoisomerase II treatment in germ cell tumors [114]. In 1992, a first report was published indicating a higher leukemia risk after epirubicin treatment [115] which had not been demonstrated for dauno-, ida- and doxorubicin so far although all drugs target DNA topoisomerase II. There are two forms of t-AML. Alkylating agents cause t-AML characterized by antecedent myelodysplasia, a mean latency period of 5–7 years and complete or partial deletion of chromosome 5 or 7. The risk is related to the cumulative alkylating agent dose. DNA topoisomerase II inhibitors (epipodophyllotoxins, anthracenedione and anthracyclines) cause leukemias with translocations of the MLL gene at chromosome band 11q23 or, less often, t(8;21), t(3;21), inv(16), t(8;16), t(15;17) or t(9;22). The mean latency period is much shorter, about 2 years. Most cases are of FAB M4 or M5 morphology. There is a correlation between DNA topoisomerase II cleavage sites and the translocation breakpoints. DNA topoisomerase II catalyzes transient double-standed DNA cleavage and rejoining. DNA topoisomerase II inhibiting agents form a complex with DNA and topoisomerase II, decrease DNA rejoining and cause chromosomal breakage. Reactive oxygen species (ROS) that are generated by the complex metabolism of the anthracyclines could create abasic sites, i.e., potent position-specific enhancers of DNA topoisomerase II cleavage [116]. The risks of chemotherapy-induced acute myeloid leukemia and myelodysplasia are dependent on the specific alkylating drug, on the the use of DNA topoisomerase II inhibitors, the cumulative dose, the schedule, and the duration of treatment [117]. T-AML and MDS resulting from treatment with anthracyclines respond less well to either chemotherapy or blood stem cell transplantation than their *de novo* counterparts. Adjuvant therapy in breast cancer patients using CMF schemes featured a small risk, enhanced only by combinations with radiotherapy [118, 119]. The cumulative incidence of t-AML/MDS of standard AC (60/600 mg/m^2) is known from NSABP B22 and B25 trials and was 0.21% [120]. Epirubicin is in use at higher dosages in combination with cyclophosfamide and 5-Fluoruracil as CEF (CYC, EPI, 5-FU; CYC 75 mg/m^2 d 1–4, EPI 60 mg/m^2 d1+8 and 5-FU 500 mg/m^2 d1+8 q4 wk) and EC (CYC 830 mg/m^2 d1, EPI 100 mg/m^2 d1 q3 wk). The cumulative incidence rates of t-AML have been published recently and are 1,7% for CEF and 1,2% for EC [121, 122]. Mitoxantrone, an anthracenedione that poisons the DNA topoisomerase II, has been associated with a 4-year 3.9% cumulative risk of leukemia [123]. It becomes clear that the use of anthracyclines and anthracenedione, both DNA topoisomerase II inhibitors, in the adjuvant setting has some severe disadvantages. The use of 2-times higher doses/treatment cycle in case of EPI compared to DOX results in an increase of the risk of developing a treatment-related AML from 0,2 in case of DOX to 1,2 or 1,7 for EPI which is a significant increase by 6 to 8.5 of the leukemogenic risk. Such an increase in the incidence of t-AML, if induced by the anthracycline epirubicin, is not acceptable. The same holds for the use of mitoxantrone in the adjuvant setting. The combination chemotherapies AC (60/600 mg/m^2) and FAC (500/50/500 mg/m^2) feature a risk of only 0,21 (117) and 0,19 [124] which is about the risk of the general population. Because EC or CEF have not shown better results than AC and FAC (direct comparisons were not performed) both regimens remain the standard therapy for many breast cancer patients in terms of efficacy who have to be treated for the risk reduction of recurrences.

Dosage and administration

There is considerable room for discussions of the optimal dose, the optimal dose density (how much and how often) and how to administer this dose. Dauno- and doxorubicin are the 'old' drugs and there is cumulative evidence that the optimal dose of single agent doxorubicin is 75 mg/m^2, and 50–60 mg/m^2 in combinations. Increasing the dose above 75 mg/m^2 has not shown any advantage in terms of better efficacy but increases the risk of more and prolonged toxicities. The optimal administration mode seems to be an infusion of 1–2 h (or even longer) instead of a bolus injection which had been the standard application procedure. The reason for the suggestion of an infusion instead of a bolus is simply the reduction of the peak plasma levels which is thought to be one of the important factors of the development of congestive heart failure. A bolus administration leads to a 'storm' of reactive oxygen species (ROS) in the cardiac tissue with consecutive damage in cardiac cells. Simply by prolonging infusion times, lower peak plasma levels with less damage to the heart will result.

In case of severe elevated liver enzymes and bilirubin the administration of a full dose of doxorubicin cannot be recommended because metabolism and excretion via the bile is severely hampered. There is no really good algorithm how to manage such a patient. For example, for a young breast cancer women with a complete diffuse metastatic infiltration of the liver at the time of diagnosis it is feasible to start with weekly DOX application as first line therapy with flat doses, e.g., 20 mg. If the patient tolerates this well, the drug dose can be escalated to 25 and 30 mg depending on the course of the liver enzymes. If therapy with DOX is satisfactory, a normalization will occur and a switch from a weekly schedule to a 3-week schedule can be considered. The normalization of the drug dose by use of body surface area is a matter of debate [125]. The reason why all anthracyclines doses including DOX are normalized by use of BSA, can only be seen in a historical context. The first studies in man were performed in this way and there was no good reason to stop this procedure. Only in recent years some researchers have focused on this question: is BSA-normalization of the drug dose necessary? [126]. Indeed, although not accepted by all, there is much evidence that it makes no sense to use BSA formula for a therapy plan.

Different doxorubicin administration modes have been described, most of them are still investigational. In Table 3 different application modes and their advantages/disadvantages are described.

Daunorubicin remains the backbone of the treatment of acute leukemias. The most common dose is between 30–60 mg/m^2 iv on three consecutive days in cases of induction therapy for AML. The German AML cooperative group used 3×60 mg/m^2 DNR together with 6-thioguanin and cytosine arabinoside (TAD) for the remission induction therapy in younger patients (<60 years) [127]. Patients older than 60 years often received lower DNR dosages although a dose reduction leads to inferior results with lower complete remission rates and less disease free survival at 5 years [128].

Epirubicin has become significant for the adjuvant and palliative treatment of breast cancer patients driven by its reduced cardiac toxicity profile when comparing equimolar dosages. The drug dose varies between 60 and 120 mg/m^2 combined with 5-FU and cyclophosfamide (FEC or CEF) or only with cyclophosfamide (EC) or as single agent. FEC (E 50 mg/m^2) *versus* FEC (E 75 mg/m^2) as first line treatment in metastatic breast cancer failed to show a benefit for the higher EPI dose. Overall response rate as well as overall survival time were similar [129]. The control arm with single agent EPI 75 mg/m^2 resulted in a 31% objective response rate. A comparison of 50 *versus* 100 mg/m^2 EPI showed a higher response rate in the case of 100 mg/m^2 but no difference in survival. The objective response rate was 23% and 41%, respectively [131]. A comparison of 60 *versus* 120 mg/m^2 EPI each combined with cyclophosphamide (600 mg/m^2) showed no difference in survival [132]. Two prospective randomized trials comparing EPI and DOX in first and second line in metastatic breast cancer failed to show any improvement in response rate, response duration and overall survival. 60 mg/m^2 DOX was compared with 90 mg/m^2 EPI as first line therapy in metastatic breast cancer. The objective

Table 3. Different application modes

Application Mode	Indication	Advantages	Disadvantages
iv Bolus q3wk	Standard method for most anthracycline regimen	Lowest amount of patient visits (in-/outpatient), very extensive data from studies	High peak plasma levels with highest risk for cardiac damage
Iv Bolus q2wk	Investigational Within studies e.g., adj. breast cancer	Increased dose density	Only possible with G-CSF support
iv Infusion (1–2 h) q3wk	Standard method in some centers	Lower peak plasma levels, less cardiotoxic	Requires a longer control of the intravenous line, longer stay in hospital or outpatient
iv Infusion (1–2 h) weekly	Investigational, standard method in some centers	Lower peak plasma levels, better toxicity control in difficult situations (e.g., massive tumor infiltration in the liver)	Requires more frequent infusions with higher costs, only manageable for single agent therapy
Continuous iv Infusion	Standard in Multiple Myeloma Investigational in all other tumor types	Very low plasma levels, cardiac toxicity reduced, metronomic application (proliferating endothelial cells are damaged)	Requires central venous cathether (e.g., Port-a Cath), high costs, exact monitoring necessary paravasation), more stomatitis
Intra-arterial Infusion	Investigational Localized hepatic or regional limb metastases	Systemic side effects reduced, allows maximal intratumoral drug levels	High rate of local drug and catheter-related complications, requires expert management of catheters (surgeon or radiologists), no cancer control benefit
Intraperitoneal Instillation	Investigational Peritoneal carcinosis e.g., ovarian cancer	Very high ip drug levels, less systemic toxicity	Abdominal pain, chemical-induced peritonitis, local catheter problems, no cancer control benefit
Intravesical Instillation	Standard Superficial flat bladder cancer (Tis, T1a)	Local disease control, no systemic side effects	Not effective in larger tumors, chemical-induced cystitis, bladder cramps

response rate were entirely the same with 47% and 49%, respectively [130]. In second line (after CMF) 60 mg/m^2 DOX was compared to 85 mg/m^2 EPI and both response rates reached 25% [133]. In summary, EPI is not able to induce better tumor response than DOX and the dose range where significant anti-cancer activity was observed based on large randomized trials is between 60 and 120 mg/m^2. For epirubicin it was shown that there is evidence against using body-surface area for dose calculation [134] and suggests to use a flat dose (e.g., 100 or 150 mg).

The use of idarubicin plus cytosine arabinoside has been evaluated as induc-tion therapy for acute myeloid leukemia. In a prospective randomized trial, idarubicin had an efficacy superior to that of daunorubicin [12], which was confirmed in a meta analysis using all available data from randomized trials [135]. Idarubicin was used in these trials with a dose varying between 8–13 mg/m^2 × 3d and daunorubicin was used for comparison with a dose of 45 and 50 mg/m^2 × 3d. One of the major drawbacks of these comparisons is the fact that a correct comparison should have included the use of 60 mg/m^2 × 3d daunorubicin (or even higher) because these two drugs (DNR and IDA) were not compared at equitoxic doses. During the consolidation phase of the randomized studies, when patients received cytarabine and either idarubicin or daunorubicin at these doses for 2 days, idarubicin resulted in sig-nificantly greater myelosuppression. Therefore, it is not clear that any observed improvement with IDA represents an inherent advantage of the drug, rather than a failure to compare drugs at biologic dose equivalence. No prospective randomized trial has been reported comparing DNR at 45 and 60 mg/m^2 (or even higher), nor have studies compared idarubicin 12 mg/m^2 to daunorubicin at 60 mg/m^2 (which would be a comparison at equitoxic doses). The latest randomized trial in patients with acute myeloid leukemia failed to demonstrate any advantage for one of the used drugs (IDA, DNR and MITOX). No difference in the disease-free, overall survival or toxicity was found [136]. Idarubicin within the dose range 8–13 mg/m^2 is an active drug for the treatment of acute leukemia alone as well as in combination regimen (e.g., AIDA) but the age of the patient and karyotype of the leukemia are of more importance than the used anthracycline [137].

Therapeutic use

All anthracyclines are approved drugs in many countries. The number of indi-cations is highest for doxorubicin. This drug has been evaluated in nearly all tumor types. Doxorubicin is a registered drug (US and EU) for the treatment of:
• breast cancer
• ovarian cancer
• transitional cell bladder cancer
• bronchogenic lung cancer
• thyroid cancer

- gastric cancer
- soft tissue sarcoma
- osteogenic sarcomas
- neuroblastoma
- Wilms' tumor
- malignant lymphoma (Hodgkin's and non-Hodgkin's)
- acute myeloblastic leukemia
- acute lymphoblastic leukemia
- Kaposi's sarcoma related to acquired immunodeficiency syndrome (AIDS)

Epirubicin is an approved drug for the treatment of several tumor types. In the US the approval is restricted to breast cancer adjuvant therapy, whereas outside the US the spectrum of indication is much broader:
- breast cancer – adjuvant (US, EU) and palliative (EU)
- small cell lung cancer
- ovarian cancer
- gastric cancer
- rectal cancer
- pancreatic cancer
- non-Hodgkin's lymphoma
- soft tissue sarcoma

Daunorubicin is approved for the treatment of:
- acute myeloid leukemia
- lymphatic leukemias

Idarubicin is approved for the treatment of:
- acute myeloid leukemia

For further readings on the differential use of these drugs the oncology text books (e.g., *Cancer Principals and Practice of Oncology*, *Cancer Medicine*, *Oxford Textbook of Oncology*) are highly recommended.

Liposomal anthracyclines

Introduction

Liposomes as drug targeting systems [138, 139] have been under discussion since the 1970s [140, 141]. Thus, it is striking that there are only three liposomal formulations of cytostatic drugs for iv applications, which are registered and in clinical use. And it is further striking that these three formulations are all anthracycline liposomes (Doxil®/Caelyx®; Myocet® (both liposomal Doxorubicin) and Daunoxome®, liposomal Daunoxome) [142]. But this is not by chance: Anthracyclines comprise a group of drugs which – in contrast to most other drugs – can be entrapped within liposomes in a rational way with high trapping efficiencies and the liposomal products can have long shelf-lives. In other words, anthracycline liposomes are on the market because it is possible to produce them with reasonable effort, even in bulk quantities. But this is not *per se* an advantage. Conventional anthracyclines have been widely used in clinical practice for a long time with good clinical results but severe side effects. Entrapping of anthracyclines into liposomes helps to reduce some of the most important side effects, e.g., cardiotoxicity. The principles of anthracycline liposome preparation and how liposomes can improve the therapeutic index of these agents will be described.

Anthracycline liposomes can be easily prepared by 'remote loading' technique

The majority of cytostatic drugs are water soluble and able to diffuse through phospholipid bilayers. Diffusion is necessary for many of the drugs to reach the inner volume of tumor cells. However, this prerequisite for cytotostatic activity reduces the lifetime of drug-containing liposomes since the drugs are also able to diffuse out of the liposomes [143]. Anthracyclines are no exception to this rule, but they share one helpful feature: an amino group in the sugar moiety which has an pK-value of around 8 (e.g., 8.22 for DOX). At a slightly basic pH-value of 8, anthracyclines represent an equilibrium of about 50% protonated and 50% non-protonated molecules. In contrast, at pH 4 almost all of the anthracycline molecules are protonated. The *remote loading technique* [144] is based on this difference since protonated (charged) anthracyclines are not able to pass the liposome membranes. For the procedure of remote loading, empty liposome dispersions have to be prepared which are mildly acidic inside the liposomes (pH 4, e.g., citrate buffer) and roughly neutral in the exterior (pH 8). Then the anthracyclines have only to be added and the neutral anthracycline molecules now diffuse through the liposome membrane. Once inside the liposomes, they immediately become protonated. The process of diffusion was accelerated by a slight increase in temperature. Since a protonated anthracycline molecule is no longer able to pass the liposome membranes, the molecule is now trapped. Nearly 100% of anthracycline molecules can be entrapped inside liposomes by this technology. The process is shown in Figure 7. In analogy to the *remote loading* using a pH-gradient which is used for Myocet and Daunoxome preparation, a transmembrane ammonium sulfate gradient [145] can also be used for efficient and stable entrapment of anthracyclines (Caelyx/Doxil). For two of the anthracycline liposomes on the market (Caelyx/Doxil and Daunoxome), drug loading is carried out by the manufacturers (ready to use). For Myocet, the loading procedure has to be performed at the clinical pharmacy, which illustrates the ease of this process, but the preparation is still more time-consuming than a ready to use preparation.

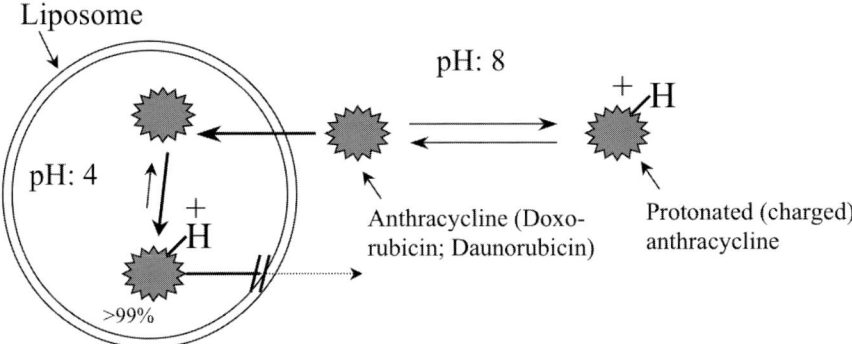

Figure 7. The remote loading technique for the entrapment of anthracyclines in liposomes.

How liposomes may improve anthracycline pharmacokinetics and anticancer activity

If drug molecules are entrapped within liposomes, the systemic environment does not recognize the free drug. Instead, it only recognizes the liposome which can be the better alternative. Especially if a drug molecule has a short half-life and therefore the drug-tumor exposure is too short for adequate anticancer activity, liposomal entrapment is a suitable way to increase its half-life. At first glance, this seems to be the case for anthracyclines because $t_{1/2}$ for, e.g., doxorubicin is only 1.3 h and for the liposomal formulations $t_{1/2}$ ranges from approximately 3–55 h. But the short half-life of free anthracyclines does not mean that the drug molecules are secreted or metabolized and no longer exist in the body. Instead, the drug molecules have penetrated into deeper compartments and the half-life of the drug serum elimination might not correlate with the half-life of the anthracyclines body elimination (estimated at about 7 days) as well as its tumor exposure time. Thus, it is questionable if liposomal entrapment of anthracyclines significantly increases long-term tumor-AUC or drug-tumor exposure time and results in higher antitumor activities. As discussed below, a series of clinical studies with different liposomal anthracyclines have not shown a higher response rate nor an increase of survival time.

Despite their ability to increase a drugs' half-life, an important feature of liposomes is their ability to accumulate in tumors due to the *enhanced permeability and retention effect* [9, 10] (EPR effect; Fig. 7). This effect is mainly based on differences between the vasculature in tumors and healthy organs/tissues. Blood vessels in tumors are more leaky. Furthermore, tumor cells are very often not as densely packed as cells in healthy tissues. The lymphatic system, important for removing substances (and also nanoparticles such as liposomes) from the tissues is very often only marginally expressed [148]. Thus, liposomes up to a diameter of 400–600 nm are able to diffuse out of the leaky blood vessels and can accumulate in the tumor tissues, but not in healthy tissues (*passive tumor targeting*) [149, 159]. Indeed, it has been shown in pre-clinical experiments with tumor bearing mice that liposomal anthracyclines have the potential to accumulate within the tumor tissue: For example, in P1798 lymphosaroma bearing mice, a 10 times higher accumulation of DNX in the tumors after application of Daunoxome in comparison to conventional DNX has been observed. Furthermore, some hints of improved anthracycline tumor accumulation after application of liposomal anthracyclines in patients have been reported as well [151]. However, *passive tumor targeting* not only has an important implication for accumulation of drug molecules within the tumor tissue. It also results in a protection of healthy tissues from the drug (Fig. 7). From clinical experience with liposomal anthracyclines, the latter seems to be more important for the therapy of cancer patients: Using liposomal anthracyclines, a much better safety profile could be observed [152] but in many cases no improved anticancer activity (see below).

Passive tumor targeting of a liposomal formulation can be controlled by the liposomes half-life: To achieve accumulation of liposomes in tumors and to achieve a reduction of the burdening of healthy tissues, liposomes in the blood stream have to be stable. This is because their chance to diffuse into the tumor is higher the longer the liposomes remain in the blood stream [153]. In addition, the more stable the drug containing liposomes are from degradation, the more the healthy tissue will be protected from the drug molecules. But the half-life of liposomes in the blood stream not only depends on their stability in serum [154]. It also depends on their uptake by cells of the monocyte phagocyte system (MPS, e.g., liver, spleen and circulation macrophages) [155]. Uptake of liposomes by MPS-cells is triggered by the binding (opsonization) of serum proteins like complement factors or antibodies (opsonines). Once opsonized, liposomes can be rapidly recognized and phagocyted by MPS-cells. Both the stability of liposomes in serum and the prevention of the liposomes uptake by MPS-cells can be improved by different ways [156–158]:

- **Reducing the liposome size:** Smaller liposomes are less vulnerable to opsonization by serum proteins, in particular by the complement system. Decreased opsonization results in a lower recognition and therefore a lower clearing by the MPS [159–165].
- **Optimization of the liposome composition:** Liposomes for iv injection usually consist of glycerophospholipids with long hydrogenated fatty acid esters (synthetic phospholipids like distearoylphosphatidylcholine (DSPC) or fully hydrated phosphatidylcholines from egg or soy). As a consequence their liposome membranes are more rigid and more stable against lipid exchange by serum proteins than membranes consisting of naturally glycerophospholipids with fatty acids of different length and saturation (e.g., not hydrated egg- or soy-lecithin) [165–167]. In addition, a rigid membrane decreases the efflux of drugs from liposomes and stabilizes the liposomes themselves. The addition of cholesterol to the liposomal bilayers stabilizes membranes by reducing membrane fluidity and by preventing membrane crystallization [168].

- **Sterical stabilization of liposomes:** Components for *sterical stabilization* of liposomes (Stealth®-components) such as polyethyleneglycol-phosphatidylethanolamine (PEG-PE; pegylation of liposomes) lower the recognition and uptake by the MPS by increasing the hydrodynamic circumference of the liposomes (Fig. 3) [144, 169–175].
- **Increasing the liposome amount:** The higher the number of liposomes, the longer the MPS needs to eliminate the liposomes from the systemic circulation. Thus, the half-life of liposomes depends on the amount of administered liposomes. This could be demonstrated in preclinical studies with Vincristine and Gemcitabine containing liposomes [176, 177]: Although liposomes were used which were not protected by stealth components, half-lives of more than 13 h could be achieved in mice by using lipid doses of more than 2 mmol/kg.

Description of commercially available anthracycline-liposomes

Probably due to patent reasons, the compositions of Caelyx/Doxil, Daunoxome and Myocet differ over a wide range (Tab. 1). The only similarity of the three formulations is the use of a neutral blend of lipids – lecithin and cholesterol – but already the type of lecithin differs. Calyx/Doxil and Daunoxome were designed as long circulating liposomes. For Caelyx/Doxil, a stealth component (MPEG-DSPE) was successfully used to increase the liposomes half-life *in vivo* by a factor of about 37 and serum-AUC by a factor of 1,200 compared to free DOX respectively. In contrast, the Daunoxome-liposomes are not protected by a stealth component. Instead, this formulation has a very rigid membrane composition (fully synthetic, long chain lecithin with defined hydrocarbon chains (di-C18)), a very small vesicle size and a 2.5-fold higher lipid content in comparison to Caelyx/Doxil. These modifications increase the half-life of the liposomes (Factor 3.8) in contrast to conventional daunorubicin. This increase was not as high as for Caelyx/Doxil which is protected from MPS-uptake by using a stealth component. Nevertheless, tumor accumulation of Daunoxome is similar to that of Caelyx/Doxil as shown in preclinical experiments. This surprising effect can be explained by the very small size of Daunoxome-liposomes and the higher lipid content of the formulation (Tab. 1). Small particles are able to diffuse much faster through leaky tumor vasculature. Furthermore, a higher lipid content (higher number of vesicles) resulted in a faster saturation of the MPS-cells and thus in an increased half-life of the liposomes. In other words, even if no stealth component is used, an intelligently designed liposome can overcome this drawback. This might be of importance because PEG-stealth components induce a new side effect, named hand-food-syndrome (see below).

A comparison of Myocet with the two other anthracycline formulations is more difficult. Myocet vesicles are rather large (180 nm) and therefore neither optimal for MPS-escape nor for tumor accumulation. Recognition of the large Myocet vesicles by the MPS might be (to a minor) part compensated by their lipid composition – hydrated egg-lecithin and cholesterol in a molar ratio of 55:45, which is the maximum possible amount of cholesterol and results in very rigid membranes. Taken together, Myocet is not designed for passive tumor targeting but rather for the protection of healthy tissues such as the heart.

Clinical aspects of liposomal anthracyclines

Pharmacokinetic data

From the early 1990s onwards, beginning with the liposomal doxorubicin in its pegylated formulation (Doxil®/Caelyx®) and followed by the liposomal daunorubicin (DaunoXome®), these compounds have undergone extensive clinical testing. The human pharmacokinetic data from early Phase I trials revealed what could be expected from the respective liposome design (Tab. 1). The elimination half-life of the PEG-protected Caelyx/Doxil liposomes was much longer than that of the conventional drug and the AUC was concomitantly raised impressively from 0.489 µg/ml·h for free DOX to 590 µg/ml·h (both: 20 mg/m^2) [178]. The half-lives of Daunoxome and Myocet were significantly lower than for Caelyx/Doxil and as a consequence the AUCs were much lower.

Antineoplastic activity

The first tumor treated was the *AIDS-related-Kaposi-sarcoma*. This tumor was chosen due to its special structure (highly vascularized) and the ability of epidermal cells to take up liposomes. Thus, a great benefit was expected by taking advantage of the passive tumor targeting effect. In two major Phase III studies between 1993 and 1995 Stewart et al. [179] and Northfelt et al. [180] were able to show the statistically significant superiority of the liposomal doxorubicin as monotherapy *versus* the standard polychemotherapy at this time (bleomycin and vincristine or free doxorubicin, bleomycin and vincristine; overall response rates 58.7% *versus* 23.3% and 45.9% *versus* 24.8%, respectively). In 1995, the formulation was approved by the FDA in the US. A similar trial for liposomal daunorubicin did not show the same results for this substance [181], but it was given at a relatively low dosage. However in a Phase II study the efficacy of liposomal daunorubicin was shown at a higher dosage [182].

In the mid 1990s, several Phase II and III studies showed a benefit of the liposomal doxorubicin formulation for patients with advanced *ovarian carcinoma* as second or third line therapy. Even in cisplatin, taxol and sometimes also topotecan refractory carcinomas it seems possible to reach an overall response rate of 23% with a median progression free survival of 6.6 months, which was not expected for this patient subgroup [183]. The FDA approved Doxil for this indication in 1999. The fact itself that anthracyclines are efficacious drugs in ovarian cancers is known because PAC (cis-platin, doxorubicin and cyclophosphamide) was an accepted regimen before the taxane era started.

Another tumor treated with liposomal doxorubicin is breast cancer. The most extensively studied liposomal drug for the treatment of metastatic breast cancer

(MBC) is Caelyx/Doxil and the drug is now approved in Europe as monotherapy for MBC-patients with higher cardiac risks: Used as a monotherapy in Phase II studies, the results are comparable to those of the free drug, even using slightly different dosages [184, 185]. Rivera [186] reported in a review article about a Phase III study with Caelyx/Doxil which has not been published to date. Liposomal DOX was compared with conventional DOX (50 mg/m^2 once every 4 weeks *versus* 60 mg/m^2 once every 3 weeks). If such a comparison is correct it is a matter of discussion because in the case of monotherapy, 75 mg/m^2 DOX is the appropriate dose. Progression free survival was not significantly different for both treatment arms (6.9 months for patients receiving Caelyx/Doxil and 7.8 months for patients receiving conventional DOX. (p. 0.99)). Overall survival was 20.1 months and 22.0 months, respectively. Although the efficacy of the liposomal DOX was not superior over conventional DOX, the safety profile was, and the study showed a significantly lower incidence of cardiotoxicity (p < 0.001). A very similar result could be shown for Myocet which is approved in the EU for the first-line treatment of MBC (with cyclophosphamide) [187]. Patients received either Myocet 75 mg/m^2 or 75 mg/m^2 conventional DOX (both: every 3 weeks). Response rates were 26% in both groups but in the liposomal group, cardiotoxicity was reduced.

Several other tumors (e.g., myeloma, soft tissue sarcoma, lymphoma, mesothelioma, leukemia, HCC, brain tumors, lung cancer and others) have been treated with the liposomal anthracyclines. Up to now, only a small number of patients have been evaluated and it is far too early to discuss these results. The use of liposomal anthracyclines in these indications is not approved and cannot be recommended until a full evaluation with results from comparative Phase III studies will be presented.

To improve the antineoplastic activity of liposomal anthracyclines, the new drugs are under extensive investigation in combination therapies. As an example, Caelyx/Doxil was tested in Phase I and II studies in combination with gemcitabine. In a Phase II study [188], MBC-patients received Caelyx/Doxil (24 mg/m^2, day 1) plus Gemcitabine (800 mg/m^2, days 1 and 8) each 21-day cycle. The treatment was well tolerated and the overall response was 52% (3 complete and 21 partial responses). But again, such results have to be evaluated within Phase III studies which remain the basis for evidence-based medicine.

Toxicity and multidrug resistance
By adopting the liposomal formulation there was a *toxicity shift* for the anthracyclines. The former anthracycline toxicity, i.e., myelosuppression, cardiotoxicity, stomatitis, alopecia, nausea and vomiting, changed significantly to a new dose-limiting skin toxicity (palmar-plantar-dysaesthesia). The myelosuppression remained about the same, but alopecia, nausea and vomiting were moderately reduced. Another intcresting finding was that there was no apparent skin necrosis after accidental paravasation of the liposomal formulations [152]. But most important for clinical use was the observation that there was significantly less cardiac toxicity [152]. This reduction of cardiac damage

offers a new quality for the use of these substances, especially for those
patients who are at high risk to develop congestive heart failure. But, the real
problem is that the liposomal anthracyclines are not approved for curative
treatment of cancers like leukemia, lymphoma and adjuvant breast cancer. The
simple application of knowledge from palliative treatment to curative treatment is scientifically not allowed. Here, we face the problem of off-label use
which cannot be discussed in detail.

Some preclinical findings suggest an influence on *multidrug resistance
(MDR)* due to the pharmacokinetic changes of the liposomal formulations. But
up to now data have been controversial and there is not yet a clear evidence for
MDR modulation [189, 190].

Costs aspects
Today, the cost of liposomal formulations of DNR and DOX are more than 20
times higher than for the non-liposomal anthracycline, but the manufactures
claimed cost savings due to lower cost for the managing of adverse effects and
for hospitalization. However, one study comparing the total costs of the treatment of patients with recurrent epithelial ovarian cancer with liposomal
Caelyx/Doxil and topotecan showed slightly lower total treatment costs if liposomal DOX was used [191]. Cost aspects are becoming an important issue in
each healthcare system and any claimed progress in medicine has to be weight-
ed on its economical impact.

Prodrugs of anthracyclines

Introduction

Any strategy by which a cytotoxic drug is targeted to the tumor, thus increasing the therapeutic index of the drug, is a way of improving cancer chemotherapy and minimizing systemic toxicity. Low- or high-molecular weight prodrugs hold promise as tumor selective drug delivery systems. Expected advantages of such formulations are a preferable tissue distribution, a prolonged
half-life of the drug in the plasma, and a controlled drug release at the tumor
site by adjustment of the chemical properties of the bond between the drug and
the linker. In the past, the design of prodrugs with antitumor agents has
focused on strategies that allow the drug to be released by extracellular or
intracellular proteases or at the low pH values present in lysosomes and endosomes, respectively.

Generally, low-molecular weight prodrugs are designed to minimize the
inherent toxicity of the antitumor drug by suitable chemical modification that
permit the parent drug to be released efficiently at the tumor site. In addition,
the prodrug can incorporate a ligand such as a peptide that targets a tumor-
associated receptor or antigen.

Table 4. Comparison of anthracycline liposomes

Liposomal formulation	Liposome characteristics						Tricks for long circu-lation	Tumor targeting via EPR	Pharmacokinetics (Anthracyclines 20 mg/m²)[7]	
	Drug	Diameter [nm]	Composition	Stealth comp.	Lipid/Drug [mg/mg]	Loading			$T_{1/2}$ [h]	AUC [µg/ml × h]
Caelyx/ Doxil	DOX	~100	Hydrogenated Soy-PC/Chol 2:1	yes, MPEG-DSPE[1]	6.4[4]	Ammonium sulfate gradient Ready to use	• Stealth-component	yes	55 (F.: 37)[6]	590 (F.:1200)[6]
Myocet	DOX	~180	Hydrogenated Egg-PC/Chol 55:45	no	3.8[2]	pH-gradient (citrate), bedside preparation[3]	• Rigid membrane	not clear	2–3 (F.: ~1.7)[6]	30 (F.: 62)[6]
Daunoxome	DNR	~35–65	DSPC/Chol 2:1	no	17.4[5]	pH-gradient (citrate) Ready to use	• Small particles • High lipid content • Rigid membrane	yes	3.8 (F.: 2.5)[6]	57

[1] MPEG-DSPE: α-(2-[1,2-distearoyl-sn-glycero-3-phosphooxy]-ethylcarbamoyl)-ω-methoxy(polyethylenglycol)-40 sodium salt
[2] Calculated from [192] and Myocet product information
[3] Have to be prepared (loaded) at the bedside (clinical pharmacy)
[4] Essex-Hotline, Munich, Germany
[5] Daunoxome-product information
[6] t1/2b (Dox): 1.3 hours, t1/2b (DNR): 1.5 hours, AUC (DOX): 0.489 µg/ml × h
[7] For comparison reasons, all PK-values were given for a dose 20 mg/m², which is not the recommended dose

High-molecular weight prodrugs also follow an active or passive targeting approach. On the one hand, active targeting with antibodies is based on the presumption that characteristic differences exist between normal and cancer cells on their cell surface, which can be exploited for the selective delivery of antineoplastic agents to solid tumors [193, 194]. On the other hand, the pathophysiology of tumor tissue, characterized by angiogenesis, hypervasculature, a defective vascular architecture and an impaired lymphatic drainage leads to passive targeting of macromolecules with MW > 20,000 Da to solid tumors.

In the past 10 years macromolecular drug delivery strategies in oncology have gradually shifted from active targeting strategies to passive ones due to a more detailed understanding of the anatomy and physiology of solid tumors. The rationale for simply using high-molecular weight molecules as efficient carriers for the delivery of antitumor agents, even if they are not targeted towards an antigen or receptor on the surface of the tumor cell, has been strengthened by recent studies concerning the enhanced vascular permeability of circulating macromolecules for tumor tissue and their subsequent accumulation in solid tumors [195–202]. This phenomenon has been termed 'enhanced permeability and retention' in relation to passive tumor targeting (EPR effect) [195] and is depicted schematically in Figure 8. Blood vessels in most of the normal tissues have an intact endothelial layer which allows the diffusion of small molecules but not the entry of macromolecules into the tissue. In contrast, the endothelial layer of blood vessels in tumor tissue is often leaky so that small molecules as well as macromolecules have access to malignant tissue. Because tumor tissue does not generally have a lymphatic drainage system, macromolecules having a molecular weight >20,000 Da are thus retained and can accumulate in solid tumors.

A variety of low- and high-molecular weight prodrugs of anthracyclines have been developed over the past 20 years, and first candidates have recently entered clinical studies. This chapter does not include all attempts of developing anthracycline prodrugs and focuses on pertinent examples that demonstrate a clear prodrug nature, an *in vivo* proof of concept and a potential clinical future. Other examples can be found in review articles that describe targeting strategies in oncology, e.g., ADEPT or peptide targeting [203–207].

Significance of the type of chemical bond incorporated in the prodrug

In principle, prodrugs can be cleaved in the body by unspecific hydrolysis, by enzymes, by reduction or in a pH-dependent manner. The design of anthracycline prodrugs has focused on acid-sensitive and enzymatically cleavable bonds that allow the prodrug to be cleaved either extracellularly in the tumor tissue or intracellularly after cellular uptake.

In general, macromolecules are taken up by the cell either through receptor-mediated endocytosis, adsorptive endocytosis, or fluid-phase endocytosis [208]. During endocytosis a significant drop in the pH-value takes place from

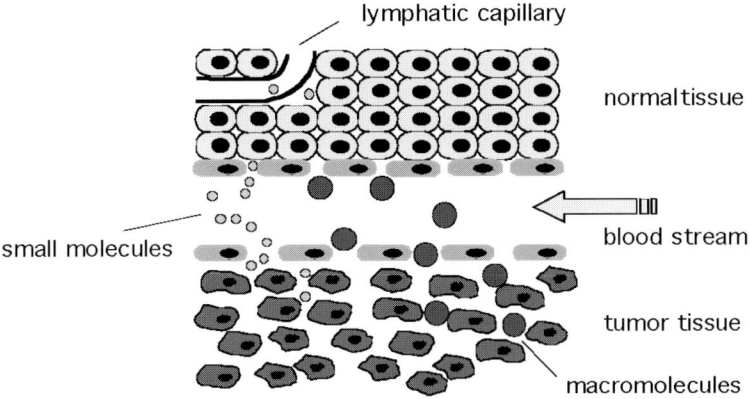

Figure 8. Schematic representation of the anatomical and physiological characteristics of normal and tumor tissue with respect to the vascular permeability and retention of small and large molecules (EPR effect).

the physiological pH (7.2–7.4) in the extracellular space to pH 6.5–5.0 in the endosomes and to around pH 4 in primary and secondary lysosomes. Additionally, a great number of lysosomal enzymes become active in the acidic environment of these vesicles, e.g., phosphatases, nucleases, proteases, esterases, and lipases.

Both the low pH-values in endosomes and lysosomes as well as the presence of lysosomal enzymes are therefore intracellular properties which can be exploited for releasing the polymer-bound drug specifically in tumor cells.

Furthermore, the microenvironment of tumors has been reported to be slightly acidic in animal models and human patients. New non-invasive techniques have demonstrated that the pH-value in tumor tissue is often 0.5–1.0 units lower than in normal tissue [209]. This pH-shift could contribute to the extracellular release of drugs bound to polymers through acid-sensitive linkers, especially if the prodrug is trapped by the tumor for longer periods of time.

Finally, extracellular proteases that are over-expressed in solid tumors serve as molecular targets for designing enzyme-specific prodrugs (see below).

Chemical considerations regarding the design of anthracycline prodrugs

From a chemical point of view, doxo- and daunorubicin are ideally suited for designing prodrugs due to the presence of two different functional groups, i.e., the 3'-amino group of the sugar moiety and the C-13-keto position (see Fig. 9).

Acid-sensitive derivatives have been developed by forming a carboxylic hydrazone bond at the C-13 carbonyl group or by attaching a cis-aconityl spacer at the 3'-NH₂-group (see Fig. 10).

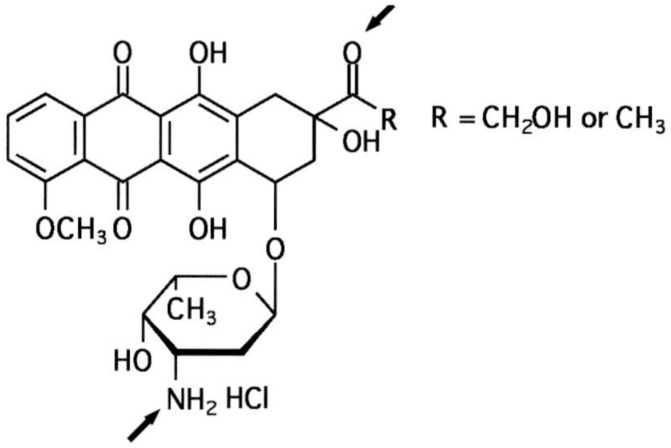

Figure 9. Sites for chemical modification of the anthracycline molecule relevant for the design of pro-drugs.

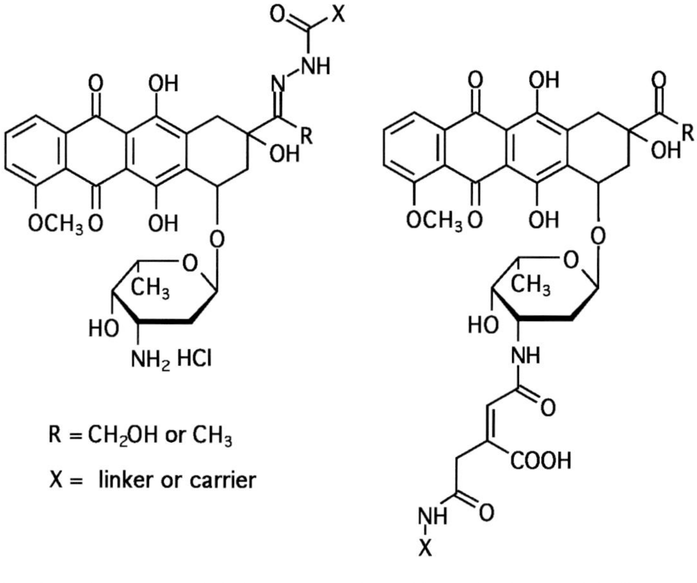

Figure 10. General structure of acid-sensitive anthracycline derivatives with hydrazone linkers (left) or *cis*-aconityl linkers (right).

Both bonds show high stability at pH values of 7.0–7.4, but release the anthracycline within a few hours at pH 5.

In enzymatically cleavable anthracycline prodrugs a suitable substrate is either attached directly to the 3'-NH$_2$ position or through a self-immolative spacer (see Fig. 11).

Peptide linkers that are bound to the 3'-NH$_2$ group are cleaved by the respective enzyme at this position with concomitant release of the native drug or within the peptide sequence with subsequent liberation of doxo- or daunoru-

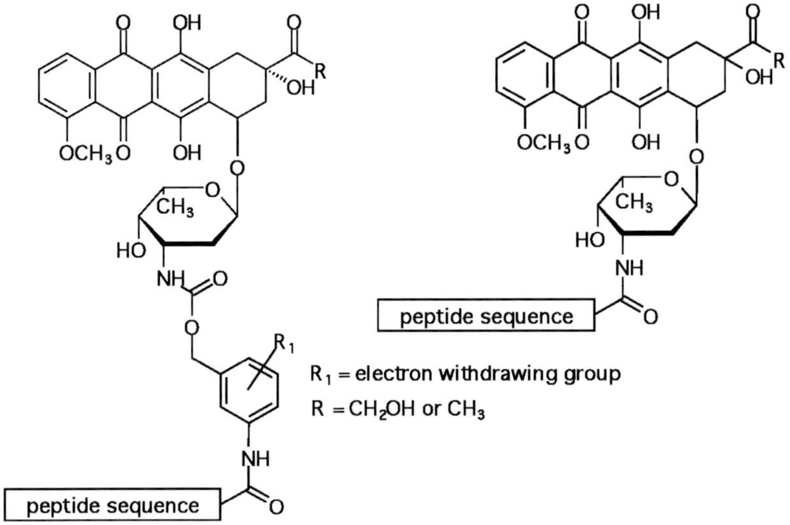

Figure 11. General structure of anthracycline derivatives with peptide linkers attached at the 3-amino position of through a self-immolative spacer molecule.

bicin peptide derivatives that are active *per se* or are degraded further to the parent compound.

Cleavage in derivatives with self-immolative spacers initially takes place at the amide bond of the aromatic linker producing a labile aromatic moiety that is hydrolyzed in a 1,4- or 1,6-elimination reaction and releases doxo- or daunorubicin.

Anthracycline prodrugs that exploit tumor-associated enzymes

Both intra- and extracellular enzymes have been envisioned as targets for specifically activating prodrugs at the tumor site. Cathepsins, especially cathepsin B, are probably the best known examples for intracellular proteases that have received considerable intention as suitable enzymes for degrading drug polymer conjugates in lysosomes [210].

During tumor invasion and progression, tumor cells also secrete a number of proteases into the extracellular space, e.g., matrix metalloproteases, plasmin, tissue-type plasminogen activator or urokinase-type plasminogen activator that degrade the extracellular matrix. Although numerous efforts have concentrated on inhibiting these proteases in order to prevent tumor growth and the formation of metastases [211, 212], research has only recently focused on drug targeting strategies in which the protease activity of these enzymes is used to release an anticancer agent from a drug carrier.

Other target enzymes that are overexpressed in tumor tissue, but do degrade the extracellular matrix, are prostate-specific antigen and β-glucuronidase.

Examples of anthracycline prodrugs that are cleaved by tumor-associated enzymes are described below.

Anthracycline prodrugs that are cleaved by cathepsins or other enzymes present in lysosomes

Pioneering work regarding the development of drug polymer conjugates containing peptide spacers, which are enzymatically degradable in lysosomes, dates back to the early 1980s. In a series of experiments, Trouet et al. bound daunorubicin to succinylated albumin through various peptide spacer arms, and the resulting conjugates differed significantly in their antitumor activity against L1210 leukemia depending on the cleavability of the oligopeptide by lysosomal hydrolases [213].

In more recent work the groups of Kopecek and Duncan have developed HPMA copolymer conjugates [HPMA = *N*-(2-hydroxypropyl)methacrylamide] containing doxorubicin bound to the polymer backbone through different peptidyl side chains, which were designed to release the drug on exposure to lysosomal thiol-proteases [214, 215]. In the tailor-made HPMA-doxorubicin conjugate (PK1) approximately 8% w/w of doxorubicin is linked to the polymer through a Gly-Phe-Leu-Gly peptide spacer that is cleaved by cathepsin B and releases doxorubicin (see Fig. 12).

PK1 has shown promising antitumor activity in solid tumor models, and the *in vivo* activity of this conjugate is correlated with the amounts of cathepsin B levels found in tumor cells and tumor tissue [215, 216].

Furthermore, sugar-modified HPMA-doxorubicin conjugates have been developed that bind to the asialoglycoprotein receptor of liver cells (see Fig. 13) with the aim of improving the treatment of hepatocellular carcinoma and liver metastases.

The conjugate, known as PK2, incorporates the same tetrapeptide linker as in PK1 as well as *N*-linked galactosamine as the receptor ligand; preclinical studies have shown that PK2 delivers doxorubicin preferentially to the liver [217]. Both PK1 and PK2 have been studied in Phase I/II studies (see below).

Doxorubicin prodrugs that are cleaved by MMP-2 and MMP-9

A first example of exploiting the activity of the matrix metalloproteases MMP-2 and MMP-9 for cleaving the anticancer agent doxorubicin from albumin has recently been reported by Kratz et al. [218, 219]. Matrix metalloproteases make up a family of approximately 20 proteases that play a key role in the degradation of collagens which is a necessary step for angiogenesis, formation of metastases and tumor progression.

Especially MMP-2 plays a critical role in the degradation of basement membranes and the extracellular matrix. Consequently, a drug targeting strategy in which the protease activity of MMP-2 is exploited to release an anticancer agent from a macromolecular carrier, i.e., circulating albumin, was

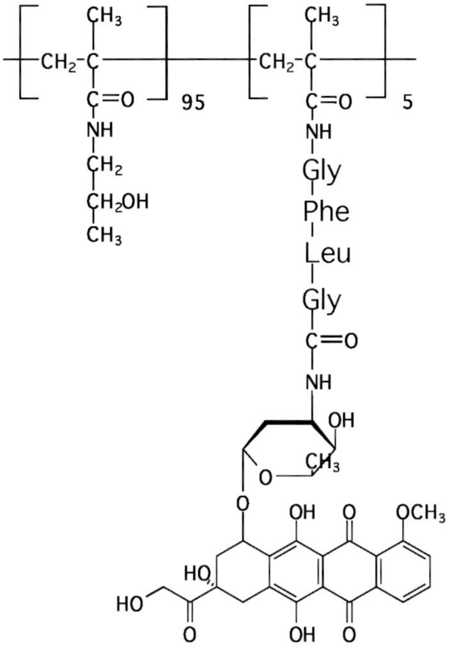

Figure 12. Structure of PK1, a HPMA-doxorubicin conjugate that is cleaved by cathespin B.

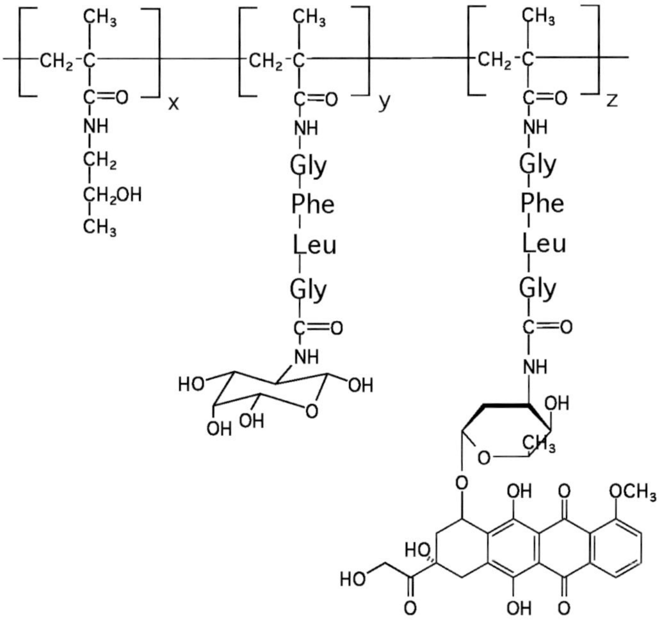

Figure 13. Structure of PK2, a HPMA-doxorubicin conjugate that is additionally tethered with N-linked galactosamine molecules.

assessed [219]. For this purpose, a water-soluble maleimide derivative of dox-orubicin incorporating a MMP-2 specific peptide sequence [Gly–Pro–Leu-Gly–Ile–Ala–Gly–Gln] was developed that binds rapidly and selectively to the cysteine-34 position of circulating albumin (Fig. 14).

The albumin-bound form of the prodrug was efficiently and specifically cleaved by MMP-2 liberating a doxorubicin tetrapeptide [Ile–Ala–Gly–Gln–DOXO]. *In vivo*, the MMP-2 specific prodrug was superior to the parent compound doxorubicin in the A375 human melanoma xenograft which is characterized by a high expression of MMP-2. A noteworthy finding was the fact that the doxorubicin tetrapeptide was subsequently degraded to doxorubicin in homogenates of tumor tissue.

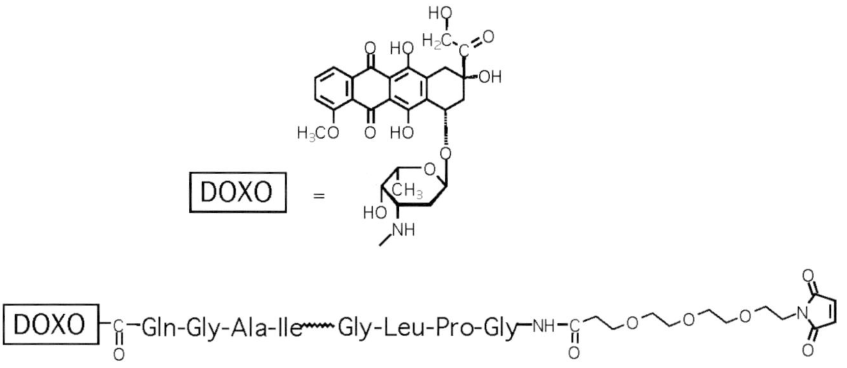

Figure 14. Structure of an albumin-binding prodrug of doxorubicin that is cleaved by MMP-2.

Doxorubicin prodrugs that are cleaved by prostate-specific antigen (PSA)

PSA is a serine protease that is especially attractive as a target protease because it is solely expressed in prostate tissue and prostate carcinoma with high levels up to mg/g present in human prostate carcinoma [220, 221]. Two low-molecular weight doxorubicin prodrugs have been developed that aim to exploit PSA as the protease target [222–225]. These doxorubicin derivatives contain the peptide sequences Mu-His-Ser-Ser-Lys-Leu-Gln-Leu-OH (Mu = morpholinocarbonyl) and *N*-glutaryl-(hydoxypropyl)-Ala-Ser-cyclo-hexaglycyl-Gln-Ser-Leu-OH (abbreviated L-377,202) bound to the amino position of doxorubicin. Both low-molecular weight prodrugs were designed to release *N*-[L-leucyl]doxorubicin following cleavage by PSA.

The MTD of both prodrugs was approximately 5-to-7-fold higher than for free doxorubicin, and at theses doses they demonstrated good antitumor activity in PSA-positive animal models (LNCAP, CWR22, PC 82) [223, 224]. A Phase I study has been carried out with L-377,202 (see below).

Anthracycline prodrugs that are cleaved by β-glucuronidase
β-glucuronidase is an enzyme that is found in elevated levels in necrotic areas of the tumor [226, 227]. A great number of doxo- and daunorubicin prodrugs of the general formulas depicted in Figure 15 have been synthesized in order to find prodrugs with optimal substrate-specificity for this enzyme [228–230].

A self-immolative spacer proved to be crucial for rapid cleavage of the pro-drugs by β-glucuronidase with concomitant release of free doxorubicin. The best studied representative within this family of prodrugs is HMR 1826 (see Fig. 16) which is stable at physiological pH, is cleaved efficiently by β-glu-curonidase to doxorubicin and is considerably less toxic than doxorubicin [228–230]. HMR 1826 has shown superior *in vivo* efficacy in several animal tumor models, albeit at ~10- to 25-fold higher doses compared to the parent compound.

Doxorubicin prodrugs that are cleaved extracellularly by unidentified peptidases
Trouet et al. have recently developed a doxorubicin prodrug, *N*-succinyl[β-alanyl-L-leucyl-L-alanyl-L-leucyl)doxorubicin, that is cleaved extracellularly to *N*-[L-leucyl]doxorubicin by unidentified peptidases [231]. *N*-[L-leucyl]dox-orubicin rapidly enters tumor cells where it can be cleaved to doxorubicin.

The prodrug can be administered at a 10-fold dose of the LD_{50} of doxoru-bicin in mice and has shown superior antitumor effects in a breast carcinoma model when compared to doxorubicin at equitoxic doses.

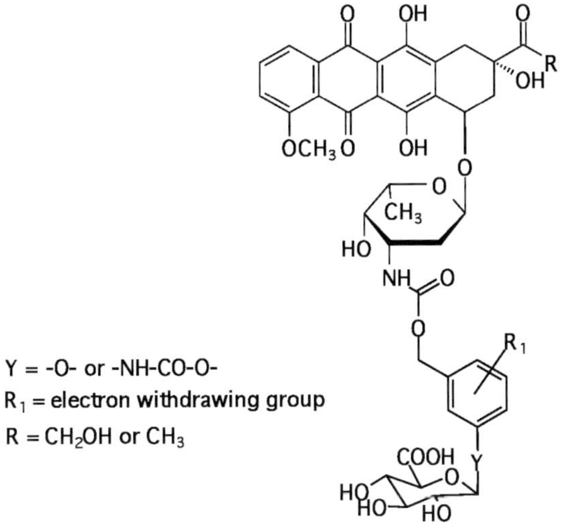

Y = -O- or -NH-CO-O-
R_1 = electron withdrawing group
R = CH$_2$OH or CH$_3$

Figure 15. General structure of anthracycline prodrugs that were designed as substrates for β-glu-curonidase.

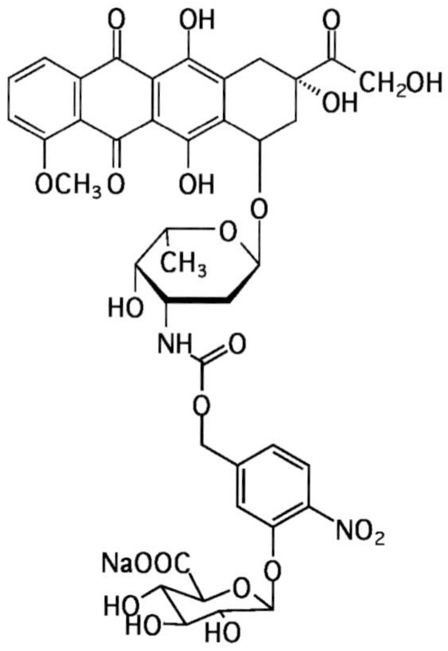

Figure 16. Structure of HMR 1826, a doxorubicin that is cleaved efficiently by β-glucuronidase.

Acid-sensitive anthracycline prodrugs

Doxorubicin prodrugs with cis-aconityl spacer molecules

The cis-aconityl spacer molecule was investigated from the early 1980s onwards with the aim of developing acid-sensitive polymer conjugates with amino-bearing drugs. Various doxo- and daunorubicin conjugates with synthetic polymers or monoclonal antibodies have meanwhile been prepared [207]. Most of the studies performed with these conjugates have shown that the release of the polymer-bound anthracycline is pH-dependent and that they exhibit *in vitro* cytotoxicity in the low micromolar range. Enhanced antitumor efficacy in murine mouse models compared to the parent compound has been demonstrated for selected acid-sensitive conjugates with polylysine [232] and monoclonal antibodies directed against antigens on leukemia and melanoma cells [233, 234].

Doxorubicin prodrugs with carboxylic hydrazone linkers

A number of doxorubicin derivatives containing an acid-sensitive hydrazone linker have been developed in the past 15 years [207]. These derivatives have been coupled to macromolecular carriers such as monoclonal antibodies, transferrin and albumin and representative examples are described below.

Doxorubicin hydrazone conjugates with monoclonal antibodies

In the late 1980s, a pharmaceutical research group at Bristol-Myers Squibb synthesized a 6-maleimidodocaproyl and a 3-(2'-pyridinyldithio)propanoyl hydrazone derivative of doxorubicin (see Fig. 17).

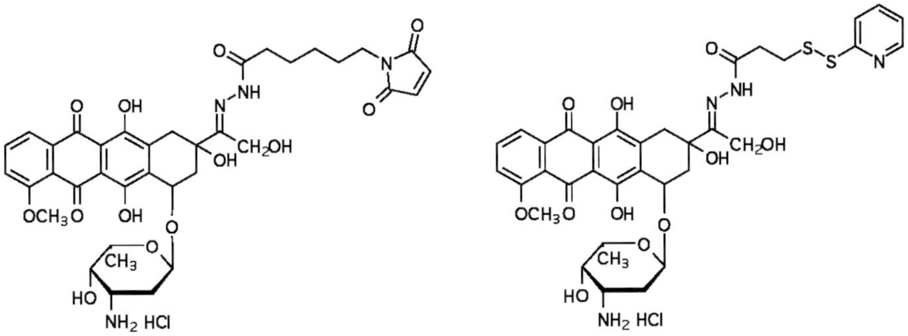

Figure 17. Structure of a (6-maleimidodocaproyl) and a 3-(2'-pyridinyldithio)propanoyl hydrazone derivative of doxorubicin.

Both derivatives were coupled to thiol-bearing monoclonal antibodies that bind to tumor-associated antigens with subsequent internalization of the antibody conjugate allowing a release of doxorubicin in the acidic pH of endosomes and lysosomes. Such designed antibody conjugates have shown high *in vitro* and *in vivo* activity [235–239].

Due to the high plasma stability of the resulting thioether bond that is formed after reaction of the maleimide with thiol groups, (6-maleimidocaproyl)hydrazone of doxorubicin was selected for developing a clinical candidate with the chimeric human/mouse monoclonal antibody that is specific for Lewis-Y, an antigen that is abundantly expressed on the surface of several human carcinomas [236]. In this conjugate, known as BR96-doxorubicin immunoconjugate, approximately eight molecules of (6-maleimidocaproyl) hydrazone of doxorubicin are coupled to the antibody. Therapy with the BR96-doxorubicin induced complete remissions in a number of xenograft tumor models and was superior to unbound doxorubicin [237–239]. Phase I/II have been performed with this immunoconjugate (see below).

Doxorubicin hydrazone conjugates with transferrin and albumin
Acid-sensitive anthracycline conjugates with serum albumin and transferrin have shown high antiproliferative activity *in vitro*, and selected conjugates that incorporate a phenylacetyl hydrazone linker, show superior antitumor efficacy in a number of animal tumor models when compared to the parent compound [240–243].

Interestingly, a comparison of analogous transferrin and albumin doxorubicin conjugates showed a very similar picture of *in vitro* as well as *in vivo* activity, i.e., inhibitory effects did not depend on the carrier protein but rather on the chemical link realized between the drug and the protein [243].

As a consequence, a therapeutic approach was investigated in which doxorubicin prodrugs bind rapidly and preferentially to circulating albumin after intravenous administration [244, 245]. Such doxorubicin prodrugs were developed to meet two features:

1. *In situ* binding of the prodrug to the cysteine-34 position of circulating albumin after intravenous administration due to the thiol-reactive maleimide
   group in the molecule.
2. Release of albumin-bound doxorubicin at the tumor site due to the incorporation of an acid-sensitive carboxylic hydrazone bond between the drug and
   the carrier.

Proof of concept was obtained with two acid-sensitive doxorubicin prodrugs, i.e., a (4-maleimidophenylacetyl)hydrazone and a (6-maleimido-
caproyl)hydrazone derivative of doxorubicin that are rapidly and selectively
bound to circulating albumin and are distinctly superior to the parent compound doxorubicin in murine tumor models [244, 245].

The (6-maleimidocaproyl)hydrazone derivative of doxorubicin (abbreviated
DOXO-EMCH) was selected as the investigational product for clinical evaluation due to:

- superior efficacy of DOXO-EMCH compared to free doxorubicin, the clinical standard, in a murine renal cell carcinoma model (RENCA) and in two
  mamma carcinoma xenograft models in nude mice (MDA-MB 435,
  MCF-7). Complete remissions were achieved with DOXO-EMCH in the
  RENCA and MDA-MB 435 model in contrast to therapy with doxorubicin.
- substantial increase of the maximum tolerated dose (MTD) of DOXO-
  EMCH in mice, rats and dogs when compared to conventional doxorubicin.
- rapid and selective binding to circulating albumin.
- high plasma stability.
- five to seven carbon atoms is the optimal length of an aliphatic maleimide
  spacer according to molecular modeling of the covalent interaction of
  maleimide spacers with the cysteine-34 position of human serum albumin.

Coincidentally, the (6-maleimidocaproyl)hydrazone derivative of doxorubicin, DOXO-EMCH, which has been evaluated in a Phase I study, is the identical molecule that was used for the preparation of the clinically tested BR96-
doxorubicin immunoconjugate.

Doxorubicin prodrugs in clinical studies

To the best of our knowledge, four macromolecular prodrugs of doxorubicin
(PK1, PK2, BR-96-doxorubicin immunoconjugate and DOXO-EMCH) and
two low-molecular weight prodrugs (L-377,202 and *N*-L-Leucyldoxorubicin)
have or are being evaluated in clinical trials.

N-Leucyldoxorubicin
In 1992 a Phase I study with *N*-Leucyldoxorubicin was reported, a prodrug
that was developed with the intention of reducing cardiotoxicity [246]. The
maximum tolerated dose of 225 mg/m^2 associated with bone marrow toxicity

was established. Doxorubicin was rapidly formed from *N*-Leucyldoxorubicin within a few minutes after administration.

L-377,202, a PSA-activated doxorubicin prodrug

L-377,202, a novel peptide doxorubicin conjugate that is cleaved by prostate-specific antigen, has been evaluated in a Phase I study. 19 patients with advanced hormone-refractory prostate cancer were treated intravenously with L-377202 at escalating dose levels of 20 to 315 mg/m^2 of L-377,202 [247]. Dose-limiting grade 4 neutropenia was noted in two of the patients receiving 315 mg/m^2. The recommended dose for Phase II studies was 225 mg/m^2. PK studies demonstrated that L-377202 was cleaved to Leucyldoxorubicin and doxorubicin. The two patients at 315 mg/m^2 had a greater than 75% decrease in PSA, and one patient had a stabilized PSA level. No response was noted at dose levels less than 225 mg/m^2 which was established as the MTD in this study and corresponds to approximately 90 mg/m^2 doxorubicin equivalents.

PK1 (doxorubicin-HPMA-copolymer)

PK1 is a doxorubicin-HPMA-copolymer-conjugate which is stable in the blood stream and releases doxorubicin in the lysosomal compartments of tumor cells. A Phase I study was carried out with 36 patients in Great Britain, which revealed that the maximum tolerated dose (MTD) was 320 mg/m^2 doxorubicin equivalents (intravenous application every 3 weeks) [248]; the dose-limiting factor observed in this study was bone marrow toxicity and mucositis. Other side-effects (e.g., nausea, diarrhea) were moderate (CTC-Grade 1). A noteworthy finding of this study was that no cardiotoxicity was observed even at these high doses. Two partial and two minor responses were seen in four patients with lung, breast and colorectal cancer. The recommended dose for Phase II studies was 280 mg/m^2 every 3 weeks. Phase II studies are ongoing [210].

PK2 (N-galactosamine linked doxorubicin-HPMA-copolymer)

PK2 is the first clinically tested drug polymer conjugate that additionally incorporates a targeting ligand, i.e., a galactosamine. 31 patients with primary or metastatic liver cancer were evaluated in a Phase I study [249, 250]. The MTD of PK2 was 160 mg/m^2 doxorubicin equivalents and was associated with severe fatigue, neutropenia and mucositis; 120 mg/m^2 was recommended as the dose for Phase II studies. Two partial responses and one minor response were achieved in this study.

BR96-doxorubicin immunoconjugate, an acid-sensitive immunoconjugate of doxorubicin

The BR96-doxorubicin immunoconjugate (BR96-DOX) has been evaluated in Phase I and II studies [251–254]. In a first Phase I study, the immunoconjugate was administered to 62 patients as an intravenous infusion every 21 days [251, 252]. Doses of BR96-DOX ranged from 66 to 875 mg/m^2, which is equivalent to 2 to 25 mg/m^2 of free doxorubicin. Two patients exhibited partial responses, one with breast and the other with gastric carcinoma.

In a second Phase I dose-escalation study, 34 patients with Le^y expressing tumors were treated with BR96-DOX administered as a weekly infusion of 100–500 mg/m^2 of BR96-DOX (equivalent to 3–15 mg/m^2 doxorubicin) [253]. Although antibody localization studies demonstrated binding of the immunoconjugate at the tumor site, no objective responses were observed. In both studies, BR96-DOX showed dose-limiting gastrointestinal (GI) toxicity at the highest doses. Recently, a randomized Phase II study was performed to evaluate the activity of BR96-DOX against metastatic breast cancer in patients with confirmed sensitivity to single-agent doxorubicin [254]. Patients received either 700 mg/m^2 of BR96-DOX (equivalent to 20 mg/m^2 DOX) or 60 mg/m^2 doxorubicin every 3 weeks. There was one partial response in the 14 patients receiving BR96-DOX but one complete and three partial responses in the 9 patients treated with doxorubicin alone. The cross-reactivity of BR96-DOX with normal gastrointestinal tissue led to prominent toxicities and probably impaired the delivery of the immunoconjugate to the tumor sites.

The low clinical response rates observed in these studies suggest that the dose which could be safely administered every 3 weeks was insufficient for maintaining the intratumoral concentration of doxorubicin required to achieve tumor regression.

DOXO-EMCH, the first albumin-binding doxorubicin prodrug to enter clinical trials

In a recently completed phase I study with DOXO-EMCH, the albumin-binding prodrug showed a good safety profile and antitumor efficacy. 41 patients with advanced cancer disease were treated with 2–6 intravenous cycles of DOXO-EMCH once every 3 weeks at a dose level of 20–340 mg/m^2 doxorubicin equivalents. Treatment with DOXO-EMCH was well tolerated up to 200 mg/m^2 without manifestation of drug-related side effects. Myelosuppression (grade 1–2), mucositis (grade 1–2) were the predominant adverse effects at dose levels of 260 mg/m^2 and myelosuppression (grade 1–3) as well as mucositis (grade 1–3) were dose-limiting at 340 mg/m^2. No acute cardiac toxicity was observed. Of 35 evaluable patients, 34% had progressive disease, 51% had disease stabilization, 6% had a minor response (sarcoma and parotis), 6% had a partial remission (mamma carcinoma and sarcoma) and 3% had a complete remission (small cell lung cancer). The recommended dose for phase II studies is 260 mg/m^2 which is approximately a 4-fold increase compared to standard treatment with doxorubicin (60 mg/m^2).

Perspectives

Will we ever find a better anthracycline [255]? DNR and DOX were the first anthracyclines that were clinically extremely useful for the treatment of leukemias and solid tumors. This was more than 40 years ago. Despite several decades of intense worldwide research by investigators in pharmaceutical companies and university institutes [256], better anthracyclines have not been approved although anthracyclines with a better preclinical profile were often described. Some analogs have some modestly reduced acute and/or chronic toxicity but they are not more effective against cancer than their parents drugs. It is possible that the best anthracyclines ever found are the first two discovered 50 years ago. Nature had a million years to develop and optimize these compounds, thus it seems possible that we still have the best anthracyclines in daily use. It is possibly hard to accept that a natural product cannot be improved by technology and human intelligence.

Our understanding of how anthracyclines act in the tumor cell is by far not complete. There is emerging evidence that different signaling pathways in the cell are affected. DNR activates the classical Raf-1/MEK/ERK pathway [257]. Raf-1 activation is mediated by complex signaling pathways that involves phosphatidylcholin-derived diacylglycerol and phosphoinositide 3 kinase lipid products that converge toward protein kinase C. Raf-1 activation itself mediates drug resistance and there are hints that increased activation of Raf-1 may upregulate transcription of P-gp because Raf-1 regulates the expression of mdr-1 (multi-drug-resistance gene) [258]. Drug resistance remains one of the most important causes of suboptimal results in cancer therapy. ATP-binding cassette (ABC) transporters are a family of transporter proteins that contribute to drug resistance via ATP-dependent drug efflux pumps. P-glycoprotein (P-gp), encoded by the MDR-1 gene, is an important ABC transporter and confers resistance to different anticancer agents, e.g., all anthracyclines, toxoids, podophyllotoxins and vinca alkaloids [259]. Moreover, anticancer therapy with cytotoxic drugs is involved in apoptosis, i.e., programmed cell death [260]. It has been shown that low levels of reactive oxygen species (ROS) induce apoptosis [261]. ROS generation after anthracycline administration has been extensively studied and is one of the major chemical reaction pathways of anthracyclines during metabolism. But even without ROS, anticancer agents are able to trigger apoptosis [262]. This was shown recently for a marine cytotoxic alkaloid, a DNA intercalating agent [263] and allows to suggest – in addition to all pathways known for the anthracyclines – that anthracyclines are able to produce apoptotic signals (from the nucleus by intercalation, from the nucleus by inhibition of the topoisomerase-II, from the intracellular space by ROS, etc.) leading to an activation of the intrinsic apoptotic pathway which is the mitochondria-apoptosome-mediated apoptotic pathway leading to cell death which is illustrated in Figure 18.

Signal transduction via the extrinsic pathway and use of the CD95 receptor (death receptors: FASL, DR4/5, TNFR) which is activated in case of binding with a death ligand is not involved.

Finally, angiogenesis research is now 30 years old and the first specific antiangiogenic drugs are entering the clinic. It has been known for years that angiogenesis is inhibited by anthracyclines [264] and recently metronomic

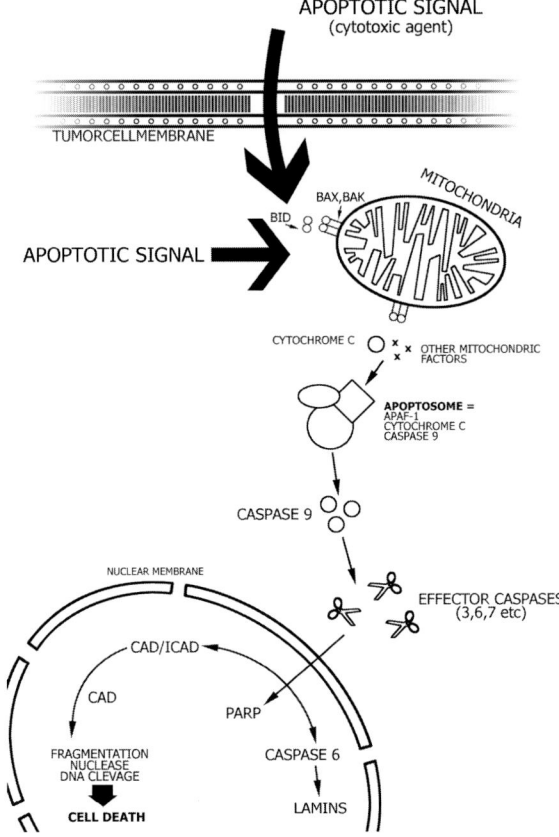

Figure 18. Mitochondrial pathway for apoptosis induction by apoptotic signals.

regimens were discussed which means that antiangiogenic effects of traditional cytotoxics might be exploited by schedules providing chronic exposure to
low drug concentrations [265]. Endothelial cell proliferation which is necessary for the development of new vessels can be blocked with very low cytotoxic drug levels because endothelial cells are more sensitive than others,
including cancer cells [266]. It will be worthwhile to follow the literature in
this emerging field of interest during the following years to see if metronomic
anticancer therapy will attain clinical practice. Liposomal formulations as well
as albumin-conjugates offer new chances to optimize anthracycline (DOX)
anticancer therapy by changing the pharmacokinetics and consecutively the
pharmacodynamics. The real clinical potential of these drug targeting technologies applied for anthracycline cancer treatment has to be explored broadly and the experienced oncologist has to be convinced that a real progress can
be verified.

References

1 Ghione M (1975) Development of adriamycin (NSC-123127). *Cancer Chemotherapy Reports* 6: 83–89

2 Di Marco A, Gaetani M, Dorigotti L et al. (1963) Studi sperimentali sull àttivita antineoplastica del nuovo antibiotico daunomicina. *Tumori* 49: 203–217

3 Di Marco A, Gaetani M, Scarpinato B (1969) Adriamycin (NSC-123127): a new antibiotic with antitumor activity. *Cancer Chemotherapy Reports* 53: 33–37

4 Bonadonna G, Monfardini S, de Lena M et al. (1969) Clinical evaluation of adriamycin, a new antitumor antibiotic. *Br Med J* 3: 503–506

5 Bonadonna G (1984) *Advances in anthracycline chemotherapy: Epirubicin.* Masson Italia Editori, Milano

6 Murdock KC, Wallace RE, Durr FE et al. (1979) Antitumor agents: I. 1,4-Bis((aminoalkyl)amino)-9,10-anthracenediones. *J Med Chem* 22: 1024–1030

7 Showalter HDH, Johnson JL, Werbel LM et al. (1984) 5-[(Aminoalkyl)amino]-substituted anthrax [1,9-cd]pyrazol-6(2H)-ones as novel anticancer agents. *J Med Chem* 27: 253–255

8 Lown JW (1988) *Anthracycline and anthracenediones-based anticancer agents.* Elsevier, Amsterdam

9 Monneret C (2001) Recent developments in the field of antitumor anthracyclines. *Eur J Med Chem* 36: 483–493

10 Hill BT, Whelan RDH (1982) A comparison of the lethal and kinetic effects of doxorubicin and 4'-epi-doxorubicin *in vitro*. *Tumori* 68: 29–37

11 Plosker GL, Faulds D (1993) Epirubicin, a review of its pharmacodynamic and pharmacokinetic proerties, and therapeutic use in cancer therapy. *Drugs* 45: 788–856

12 Hollingshead LM, Faulds D (1991) Idarubicin, a review of its pharmacodynamic and pharmacokinetic proerties, and therapeutic use in cancer therapy. *Drugs* 42: 690–719

13 Casazza AM, Giuliani FC (1984) Preclinical properties of epirubicin. In: G Bonnadonna (ed.): In: *Advances in Anthracycline chemotherapy*. Masson, Milano

14 Arcamone F, Berbardi L, Patelli B et al. (1979) Synthesis and antitumour activity of new daunoribicin und doxorubicin analogues. *Experientia* 34: 1255–1257

15 Ames MM, Spreafico F (1992) Selected pharmacologic characteristics of idarubicin and idarubicinol. *Leukemia* 6: 70–75

16 Dalmark M, Strom HH (1981) A fickian diffusion transport process with features of transport catalysis. *J Gen Physiol* 78: 349–364

17 Gianni L, Corden B, Myers C (1983) The biochemical basis of anthracycline toxicity and antitumor action. *Rev Biochem Toxicol* 5: 1–82

18 Kartner N, Riordan JR, Ling V (1983) Cell-surface P-glycoprotein associated with multidrug resistance in mammalian cell lines. *Science* 221: 1285–1288

19 Tritton T, Yee G (1982) The anticancer agent adriamycin can be actively cytotoxic without entering the cell. *Science* 217: 248–250

20 Pollard TD, Earnshaw WC (2002) *Cell biology*. Saunders, Philadelphia

21 Malisza KL, McIntosh AR, Sceinson E et al (1996) Semiquinone free radical formation by daunorubicin aglycone incorporated into the cellular membranes of intact Chinese hamster ovary cells. *Free Rad Res* 24: 9–18

22 Leonhard GA, Brown T, Hunter WT (1992) Anthracycline binding to DNA. *Eur J Biochem* 204: 69–74

23 Dessypris EN, Brenner DE, Hande KR (1986) Toxicity of doxorubicin metabolites to human erythroid and myeloid progenitors *in vitro*. *Cancer Treatment Reports* 70: 487–490

24 Quigley JG, Wang AH, Ughetto G et al. (1980) Molecular structure of an anticancer-drug-DNA complex: daunomycin plus d(CpGpTpApCpG). *Proc Natl Acad Sci USA* 77: 7204–7205

25 Ross WA, Glaubiger DL, Kohn KW (1978) Protein-associated DNA breaks in cells treated with adriamycin and ellipicine. *Biochim Biophys* 519: 23–30

26 Tewey KM, Chen GI, Nelson EM et al. (1984) Intercalative antitumor drugs interfere with the breakage-reunion reaction of mammalian DNA topoisomerase. *J Biol Chem* 259: 9182–9187

27 Glisson B, Gupta R, Hodges P et al. (1986) Cross-resistance to intercalating agents in an epipodophyllotoxin-resitant Chinese hamster ovary cell line: evidence for a common intracellular target. *Cancer Res* 46: 1939–1942

28 Abdella BRJ, Fisher J (1985) A chemical perspective on the anthracycline antitumor antibiotics. *Environ Health Perspect* 64: 3–18

29 Bachur NR, Gordon SL, Gee MV (1977) A general mechanism for microsomal activation of quinone anticancer agents to free radicals. *Cancer Res* 38: 1745–1752

30 Doroshow JH, Locker GY, Meyers CE (1980) The enzymatic defenses of the mouse heart against reactive oxygen metabolites. *J Clin Invest* 65: 128–135

31 Burton KP, McCord JM, Ghai G (1984) Myocardial alterations due to free-radical generation. *Heart Circ Physiol* 15: H776–H783

32 Laurent G, Jaffrezou JP (2001) Signaling pathways activated by daunorubicin. *Blood* 98: 913–924

33 Mross K (1993) *Klinische und pharmakologische Untersuchungen zur Pharmakokinetik, Metabolisierung, Pharmakodynamik und Toxizität von Anthrazyklinen.* Zuckschwerdt, München

34 Andriollo M, Favier A, Guiraud P (2003) Adriamycin activates NF-κB in human carcinoma cells by IκBα degradadtion. *Archives Biochem Biophys* 413: 75–82

35 Weistein-Oppenheimer CR, Hendiquez-Roldan CF, Davis JM et al. (2001) Role of the Raf signal transduction cascade in the *in vitro* resistance to the anticancer drug doxorubicin. *Clin Cancer Res* 7: 2898–2907

36 Mansat-De Mas V, Hernandez H, Plo I et al. (2003) Protein kinase Cξ mediated Raf-1/extracellular-regulated kinase activation by daunorubicin. *Blood* 101: 1543–1550

37 Maessen PA, Mross K, Pinedo HM et al (1988) A new method for the determination of doxorubicin, 4'-epi-doxorubicin and all known metabolites in tissue. *J Chromat Biomed Appl* 424: 103–110

38 Maessen PA, Mross K, Oinedo HM et al (1987) Improved method for the determination of 4'-epi-doxorubicin and seven metabolites in plasma by HPLC. *J Chromat Biomed Appl* 417: 339–346

39 Alberts DS, Bachur NR, Holtzman JL (1971) The pharmacokinetics of daunomycin in man. *Clin Pharmacol Ther* 12: 96–104

40 Riggs CE, Bachur NR (1983) Clinical pharmacokinetics of anthracycline antibiotics. In: MM Ames, G Powlis, JS Kovach (eds): In: *Pharmacokinetics of anticancer agents in humans*. Elsevier Science Publisher, Amsterdam

41 Benjamin RS, Riggs CE, Bachur NR (1977) Plasma pharmacokinetcs of adriamycin and its metabolites in humans with normal hepatic and renal function. *Cancer Res* 37: 1416–1420

42 Brenner DE, Grosh WW, Noone R et al. (1984) Human plasma pharmacokinetics of doxorubicin: Comparison of bolus and infusional administration. *Cancer Treat Symp* 3: 77–83

43 Robert J, Vrignaud P, Hguyen-Ngoc T et al. (1985) Comparative pharmacokinetics and metabolism of doxorubicin and epirubicin in patients with metastatic breast cancer. *Cancer Treat Rep* 69: 633–640

44 Camaggi CM, Comparsi R, Strocchi E et al. (1988) HPLC analysis of doxorubicin, epirubicin ans fluorescent metabolites in biological fluids. *Cancer Chemother Pharmacol* 21: 216–220

45 Mross K, Maessen P, van der Vijgh WJF et al. (1988) Pharmacokinetics and metabolism of epirubicin and doxorubicin in man. *J Clin Oncol* 6: 517–526

46 Eksborg S, Stendahl U, Lönroth U (1986) Comparative pharmacokinetic study of adriamycin and 4'-epi-adriamycin after their simultaneous intravenous administration. *Eur J Clin Pharm* 30: 629–631

47 Camaggi CM, Strocchi E, Carisi P et al. (1992) Idarubicin metabolism and pharmacokinetics after intravenous and oral administration in cancer patients: a crossover study. *Cancer Chemother Pharmacol* 30: 307–316

48 Robert J, Rigal-Huguet F, Hurteloup P (1992) Comparative pharmacokinetic study of idarubicin and daunorubicin in leukaemia patients. *Hematological Oncology* 10: 111–116

49 Reid JM, Pendergrass TW, Krailo MD et al. (1990) Plasma pharmacokientics and cerebrospinal fluid concentrations of idarubicin and idarubicinol in pediatric leukemia patients: a childrens cancer study group report. *Cancer Res* 50: 6525–6528

50 Mross K, Mayer U, Hamm K et al (1990) High-performance liquid chromatography analysis of iodo-doxorubicin and fluorescent metabolites in plasma samples. *J Chromatography* 530: 192–199

51 Speth PA, Minderman H, Haanen C (1989) Idarubicin vs daunorubicin: preclinical and clinical pharmacokinetic studies. *Sm Oncol* 16: 2–9

52 Riggs CE, Linssen PCM, Serpick A et al (1987) Biliary disposition of adriamycin. *Clin Pharmacol Ther* 22: 234–241

53 Camaggi CM, Strocchi E, Comparsi R et al. (1986) Biliary excretion and pharmacokinetics of

4'-epidoxorubicin (epirubicin) in advanced cancer patients. *Cancer Chemother Pharmacol* 18: 47–50

54 Speth PAJ, Linssen PCM, Holdrinet RSG et al (1987) Plasma and cellular adriamycin concentrations in patients with myeloma treated with ninety-six-hour continuous infusion. *Clin Pharmacol Ther* 41: 661–665

55 Speth PAJ, Linssen PCM, Beex LVA et al. (1986) Cellular and plasma pharmacokinetics of weekly 20 mf 4'-epi-adriamycin bolus injection in patients with advanced breast carcinoma. *Cancer Chemother Pharmacol* 18: 78–82

56 Van der Vijgh WJF, Maessen PA, Pinedo HM (1990) Comparative metabolism and pharmacokinetics of doxorubicin and 4'-epidoxorubicin in plasma, heart and tumor-bearing mice. *Cancer Chemother Pharmacol* 26: 9–12

57 Maessen PA, Mross K, Pinedo HM et al (1987) Metabolism of epidoxorubicin in animals: absence of glucuronidation. *Cancer Chemother Pharmacol* 20: 85–87

58 Bronschud MH, Margison JM, Howell A et al. (1990) Comparative pharmacokinetics of escalating doses of doxorbicin in patients wit metastatic breasr cancer. *Cancer Chemother Pharmacol* 25: 435–439

59 Loveless H, Arena E, Felsted RL et al. (1978) Comparative mammalian metabolism of adriamycin and daunomycin. *Cancer Res* 38: 593–598

60 Cummings J, Merry S, Willmott N (1986) Disposition kinetics of adriamycin, adriamycinol and their 7-deoxyaglycones in AKR mice bearing a sub-cutaneously growing ridgway osteogenic sarcoma (ROS). *Eur J Cancer Clin Oncol* 22: 451–460

61 Cummings J, Milstead R, Cunnigham D et al (1986) Marked inter-patient variation in adriamycin biotransformation to 7-deoxyaglycones: Evidence from metabolites identified in serum. *Eur J Cancer Clin Oncol* 22: 991–1001

62 Takanashi S, Bachur NR (1976) Adriamycin metabolism in man. Evidence from urinary metabolites. *Drug Metab Disp* 4: 79–87

63 Weenen H, Lankelma J, Penders PGM et al. (1983) Pharmacokinetics of 4-epi-doxorubicin in man. *Invest New Drugs* 1: 59–64

64 Kaplan S, Sessa C, Willems Y et al. (1984) Phase I trial with 4-demethoxydaunorubicin (idarubicin) with single doses. *Investigational Drugs* 2: 281–286

65 Smith SR, Marginon JM, Lucas SB et al. (1987) Clinical pharmacology of oral and intravenous 4-demethoxy-daunorubicin. *Cancer Chemother Pharmacol* 19: 138–142

66 Benjamin RS, Wiernik PH, Bachur NR (1974) Adriamycin chemotherapy efficacy, safety and pharmacologic basis of an intermittent single high dose schedule. *Cancer* 33: 19–27

67 Chan KK, Chlebowski RT, Tong M et al. (1980) Clinical pharmacokinetics of adriamycin in hepatoma patients with cirrhosis. *Cancer Res* 40: 1263–1268

68 Müller HJ, Port RE, Grubert M et al. (2003) The influenece of liver metastasis on the pharmacokinetics of doxorubicin – a population based pharmacokinetic project of the "Arbeitsgruppe Pharmakologie in der Onkologie und Hämatologie" (APOH). *Int J Clin Pharm Ther* 41: 598–599

69 Camaggi CM, Strocchi E, Tamassia V et al. (1982) Pharmacokinetic studies of 4'epi-doxorubicin in cancer patienst with normal and impaired renal function and with hepatic metastasis. *Cancer Treatment Rep* 66: 1819–1824

70 Twelves CJ, Dobbs NA, Michael Y et al. (1992) Clinical pharmacokinetics of epirubicin: the importance of liver biochemistry. *Br J Cancer* 66: 765–769

71 Dobbs NA, Twelves CJ (1998) Anthracycline doses in patients with liver dysunction: do UK oncologists follow current recommendations? *Brit J Cancer* 77: 1145–1148

72 Ralph LD, Thomson AH, Dobbs NA et al (2003) A population model of epirubicin pharmacokinetics and application to dosage guidelines. *Cancer Chemother Pharmacol* 52: 34–40

73 Dobbs NA, Twelves CJ, Gregory W et al. (2003) Epirubicin in patients with liver dysfunction: development and evaluation of a novel dose modification scheme. *Eur J Cancer* 39: 557–559

74 Columbo T, Donelli MG, Urso R et al. (1989) Doxorubicin toxicity and pharmacokinetics in old and young rats. *Experimental Gerontology* 24: 159–171

75 Li J, Gwilt (2003) The effect of age on the early disposition of doxorubicin. *Cancer Chemother Pharmacol* 51: 395–402

76 Wade JR, Kelman AW, Kerr DJ et al. (1992) Variability in he pharmacokinetics of epirubicin: a population analysis. *Cancer Chemother Pharmacol* 29: 391–395

77 Mross K, Hamm K, Hossfeld DK (1993) The effects of verapamil on the pharmacokinetics and metabolism of epirubicin. *Cancer Chemother Pharmacol* 31: 369–375

78 Cusack BJ, Tesnohlidek DA, Loseke VL et al. (1988) Effect of phenytoin on the pharmacokinetics of doxorubicin and doxorubicinol in the rabbit. *Cancer Chemother Pharmacol* 22: 294–298

79 Brenner DE (1987) Approaches to the problem of individual doxorubicin dosing schedules. *Path Biol* 35: 31–39

80 Minderman H, Linssen PCM, Wessels JMC et al (1991) Doxorubicin toxicity in relation to the proliferative state of human hematopoietic cells. *Exp Hematol* 19: 110–114

81 Drewinko B, Patchen M, Yang LY et al (1981) Differential killing efficacy of twenty antitumor drugs on proliferating and nonproliferating human tumor cells. *Cancer Res* 41: 2328–2333

82 Raijmakers R, Speth P, de Witte T et al. (1987) Infusion-rate independent cellular adrimycin concentrations and cytotoxicity to human bone marrow clonogenic cell (CFU-GM). *Br J Cancer* 56: 123–126

83 Bristow MR, Thomson PD, Martin RP (1978) Early anthracycline cardiotoxicity. *Am J Med* 65: 823–832

84 Billingham ME, Mason GW, Bristow MR et al. (1978) Anthracycline cardiomyopathy monitored by morphologic changes. *Cancer Treat Rep* 62: 865–872

85 von Hoff DD, Rosencweig M, Layard M et al. (1977) Daunomycin-induced cardiotoxicity in children and adults. *Am J Med* 62: 200–205

86 Minow RA, Benjamin RS, Gottlieb JA (1975) Adriamycin (NSC-123127) cardiomyopathy-an overview with determination of risk factors. *Cancer Chemother Rep* 6: 195–201

87 Rinehart JJ, Lewis RP, Balcerzak SP (1974) Adriamycin cardiotoxicity in man. *Ann Intern Med* 81: 475–478

88 von Hoff DD, Layard (1981) Risk factors for development of daunorubicin cardiotoxicity. *Cancer Treat Rep* 65: 19–23

89 Ryberg M, Nielsen D, Skovsgaard T et al. (1998) Epirubicin cardiotoxicity: an analysis of 469 patients with metastatic breast cancer. *J Clin Oncol* 16: 3502–3508

90 Jensen BV, Skovsgaard T, Nielsen SL et al. (2002) Functional monitoring of anthracycline cardiotoxicity: a prospective, blinded, long-term observational study of outcome in 120 patients. *Ann Oncol* 13: 699–709

91 Swain SM, Hhaley FS, Ewer MS (2003) Congestive heart failure in patients treated with doxorubicin. *Cancer* 97: 2869–2879

92 Schwartz R, McKenzie W, Alexander J et al. (1987) Congestive heart failure and left ventricular dysfunction complicating doxorubicin therapy. Seven years experience using serial radionuclide angiocardigraphy. *Am J Med* 82: 1109–1118

93 Sorenson K, Levitt GA, Bull C et al. (2003) Late anthracycline cardiotoxicity after childhood cancer – a prospective longitudinal study. *Cancer* 97: 1991–1998

94 Billingham ME, Bristow MR (1984) Evaluation of anthracycline cardiotoxicity: predictive ability and functional correlation of endomyocardial biopsy. *Cancer Treat Symp* 3: 71–76

95 Bristow MR, Mason JW, Billingham ME et al. (1981) Dose-effect and structure-function relationship in doxorubicin cardiomyopathy. *Am Heart J* 102: 709–718

96 van Hoff D, Layard M, Basa P et al. (1979) Risk factors for doxorubicin-induced congestive heart failure. *Ann Intern Med* 91: 710–717

97 Lum BL, Svec JM, Torti FM (1985) Doxorubicin: alteration of dose scheduling as a means of reducing cardiotoxicity. *Drug Intelligence Clin Pharm* 19: 259–264

98 Hortobagyi GN, Frye D, Buzdar AU et al. (1989) Decreased cardiac toxicity of doxorubicin administered by continuous intravenous infusion in combination chemotherapy for metastatic breast cancer. *Cancer* 63: 37–45

99 Kalyanaraman B, Joseph J, Kalivendi S et al. (2002) Doxorubicin-induced apoptosis: implications in cardiotoxicity. *Mol Cell Biochem* 234: 119–124

100 Myers CE, McGuire WP, Liss RH et al. (1977) Adriamycin: the role of lipid peroxidation in cardiac toxicity and tumor response. *Science* 197: 165–167

101 Doroshow JH, Locker GY, Ifrim I et al (1981) Prevention of doxorubicin cardiac toxicity in the mouse by N-acetylcysteine. *J Clin Invest* 68: 1053–1064

102 Speyer JL, Green MD, Kramer E et al. (1988) Protective effect of the bispiper azinrdione ICRF-187 against doxorubicin-induced cardiac toxicity in women with advanced breast cancer. *New Eng J Med* 319: 745–752

103 Pollakis G, Goormaghtigh E, Ruysschaert JM (1983) Role of the quinone structure in the mitochondrail damage induced by antitumor anthracyclines. Comparisons of adriamycin and 5-iminodaunorubicin. *FEBS Letters* 155: 267–272

104 Nielsen D, Jensen JB, Dombernowsky P et al. (1990) Epirubicin cardiotoxicity: a study of 135 patients with advanced breast cancer. *J Clin Oncol* 8: 1806–1810

105 Dunagin WG (1982) Clinical toxicity of chemotherapy agents: dermatologic toxicity. *Sem Oncol* 9: 14–21

106 Desai MH, Teres D (1982) Prevention of doxorubicin-induced skin ulcers in the rat and pig with dimethylsulfoxide (DMSO). *Cancer Treat Rep* 66: 1371

107 Lawrence HJ, Walsh D, Zapatowski KA (1989) Topical dimethylsulfoxide may prevent tissue damage from anthracycline extravasation. *Cancer Chemother Pharmacol* 23: 316–318

108 Mader I, Fürst-Weger PR, Mader RM et al. (2002) Paravasation von Zytostika, Springer Wien, New York

109 Olver N, Aisner J, Hament A et al. (1988) A prospective study of topical dimethyl sulfoxide for treating anthracycline extravasation. *J Clin Oncol* 6: 1732–1735

110 Dührsen U, Heinrichs V, Beeken WD et al. (1997) Local and systemic sequelae of mediastinal daunorubicin extravasation in a patient with acute myelomonocytic leukemia. *Ann Oncology* 8: 1167–1168

111 Morris D, Aisner J, Wiernik PH (1977) Horizontal pigmented banding of the nails in association with adriamycin chemotherapy. *Cancer Treat Rep* 61: 499–501

112 Priestman TJ, James KW (1975) Adriamycin and longitudinal banding of fingernails. *Lancet* 7920: 1337–1338

113 Pedersen-Bjergaard J, Specjt L, Larson SO et al. (1987) Risk of therapy – related leukemia and preleukemia after Hodgkin's disease: Relation to age, cumulative dose of alkylating agents, and time from chemotherapy. *Lancet* 2: 83–88

114 Pedersen-Bjergaard J, Daugaard G, Hansen SW et al. (1991) Increased risk of myelodysplasia and leukaemia after etoposide, cisplatin, and bleomycin for germ-cell tumours. *Lancet* 338: 359–363

115 Pedersen-Bjergaard J, Sigsgard TC, Nielsen D et al. (1992) Acute monocytic or myelomonocytic leukaemia with balanced chromosome translocations to band 11q23 after therapy with 4-epi-doxorubicin and cisplatin or cyclophosphamide for breast cancer. *J Clin Oncol* 10: 1444–1451

116 Felix CA (1998) Secondary leukemias induced by topoisomerase-targeted drugs. *BBA-Gene Structure and Expression* 1400: 233–255

117 Alastair JJ, Wood MD (2001) Side effects of adjuvant treatment of breast cancer. *New Eng J Med* 344: 1997–2008

118 Tallmann MS, Gray R, Bennett JM et al. (1995) Leukemogenic potential of adjuvant chemotherapy for early-stage breast cancer: the Eastern Cooperative Oncology Group experience. *J Clin Oncol* 12: 1557–1563

119 Curtis RE, Boice JD, Stovall M et al. (1992) Risk of leukemia after chemotherapy and radiation treatment for breast cancer. *New Eng J Med* 326: 1745–1751

120 Smith RE, Bryant J, DeCillis A et al (2003) Acute myeloid leukaemia and myelodysplastic syndrome after doxorubicin-cyclophosphamide adjuvant therapy for operable breast cancer: The National Surgical Adjuvant Breast and Bowel Project experience. *J Clin Oncol* 21: 1195–1204

121 Crump M, Tu D, Shephard L et al. (2003) Risk of acute leukemia following epirubicin-based adjuvant chemotherapy: A report from the National Cancer Institute of Canada Clinical Trial Group. *J Clin Oncol* 21: 3066–3071

122 Bernard-Marty C, Mano M, Paesmans M et al. (2003) Second malignancies following adjuvant chemotherapy: 6-year results from a Belgian randomised study comparing cyclophosphamide, methotrexate and 5-fluoruracil (CMF) with an anthracycline-based regimen in adjuvant treatment of node-positive breast cancer patients. *Ann Oncol* 14: 693–698

123 Chaplin G, Milan C, Sgro C et al. (2000) Increased risk of acute leukaemia after adjuvant chemotherapy for breast cancer: a population-based study. *J Clin Oncol* 18: 2836–2842

124 Budman DR, Berry DA, Cirrincione CT et al. (1998) Dose and dose intensity as determinants of outcome in the adjuvant treatment of breast cancer. *J Natl Cancer Inst* 90: 1205–1211

125 Bake SD, Verweij J, Rowinsky EK et al. (2002) Role of body surface area in dosing of investigational anticancer agents in adults. *J Nat Cancer Inst* 94: 1883–1888

126 Reilly JJ, Workman P (1993) Normalization of anti-cancer drug dosage using body surface area: is it worthwhile? *Cancer Chemother Pharmacol* 32: 411–418

127 Büchner T, Urbanitz D, Hiddemann W et al. (1985) Intensified induction and consolidation with or without maintenance chemotherapy for acute myeloid leukaemia (AML): two multicenter studies of the german AML cooperative study group. *J Clin Oncol* 3: 1583–1580

128 Hiddemann W, Kern W, Schoch C et al. (1999) Management of acute myeloid leukemia in elderly patients. *J Clin Oncol* 17: 3569–3576

129 French Epirubicin Study Group (1991) A prospective randomized trial comparing epirubicin monochemotherapy to two fluoruracil, cyclophosphamide, and epirubicin regimens differing in epirubicin dose in advanced breast cancer patients. *J Clin Oncol* 9: 305–312

130 Perez DJ, Harvey VJ, Robinson BA et al. (1991) A randomized comparison of single-agent doxorubicin and epirubicin as first-line cytotoxic therapy in advanced breast cancer. *J Clin Oncol* 9: 2148–2152

131 Habeshaw T, Jones JPR, Stallard S et al. (1991) Epirubicin at two dose levels with prednisolone as treatment for advanced breast cancer: the results of a randomized trial. *J Clin Oncol* 9: 295–304

132 Marschner N, Kreiberg R, Souchon R et al. (1994) Evaluation of the importance and relevance of dose intensity using epirubicin and cyclophosphamide in metastatic cancer: analysis of a prospective randomized trial. *Sem Oncol* 21 (Suppl 1): 10–16

133 Jain KK, Casper ES, Geller et al. (1985) A prospective randomised comparison of epirubicin and doxorubicin in patients with advanced breast cancer. *J Clin Oncol* 3: 818–826

134 Gurney H, Ackland S, Gebski V, Farell G (1998) Factors affecting epirubicin pharmacokinetics and toxicity: evidence against using body-surface area for dose calculation. *J Clin Oncol* 16: 2299–2304

135 AML Collaborative Group (1998) A systematic collaborative overview of randomized trials comparing idarubicin with daunorubicin (or other anthracyclines) as induction therapy for acute myeloid leukemia. *Br J Haematol* 103: 100–109

136 Rowe JM, Neuberg D, Friedenberg W et al. (2003) A Phase III study of three induction regimens and of priming with GM-CSF in older adults with acute myeloid leukaemia: a trial by the Eastern Cooperative Oncology Group. *Blood* Sept 25 (Epub ahead of print)

137 Flasshove M, Meussers P, Schutte J et al. (2000) Long-term survival after induction therapy with idarubicin and cytosine arabinoside for *de novo* acute myeloid leukaemia. *Ann Hematol* 79: 533–542

138 Gregoriadis G (1978) Liposomes in therapeutic and preventive medicine: the development of the drug-carrier concept. *Ann NY Acad Sci* 308: 343–370

139 Gregoriadis G (1995) Engineering liposomes for drug delivery: progress and problems. *Trends Biotechnol* 13: 527–537

140 Gregoriadis G (1976) The carrier potential of liposomes in biology and medicine (second of two parts). *N Engl J Med* 295: 765–770

141 Gregoriadis G (1976) The carrier potential of liposomes in biology and medicine (first of two parts). *N Engl J Med* 295: 704–710

142 Massing U (1997) Cancer therapy with liposomal formulations of anticancer drugs. *Int J Clin Pharmacol Ther* 35: 87–90

143 Brandl M, Massing U (2003) Vesicular phospholipid gels. In: P Torchilin, V Weissig (eds): In: *Liposomes – a practical approach*, Oxford University Press, Oxford 353–374

144 Madden TC, Hanigan PR, Tai L et al. (1990) The accumulation of drugs with unilamellar vesicles exhibiting a proton gradient: a survey. *Chem Phys Lipids* 53: 37–46

145 Haran G, Cohen R, Bar LK et al. (1993) Transmembrane ammonium sulfate gradients in liposomes produce efficient and stable entrapment of amphipathic weak bases. *Biochim Biophys Acta* 1151: 201–215

146 Matsumura Y, Maeda H (1986) A new concept for macromolecular therapeutics in cancer chemotherapy: mechanism of tumoritropic accumulation of proteins and the antitumor agent smancs. *Cancer Res* 46: 6387–6392

147 Maeda H, Matsumura Y (1989) Tumoritropic and lymphotropic principles of macromolecular drugs. *Crit Rev Ther Drug Carrier Syst* 6: 193–210

148 Yuan F, Leunig M, Huang SK et al (1994) Microvascular permeability and interstitial penetration of sterically stabilized (stealth) liposomes in a human tumor xenograft. *Cancer Res* 54: 3352–3356

149 Drummond DC, Meyer O, Hong K et al (1999) Optimizing liposomes for delivery of chemotherapeutic agents to solid tumors. *Pharmacol Rev* 51: 691–743

150 Yuan F, Dellian M, Fukumura D et al (1995) Vascular permeability in a human tumor xenograft: molecular size dependence and cutoff size. *Cancer Res* 55: 3752–3756

151 Forssen EA, Ross ME (1994) Daunoxome treatment of solid tumors: preclinical and clinical

investigations. *J Liposome Res* 4: 481–512

152	Alberts DS, Garcia DJ (1997) Safety aspects of pegylated liposomal doxorubicin in patients with cancer. *Drugs* 54: 30–35

153	Gabizon A, Papahadjopoulos D (1988) Liposome formulations with prolonged circulation time in blood and enhanced uptake by tumors. *Proc Natl Acad Sci USA* 85: 6949–6953

154	Scherphof GL, Morselt H, Allen TM (1994) Intrahepatic distribution of long-circulating liposomes containing poly(ethylene glycol) distearoyl phosphatidylethanolamine. *J Liposome Res* 4: 213–228

155	Patel HM (1992) Serum opsonins and liposomes: their interaction and opsonophagocytosis. *Crit Rev Ther Drug Carrier Syst* 9: 39–90

156	Massing U, Fuxius S (2000) Liposomal formulations of anticancer drugs: selectivity and effectiveness. *Drug Resist Updat* 3: 171–177

157	Chonn A, Cullis PR (1995) Recent advances in liposomal drug-delivery systems. *Curr Opin Biotechnol* 6: 698–708

158	Lasic DD, Papahadjopoulos D (1995) Liposomes revisited. *Science* 267: 1275–1276

159	Bakker-Woudenberg IA, Lokerse AF, ten Kate MT, Storm G (1992) Enhanced localization of liposomes with prolonged blood circulation time in infected lung tissue. *Biochim Biophys Acta* 1138: 318–326

160	Senior JH (1987) Fate and behavior of liposomes *in vivo*: a review of controlling factors. *Crit Rev Ther Drug Carrier Syst* 3: 123–193

161	Harashima H, Hiraiwa T, Ochi Y et al (1995) Size dependent liposome degradation in blood: *in vivo/in vitro* correlation by kinetic modeling. *J Drug Target* 3: 253–261

162	Harashima H, Sakata K, Funato K et al (1994) Enhanced hepatic uptake of liposomes through complement activation depending on the size of liposomes. *Pharm Res* 11: 402–406

163	Harashima H, Ochi Y, Kiwada H (1994) Kinetic modelling of liposome degradation in serum: effect of size and concentration of liposomes *in vitro*. *Biopharm Drug Dispos* 15: 217–225

164	Forssen EA, Coulter DM, Proffitt RT (1992) Selective *in vivo* localization of daunorubicin small unilamellar vesicles in solid tumors. *Cancer Res* 52: 3255–3261

165	Devine DV, Wong K, Serrano K et al. (1994) Liposome-complement interactions in rat serum: implications for liposome survival studies. *Biochim Biophys Acta* 1191: 43–51

166	Moghimi SM, Patel HM (1992) Opsonophagocytosis of liposomes by peritoneal macrophages and bone marrow reticuloendothelial cells. *Biochim Biophys Acta* 1135: 269–274

167	Patel HM (1992) Influence of lipid composition on opsonophagocytosis of liposomes. *Res Immunol* 143: 242–244

168	Papahadjopoulos D, Jacobson K, Nir S et al (1973) Phase transitions in phospholipid vesicles. Fluorescence polarization and permeability measurements concerning the effect of temperature and cholesterol. *Biochim Biophys Acta* 311: 330–348

169	Allen TM, Hansen C, Martin F et al. (1991) Liposomes containing synthetic lipid derivatives of poly(ethylene glycol) show prolonged circulation half-lives *in vivo*. *Biochim Biophys Acta* 1066: 29–36

170	Blume G, Cevc G (1993) Molecular mechanism of the lipid vesicle longevity *in vivo*. *Biochim Biophys Acta* 1146: 157–168

171	Papahadjopoulos D, Allen TM, Gabizon A et al (1991) Sterically stabilized liposomes: improvements in pharmacokinetics and antitumor therapeutic efficacy. *Proc Natl Acad Sci USA* 88: 11460–11464

172	Gabizon A, Catane R, Uziely B et al (1994) Prolonged circulation time and enhanced accumulation in malignant exudates of doxorubicin encapsulated in polyethylene-glycol coated liposomes. *Cancer Res* 54: 987–992

173	Alberto A, Gabizon A (1995) Stealth liposomes and cancer targeting, a realistic compromize in drug delivery. *J Liposome Res* 5: 705–710

174	Woodle MC, Collins LR, Sponsler E et al (1992) Sterically stabilized liposomes. Reduction in electrophoretic mobility but not electrostatic surface potential. *Biophys J* 61: 902–910

175	Torchilin VP (1996) How do polymers prolong circulation time of liposomes? *J Liposome Res* 6: 99–116

176	Guthlein F, Burger AM, Brandl M et al (2002) Pharmacokinetics and antitumor activity of vincristine entrapped in vesicular phospholipid gels. *Anticancer Drugs* 13: 797–805

177	Moog R, Burger AM, Brandl M et al (2002) Change in pharmacokinetic and pharmacodynamic behavior of gemcitabine in human tumor xenografts upon entrapment in vesicular phospholipid gels. *Cancer Chemother Pharmacol* 49: 356–366

178 Martin FJ (1997) Stealth liposome technology: an overview. *Doxil Clinical Series* 1: 1–7

179 Stewart S, Jablonowski H, Goebel FD et al (1998) Randomized comparative trial of pegylated liposomal doxorubicin *versus* bleomycin and vincristine in the treatment of AIDS-related Kaposi's sarcoma. International Pegylated Liposomal Doxorubicin Study Group. *J Clin Oncol* 16: 683–691

180 Northfelt DW, Dezube BJ, Thommes JA et al (1998) Pegylated-liposomal doxorubicin *versus* doxorubicin, bleomycin, and vincristine in the treatment of AIDS-related Kaposi's sarcoma: results of a randomized Phase III clinical trial. *J Clin Oncol* 16: 2445–2451

181 Gill PS, Wernz J, Scadden DT et al (1996) Randomized Phase III trial of liposomal daunorubicin *versus* doxorubicin, bleomycin, and vincristine in AIDS-related Kaposi's sarcoma. *J Clin Oncol* 14: 2353–2364

182 Tulpule A, Yung RC, Wernz J et al (1998) Phase II trial of liposomal daunorubicin in the treatment of AIDS-related pulmonary Kaposi's sarcoma. *J Clin Oncol* 16: 3369–3374

183 Rose P, Gordon AN, Granai CO et al. (1999) Interim analysis of a non-comparative, multicenter study of Doxil/Caelyx in the treatment of patients with refractory ovarian cancer. *Proc ASCO Abstract* No 1392

184 Ranson MR, Carmichael J, O'Byrne K et al (1997) Treatment of advanced breast cancer with sterically stabilized liposomal doxorubicin: results of a multicenter Phase II trial. *J Clin Oncol* 15: 3185–3191

185 Lyass O, Uziely B, Heching NI et al (1998) Doxil(R) in metastatic breast cancer (MBC) after prior chemotherapy: therapeutic results in two consecutive studies. *Proc ASCO Abstract* No 597

186 Rivera E (2003) Liposomal anthracyclines in metastatic breast cancer: clinical update. *The Oncologist* 8: 3–9

187 Harris L, Batist G, Belt R et al (2002) Liposome-encapsulated doxorubicin compared with conventional doxorubicin in a randomized multicenter trial as first-line therapy of metastatic breast carcinoma. *Cancer* 94: 25–36

188 Rivera E, Valero V, Arun B et al (2003) Phase II study of pegylated liposomal doxorubicin in combination with gemcitabine in patients with metastatic breast cancer. *J Clin Oncol* 21: 3249–3254

189 Michieli M, Damiani D, Ermacora A et al (1999) Liposome-encapsulated daunorubicin for PGP-related multidrug resistance. *Br J Haematol* 106: 92–99

190 Verdonck LF, Lokhorst HM, Roovers DJ et al (1998) Multidrug-resistant acute leukemia cells are responsive to prolonged exposure of daunorubicin: implications for liposome-encapsulated daunorubicin. *Leuk Res* 22: 249–256

191 Ojeda B, de Sande LM, Casado A et al (2003) Cost-minimisation analysis of pegylated liposomal doxorubicin hydrochloride *versus* topotecan in the treatment of patients with recurrent epithelial ovarian cancer in Spain. *Br J Cancer* 89: 1002–1007

192 Shapiro CL, Ervin T, Welles L et al (1999) Phase II trial of high-dose liposome-encapsulated doxorubicin with granulocyte colony-stimulating factor in metastatic breast cancer. TLC D-99 Study Group. *J Clin Oncol* 17: 1435–1441

193 Pimm MV (1988) Drug monoclonal antibody conjugates for cancer therapy: potentials and limitations. *CRC Crit Rev Ther Drug Carrier Sys* 5: 189

194 Reisfeld RA, Cheresch DA (1987) Human tumour antigens. *Adv Immunol* 40: 323

195 Maeda H, Wu J, Sawa T et al (2000) Tumor vascular permeability and the EPR effect in macromolecular therapeutics, a review. *J Control Release* 65: 271

196 Maeda H, Matsumura Y (1989) Tumoritropic and lymphotropic principles of macromolecular prodrugs. *Crit Rev Ther Drug Carrier Sys* 6: 193–210

197 Seymour LW (1992) Passive tumor targeting of soluble macromolecules and drug conjugates. *Crit Rev Ther Drug Carrier Sys* 9: 135–187

198 Duncan R, Spreafico F (1994) Polymer conjugates. Pharmacokinetic considerations for design and development. *Clin Pharmacokinet* 27: 290–303

199 Matsumura Y, Maeda H (1986) A new concept for macromolecular therapeutics in cancer chemotherapy: mechanism of tumoritropic accumulation of proteins of the antitumor agent smancs. *Cancer Res* 46: 6387–6392

200 Jain RK (1987a) Transport of molecules across tumor vasculature. *Cancer Metast Rev* 6: 559

201 Jain RK (1987) Transport of molecules in the tumor interstitium: a review. *Cancer Res* 47: 3039

202 Yuan F, Deilian M, Fukumura D et al (1995) Vascular permeability in a human tumor xenograft:

molecular size dependence and cutoff size. *Cancer Res* 55: 3752–3756

203 de Groot FM, Damen EW, Scheeren HW (2001) Anticancer prodrugs for application in monotherapy: targeting hypoxia, tumor-associated enzymes, and receptors. *Current Medicinal Chemistry* 8(9): 1093–1122

204 Langer M, Beck-Sickinger AG (2001) Peptides as carrier for tumor diagnosis and treatment. *Current Medicinal Chemistry – Anti-Cancer Agents* 1(1): 71–93

205 Huang PS, Oliff A (2001) Drug-targeting strategies in cancer therapy. *Current Op Genet Dev* 11(1): 104–110

206 Soyez H, Schacht E, Vanderkerken S (1996) The crucial role of spacer groups in macromolecular design. *Adv Drug Del Rev* 21: 81

207 Kratz F, Beyer U, Schütte MT (1999) Polymer drug conjugates containing acid-cleavable bonds. *Crit Rev Ther Drug Carrier Sys* 16: 245

208 Mukherjee S, Ghosh RN, Maxfield FR (1997) Endocytosis. *Physiol Rev* 77(3): 759–803

209 Tannock IF, Rotin D (1989) Acid pH in tumors and its potential for therapeutic exploitation. *Cancer Res* 49: 4373

210 Thanou M, Duncan R (2003) Polymer-protein and polymer-drug conjugates in cancer therapy. *Current Op Invest Drugs* 4(6): 701–709

211 Heath EI, Grochow LB (2000) Clinical potential of matrix metalloprotease inhibitors in cancer therapy. *Drugs* 59: 1043

212 Rockway TW, Giranda VL (2003) Inhibitors of the proteolytic activity of urokinase type plasminogen activator. *Curr Pharma Design* 9(19): 1483–1498

213 Trouet A, Masquelier M, Baurain R et al (1982) D-D: A covalent linkage between daunorubicin and proteins that is stable in serum and reversible by lysosomal hydrolases, as required for a lysosomotropic drug carrier conjugate: *in vitro* and *in vivo* studies. *Proc Natl Acad Sci USA* 79: 626–629

214 Subr V, Strohalm J, Ulbrich K et al (1992) Polymers containing enzymatically degradable bonds. XII. Release of daunomycin and adriamycin from poly[N-(2-hydroxypropyl)methacrylamide] copolymers. *J Controlled Release* 18: 123–132

215 Duncan R, Kopeckova-Rejmanova P, Strohalm J et al (1988) Anticancer agents coupled to N-(2-hydroxypropyl)methacrylamide copolymers. II. Evaluation of daunomycin conjugates *in vivo* against L1210 leukaemia. *Br J Cancer* 57: 147–156

216 Loadman PM, Bibby MC, Double JA et al (1999) Pharmacokinetics of PK1 and doxorubicin in experimental colon tumor models with differing responses to PK1. *Clin Cancer Res* 5: 3682–3688

217 Seymour LW, Ulbrich K, Wedge SR et al (1991) N-(2-hydroxypropyl)methacrylamide copolymers targeted to the hepatocyte galactose-receptor: pharmacokinetics in DBA2 mice. *Br J Cancer* 63(6): 859–866

218 Kratz F, Drevs J, Bing G et al (2001) Development and *in vitro* efficacy of novel MMP-2 and MMP-9 specific doxorubicin albumin conjugates. *Bioor Med Chem Letters* 11(15): 2001–2006

219 Mansour AM, Drevs J, Esser N et al (2003) A new approach for the treatment of malignant melanoma: enhanced antitumor efficacy of an albumin-binding doxorubicin prodrug that is cleaved by matrix metalloproteinase 2. *Cancer Res* 63(14): 4062–4066

220 Denmeade SR, Isaacs JT (2002) A history of prostate cancer treatment. *Nature Reviews Cancer* 2(5): 389–396

221 Denmeade SR, Sokoll LJ, Chan DW et al (2001) Concentration of enzymatically active prostate-specific antigen (PSA) in the extracellular fluid of primary human prostate cancers and human prostate cancer xenograft models. *Prostate* 48(1): 1–6

222 Denmeade SR, Nagy A, Gao J et al (1998) Enzymatic activation of a doxorubicin-peptide prodrug by prostate-specific antigen. *Cancer Res* 58(12): 2537–2540

223 Khan SR, Denmeade SR (2000) *In vivo* activity of a PSA-activated doxorubicin prodrug against PSA-producing human prostate cancer xenografts. *Prostate* 45(1): 80–83

224 DeFeo-Jones D, Garsky VM, Wong BK et al (2000) A peptide-doxorubicin 'prodrug' activated by prostate-specific antigen selectively kills prostate tumor cells positive for prostate-specific antigen *in vivo*. *Nature Medicine* 6(11): 1248–1252

225 Garsky VM, Lumma PK, Feng DM et al (2001) The synthesis of a prodrug of doxorubicin designed to provide reduced systemic toxicity and greater target efficacy. *J Med Chem* 44(24): 4216–4224

226 Sperker B, Backman JT, Kroemer HK (1997) The role of beta-glucuronidase in drug disposition

and drug targeting in humans. *Clin Pharmacokinet* 33(1): 18–31

227 Bernacki RJ, Niedbala MJ, Korytnyk W (1985) Glycosidases in cancer and invasion. *Cancer Metastasis Revs* 4(1): 81–101

228 Azoulay M, Florent JC, Monneret C et al (1995) Prodrugs of anthracycline antibiotics suited for tumor-specific activation. *Anti-Cancer Drug Design* 10(6): 441–450

229 Bosslet K, Straub R, Blumrich M et al (1998) Elucidation of the mechanism enabling tumor selective prodrug monotherapy. *Cancer Res* 58(6): 1195–1201

230 Murdter TE, Sperker B, Kivisto KT et al (1997) Enhanced uptake of doxorubicin into bronchial carcinoma: beta-glucuronidase mediates release of doxorubicin from a glucuronide prodrug (HMR 1826) at the tumor site. *Cancer Res* 57(12): 2440–2445

231 Dubois V, Dasnois L, Lebtahi K et al (2002) CPI-0004Na, a new extracellularly tumor-activated prodrug of doxorubicin: *in vivo* toxicity, activity, and tissue distribution confirm tumor cell selectivity. *Cancer Res* 62*:* 2327–2331

232 Gaál D, Hudecz F (1998) Low toxicity and high antitumour activity of daunomycin by conjugation to an immunopotential amphoteric branched polypeptide. *Eur J Cancer* 34: 155

233 Dillman RO, Johnson DE, Shawler DL et al (1998) Superiority of an acid-labile daunorubicin-monoclonal antibody immunoconjugate compared to free drug. *Cancer Res* 48: 6097

234 Yang HM, Reisfeld RA (1988) Doxorubicin conjugated with a monoclonal antibody directed to a human melanoma-associated proteoglycan suppresses the growth of established tumor xenografts in nude mice. *Proc Natl Acad Sci USA* 85: 1189

235 Willner D, Trail PA, Hofstead SJ et al (1993) (6-Maleimidocaproyl)hydrazone of doxorubicin – A new derivative for the preparation of immunoconjugates of doxorubicin. *Bioconjugate Chem* 4: 521–527

236 Trail PA, Willner D, Lasch SJ et al (1992) Antigen-specific activity of carcinoma-reactive BR64-doxorubicin conjugates evaluated *in vitro* and in human tumor xenograft models. *Cancer Res* 52: 5693

237 Firestone RA, Willner D, Hofstead SJ et al (1996) Synthesis and antitumor activity of the immunoconjugate BR96-DOX. *J Contr Rel* 39: 251

238 Trail PA, Willner D, Lasch SJ et al (1993) Cure of xenografted human carcinomas by BR96-doxorubicin immunoconjugates. *Science* 261: 212–215

239 Trail PA, Willner D, Knipe J et al (1997) Effect of linker variation on the stability, potency, and efficacy of carcinoma-reactive BR64-doxorubicin immunoconjugates. *Cancer Res* 57: 100–105

240 Kratz F, Beyer U, Roth T et al (1998) Transferrin conjugates of doxorubicin: synthesis, characterization, cellular uptake, and *in vitro* efficacy. *J Pharma Sci* 87(3): 338–346

241 Kratz F, Beyer U, Collery P et al (1998) Preparation, characterization and *in vitro* efficacy of albumin conjugates of doxorubicin. *Biol Pharma Bull* 21(1): 56–61

242 Drevs J, Hofmann I, Marmé D et al (1999) *In vivo* and *in vitro* efficacy of an acid-sensitive albumin conjugate of adriamycin compared to the parent compound in murine renal cell carcinoma. *Drug Delivery* 6: 1–7

243 Kratz F, Roth T, Fichiner I et al (2000) *In vitro* and *in vivo* efficacy of acid-sensitive transferrin and albumin doxorubicin conjugates in a human xenograft panel and in the MDA-MB-435 mamma carcinoma model. *J Drug Targeting* 8(5): 305–318

244 Kratz F, Muller-Driver R, Hofmann I (2000) A novel macromolecular prodrug concept exploiting endogenous serum albumin as a drug carrier for cancer chemotherapy. *J Med Chem* 43(7): 1253–1256

245 Kratz F, Warnecke A, Scheuermann K et al (2002) Probing the cysteine-34 position of endogenous serum albumin with thiol-binding doxorubicin derivatives. Improved efficacy of an acid-sensitive doxorubicin derivative with specific albumin-binding properties compared to that of the parent compound. *J Med Chem* 45(25): 5523–5533

246 de Jong J, Geijssen GJ, Munniksma CN et al (1992) Plasma pharmacokinetics and pharmacodynamics of a new prodrug N-l-leucyldoxorubicin and its metabolites in a Phase I clinical trial. *J Clin Oncology* 10(12): 1897–1906

247 DiPaola RS, Rinehart J, Nemunaitis J et al (2002) Characterization of a novel prostate-specific antigen-activated peptide-doxorubicin conjugate in patients with prostate cancer. *J Clin Oncology* 20(7): 1874–1879

248 Vasey PA, Kaye SB, Morrison R et al (1999) Phase I clinical and pharmacokinetic study of PK1 [N-(2-hydroxypropyl)methacrylamide copolymer doxorubicin]: first member of a new class of chemotherapeutic agents-drug-polymer conjugates. Cancer Research Campaign Phase I/II

Committee. *Clin Cancer Res* 5(1): 83–94

249 Seymour LW, Ferry DR, Anderson D et al (2002) Cancer Research Campaign Phase I/II Clinical Trials committee. Hepatic drug targeting: Phase I evaluation of polymer-bound doxorubicin. *J Clin Oncology* 20(6): 1668–1676

250 Julyan PJ, Seymour LW, Ferry DR et al (1999) Preliminary clinical study of the distribution of HPMA copolymers bearing doxorubicin and galactosamine. *J Controlled Release* 57(3): 281–290

251 Sugerman S, Murray JL, Saleh M et al (1995) A Phase I study of BR96-doxorubicin (BR96-DOX) in patients with advanced carcinoma expressing the Lewis y-antigen. *Proc Am Soc Clin Oncol* 14: A1532

252 Slichenmyer WJ, Saleh MN, Bookman MA et al (1996) Phase I studies of BR96 doxorubicin in patients with advanced solid tumors that express the Lewis Y antigen. Sixth International Congress on Anti-Cancer Treatment, Paris p. 95

253 Giantonio BJ, Gilewski TA, Bookman M et al (1996) A Phase I study of weekly BR96-doxorubicin (BR96-DOX) in patients with advanced carcinoma expressing the Lewis Y (Le^Y) antigen. *Proc Ann Meet Am Soc Clin Oncol* 15: A1380

254 Tolcher AW, Sugarman S, Gelman KA et al (1999) Randomized Phase II study of BR96-doxorubicin conjugate in patients with metastatic breast cancer. *J Clin Oncol* 2: 478–484

255 Weiss RB (1992) The anthracyclines: Will we ever find a better doxorubicin? *Sem Oncol* 19: 670–686

256 Monneret C (2001) Recent developments in the field of antitumor anthracyclines. *Eur J Med Chem* 36: 483–493

257 Mansat-De Mas V, Hernandez H, Plo I et al. (2003) Protein kinase C mediated Raf-1/extracellular-regulated kinase activation by daunorubicin. *Blood* 101: 1543–1550

258 Weinstein-Oppenheimer C, Henriquez-Roldan CF, Davis JM et al. (2001) Role of the Raf signal transduction cascade in the *in vitro* resistance to the anticancer drug doxorubicin. *Clin Cancer Res* 7: 2898–2907

259 Leonard GD, Fojo T, Bates SE (2003) The role of ABC transporters inclinical practice. *The Oncologist* 8: 411–424

260 Hu W, Kavanagh JJ (2003) Anticancer therapy targeting the apoptotic pathway. *Lancet Oncology* 4: 721–729

261 Dumont A, Hehner SP, Hofmann TG et al. (1999) Peroxide-induced apoptosis is CD95-independent, requires the release of mitochondria-derived reactive oxygen species and the activation of NF-kappaB. *Oncogene* 18: 747–757

262 Senturker S, Tschirret-Guth R, Morrow J et al. (2002) Induction of apoptosis by chemotherapeutic drugs without generation of reactive oxygen species. *Arch Biochem Biophys* 397: 262–272

263 Kirschke SO (2002) Investigation of the apoptosis signal transduction mediated by the marine pyridoacridine alkaloid Ascidemin in human leukemic Jurkat T cells. Medical Thesis Ludwig-Maximilians-University, Munich

264 Maragoudakis ME, Peristeris P, Misirlis E et al. (1994) Inhibition of angiogenesis by anthracyclines and titanocene dichloride. *Ann NY Acad Sci* 731: 280–293

265 Miller KD, Sweeney CJ, Sledge GW (2001) redefining the target: chemotherapeutics as antiangiogenics. *J Clin Oncol* 19: 1195–1206

266 Wenzel DG, Cosma GN (1985) A model system for measuring comparative toxicities of cardiotoxic drugs for cultured rat heart myocytes, endothelial cells and fibroblasts. *Toxicology* 33: 117–128

Topoisomerase inhibitors

Hans Gelderblom[1] and Alex Sparreboom[2]

[1] *Leiden University Medical Center, Department of Clinical Oncology, Albinusdreef 2, 2300 RC*
 Leiden, The Netherlands
[2] *National Cancer Institute, Bethesda, Maryland, USA*

Summary

DNA topoisomerase inhibitors, known for their broad antitumour activity, represent one of the most widely used groups of anticancer agents. In spite of the early discovery and long-standing clinical use, the mechanism of action of these agents was not recognized until the 1980s [1–3].

Currently agents available for clinical use include the topoisomerase I inhibitors of the camptothecin class (topotecan and irinotecan) and the topoisomerase II inhibitors in the class of epipodophyllotoxins (etoposide and teniposide). Many new formulations and structurally-related agents are currently undergoing clinical development. This chapter highlights the most important aspects of the past, current and future development of topoisomerase I and II inhibitors, and provides an overview of pharmacology and clinical data, with a focus on recent developments.

Introduction

Topoisomerases

DNA topoisomerases are nuclear enzymes that change the topology (or conformation) of a segment of DNA by a complex catalytic cycle that involves DNA strand cleavage, strand passage and religation of the cleaved DNA [4, 5]. Thus, topoisomerases enable the DNA to be tightly packed and yet still assessable for processes for proper cellular function. In the strand-breakage reaction by a DNA topoisomerase, a tyrosyl oxygen of the enzyme attacks a DNA phosphorus, forming a covalent phosphotyrosine link and breaking a DNA phosphodiester bond at the same time [5, 6]. Rejoining of the DNA strand occurs by a second transesterification, which is basically the reverse of the first. These reactions cause transient enzyme mediated gates in the DNA for the passage of another DNA strand or double helix. There are two major classes of topoisomerases: Type I topoisomerases induce transient single-strand breaks in DNA, and type II enzymes induce double-strand breaks. The two types can be further divided into four subfamilies: IA, IB, IIA and IIB. Members of the same subfamily are structurally and mechanistically similar, whereas those of different subfamilies are distinct.

Agents targeting either topoisomerase I or II lead to elevated levels of the cleaved complex, where the topoisomerase is covalently bound to DNA. As the consequence of the formation of a cleavable complex, both the initial cleavage reaction and religation steps are inhibited. These events eventually trigger other cellular responses that can lead to cell cycle arrest and to cell death. Drugs acting in this matter have been termed topoisomerase poisons or inhibitors.

The sequential use of topoisomerase I inhibitors followed by topoisomerase II inhibitors might be attractive because of observed preclinical synergism, possibly due to an increase of topoisomerase II levels observed after inhibition of topoisomerase I and an increase in the S-phase cell population, possibly enhancing the sensitivity to topoisomerase II inhibition [7].

Topoisomerase I inhibitors

In the 1950s, during the National Cancer Institute's screening program of natural products, an alkaloid stem wood extract from the *Camptotheca acuminata*, an oriental tree that is cultivated throughout Asia, was found to be active against L1210 murine leukemia. Subsequent studies by Wall et al. [8] showed camptothecin to be the active ingredient of this extract. In the early 1970s, the parent compound 20-*S*-camptothecin underwent clinical testing. However, further clinical development was precluded due to severe and unpredictable toxicities including myelosuppression, diarrhea and hemorrhagic cystitis [9–12]. In the 1980s, topoisomerase I was identified as the major target for the antitumor effect of camptothecin [13] and overexpression of topoisomerase I levels were found in colon and ovarian cancer compared with normal tissue [14, 15]. These findings led to renewed interest in this class of agents, resulting in the development of better water soluble semi-synthetic analogs of camptothecin that were to be less toxic through their better solubility, whereas toxicity was also better predictable.

In *in vitro* studies topoisomerase I inhibitors showed more pronounced antitumor efficacy with protracted exposure at low concentration. Also in animal models, prolonged exposure at low dose resulted in less toxicity [16–22]. It should be stated though, that most animal models are poor models for toxicity with the camptothecin analogs since they are relatively resistant to the myelosuppressive effects. In order to simulate these prolonged exposures, various subsequent Phase I and II studies have focused on low dose continuous infusion of topoisomerase I inhibitors in cancer patients [23–29]. Most of the studies showed that continuous intravenous (iv) administration is feasible. Whether it is also more effective has not yet been proven. Since oral administration is a more convenient and more cost effective method for prolonged drug administration, further development of oral formulations of topoisomerase I inhibitors was given priority. Since most of the oral topoisomerase I inhibitors have relatively short half-lives, the use of protracted oral dosing is not always the same as continuous intravenous administration, although if the concept of time over threshold concentration is a valid indication of toxicity and efficacy, oral dosing can mimic continuous infusion. Despite efforts to develop oral topoisomerase I inhibitors, registration is thus far limited to iv topotecan (Hycamtin) and iv irinotecan (Camptosar). These two topoisomerase I inhibitors will be discussed first, followed by others that are still investigational.

Topotecan

Pharmacology

Topotecan (Hycamtin; 9-dimethylaminomethyl-10-hydroxycamptothecin) is a water soluble semi-synthetic analog of camptothecin [30]. The drug is poorly bound to plasma proteins. The active lactone structure can undergo a pH-dependent, reversible hydrolysis to an inactive carboxylate form (Fig. 1). At physiological pH, the equilibrium of topotecan is towards the inactive carboxylate form, whereas in acidic environment the equilibrium ratio is in the opposite direction. Lactone to carboxylate ratios was comparable after oral and intravenous administration [31]. The oral bioavailability of topotecan is 30–40% [31, 32]. Topotecan exhibits a linear pharmacokinetic behavior. The volume of distribution of topotecan lactone is approximately 70 L/m^2 after a 30 min iv administration and the terminal disposition half-life ($t_{1/2}$) is approximately 2.8 h. Elimination of the drug is mainly renally, necessitating dose reductions in patients with impaired renal function [33], whereas dose reductions in patients with impaired hepatic function and normal renal function are not necessary [34]. In patients with extensive pleural effusion or ascites treated with topotecan, plasma pharmacokinetics is unaltered and substantial penetration to third spaces has been observed [35]. Topotecan cerebrospinal fluid concentrations equivalent to 30% of those observed in plasma have been noted in children after iv topotecan administration, indicating that topotecan crosses the blood–brain barrier freely [36]. Topotecan also appears to be an effective radiosensitizer *in vitro,* as first reported by Kim et al. [37], although this concept has not been fully evaluated clinically.

   Although the intrapatient pharmacokinetic variability following iv or oral administration is limited, interpatient variability is considerable [38]. Since relationships between topotecan area under the curve (AUC) and its dose-limiting toxicity (i.e., neutropenia) have been established [39–41], considerable efforts have been put into the possibility of predicting exposure to topotecan (AUC) and topotecan clearance using individual patient characteristics (e.g.,

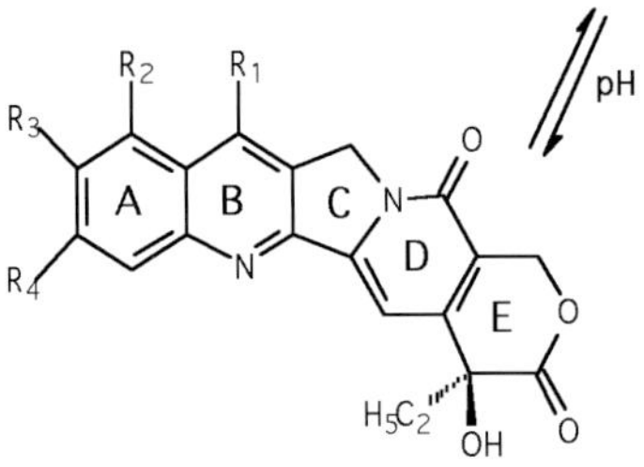

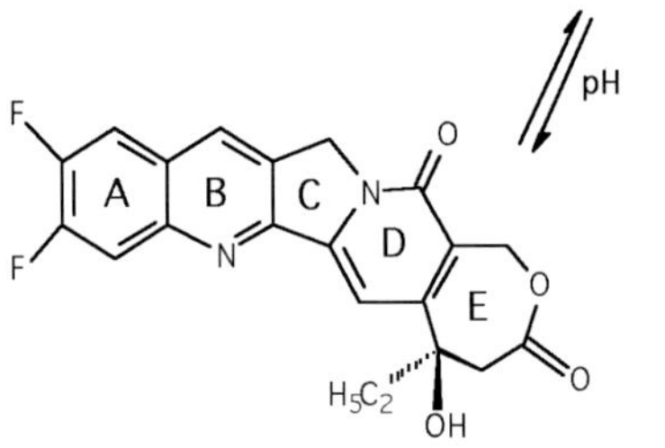

Diflomotecan

	R_1	R_2	R_3	R_4
Camptothecin	H	H	H	H
9-Nitrocamptothecin	H	NO_2	H	H
9-Aminocamptothecin	H	NH_2	H	H
Topotecan	H	$CH_2N(CH_3)_2$	H	H
Irinotecan (CPT-11)	CH_2CH_3	H	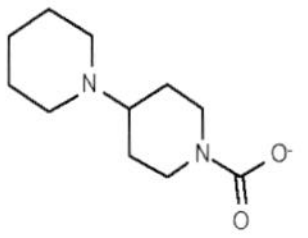 (piperidine-piperidine-carbonyloxy structure)	H
SN-38	CH_2CH_3	H	OH	H
Lurtotecan	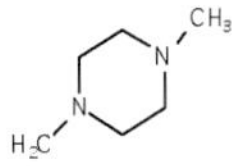 (methylpiperazine-methyl structure)	H	 (dioxane structure)	
DX-8951	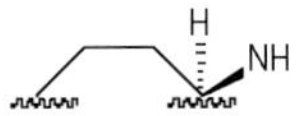 (propyl-NH structure)		CH_3	F

Figure 1. Chemical structures and pH-dependent interconversion of the lactone and carboxylate forms of topoisomerase-1 inhibitors.

body weight, serum creatinine, and sex) [42, 43], or by the use of limited sampling strategies for iv [44] and oral administration [45]. Preliminary, unpublished results from pharmacogenetic studies aimed at predicting individual exposure to oral topotecan by genotyping of genes encoding for drug transporters involved in topotecan elimination are encouraging (Gelderblom et al., ASCO 2004). All of these efforts should result in better prediction of toxicity and response to topotecan.

Clinical development

Ovarian cancer

Topotecan 1.5 mg/m^2/day, as a 30 min infusion days 1–5 every 3 weeks, was Food and Drug Administration (FDA) approved in 1996 for the treatment of ovarian cancer. This approval was based on a Phase III trial by ten Bokkel Huinink et al. [46] comparing topotecan in this schedule with paclitaxel 175 mg/m^2/day given once every 3 weeks in patients with recurrent or refractory ovarian cancer after a platinum-containing regimen. The overall response rates were 21% and 13% and the median survival was 63 and 53 weeks for patients treated with topotecan and paclitaxel, respectively. The efficacy of topotecan in ovarian cancer has further been demonstrated in paclitaxel resistant disease [47] and *versus* pegylated liposomal doxorubicin [48]. In these studies, topotecan-mediated hematological toxicity was usually predictable, manageable and of short duration. Non-hematological toxicity was generally mild, with NCI-CTC grade 3–4 nausea/vomiting and fatigue in only up to 10% of patients. A randomized trial of oral topotecan (2.3 mg/m^2/day, days 1–5 every 3 weeks) *versus* the standard intravenous scheme in relapsed epithelial ovarian cancer failed to show superiority of the more convenient oral regimen in terms of efficacy [49]. A number of studies are investigating topotecan in first-line treatment of advanced ovarian cancer either in combination, sequential or consolidation therapy.

Small-cell lung cancer
Topotecan 1.5 mg/m^2/day, days 1–5 every 3 weeks, has recently been approved for treatment of recurrent small-cell lung cancer after failure of first-line therapy. Its efficacy as a single agent in previously untreated (response rate, 39%; median survival 10 months) and platinum-sensitive relapsed patients (response rate 24–37%; median survival 6 months) was previously demonstrated in several Phase II trials and one Phase III trial [50–53]. A randomized Phase II study comparing oral *versus* iv topotecan in relapsed sensitive patients showed similar efficacy (response rate 23 *versus* 15% and median survival 32 *versus* 25 weeks), with a lower incidence of neutropenia in the oral topotecan group [54]. Results from a follow-up Phase III trial are pending, and will more clearly define the role of oral topotecan in the treatment of small-cell lung cancer. Several doublet or triplet therapies have reported promising results for first-line therapy. Consequently, a number of Phase III trials are currently investigating the role of oral and iv topotecan in combination with cisplatin, etoposide or paclitaxel in first-line therapy for small-cell lung cancer.

Hematological malignancies
In early clinical studies iv topotecan has shown significant activity against chronic myelomonocytic leukemia [55], myelodysplastic syndromes [56] and

acute leukemia [57–59]. These results were confirmed in a Phase I study with oral topotecan in hematological malignancies [60]. Modest activity was observed in Phase II studies in non-Hodgkin's lymphoma [61, 62].

Other gynecological malignancies

Topotecan also has activity in advanced cervical cancer. Single agent Phase II studies show responses in 13–19% of patients with a median survival of 6.5 months [63, 64], while studies with combination modalities (i.e., with cisplatin or paclitaxel) reported responses in 28–54% of patients with median survival of 10+ and 8.6 months [65, 66]. The combination of weekly cisplatin and oral topotecan with radiotherapy is currently being investigating in a Gynecologic Oncology Group trial. In second-line treatment for advanced endometrial cancer, weekly 72 h continuous iv topotecan showed a limited response rate of 9.1% and overall survival of 9 months [67]. However, in front-line treatment with the standard administration schedule, the response rate was 20% with an overall survival of 6.5 months [68].

Non-small cell lung cancer

Single agent topotecan has been evaluated in previously untreated patients with non-small cell lung cancer, achieving response rates of 4–25% [69, 70]. Combination therapy is currently being evaluated.

Irinotecan

Pharmacology

Irinotecan (Camptosar, CPT-11; 7-ethyl-10 [4-(pipiridino)-1-piperidino] carboxyloxy-camptothecin) also is a water-soluble analog of camptothecin. Like topotecan, it is known in two distinguishable forms, an active α-hydroxy-δ-lactone ring form and an inactive carboxylate form, for which a pH-dependent equilibrium exists, which significantly impacts on the compound's kinetic profile. The volume of distribution of irinotecan is large, suggesting extensive tissue distribution. The terminal disposition half-life of irinotecan is approximately 17 h, which is much longer than that of topotecan. Therefore, irinotecan has been studied in schedules different from those evaluated with topotecan. Irinotecan is a prodrug that is converted in the liver by carboxylesterase and/or butyrylcholinesterase [71] to SN-38 (7-ethyl-10-hydroxycamptothecin), a metabolite that is 1,000-fold more potent *in vitro* than the parent drug [72]. In animals, peripheral conversion of irinotecan to SN-38 also has been found in serum [73], small intestine [74] and possibly even within certain tumors [75]. Contrary to irinotecan, the lactone form of SN-38 predominates at physiologic pH, although with large interpatient variability [76, 77]. SN-38 undergoes further conjugation to an inactive β-glucuronide derivative (SN-38G) by the enzyme uridine diphosphate glucuronosyltransferase UGT1A1

[78]. Other known inactive irinotecan metabolites are APC (7-ethyl-10-[4-N-(5-aminopentanoicacid)-1-piperidino] carbonyloxycamptothecin) and NPC (7-ethyl-10-[4-(1-piperidino)-1-amino] carbonyloxycamptothecin) resulting from a cytochrome P-450 3A4-mediated pathway [79, 80]. As with the parent compound irinotecan, both metabolites are poor inhibitors of topoisomerase I, although a secondary conversion of NPC to SN-38 may have clinical significance (Fig. 2). A study to determine the complete metabolic fate and disposition of irinotecan in plasma, urine and feces was only able to account for half of the administered dose in urine and feces, indicating the possible existence of further unknown metabolites [81–83].

Hepatic metabolism and biliary secretion are the major pathways of irinotecan elimination in humans. Therefore, patients with hepatic dysfunction should have reductions in the administered dose of irinotecan [84]. Patients with impaired renal function do not appear to have increased sensitivity to irinotecan. In blood, irinotecan is mainly bound to and/or localized in erythrocytes, whereas SN-38 is mainly bound to albumin and lymphocytes, but also to erythrocytes and neutrophils.

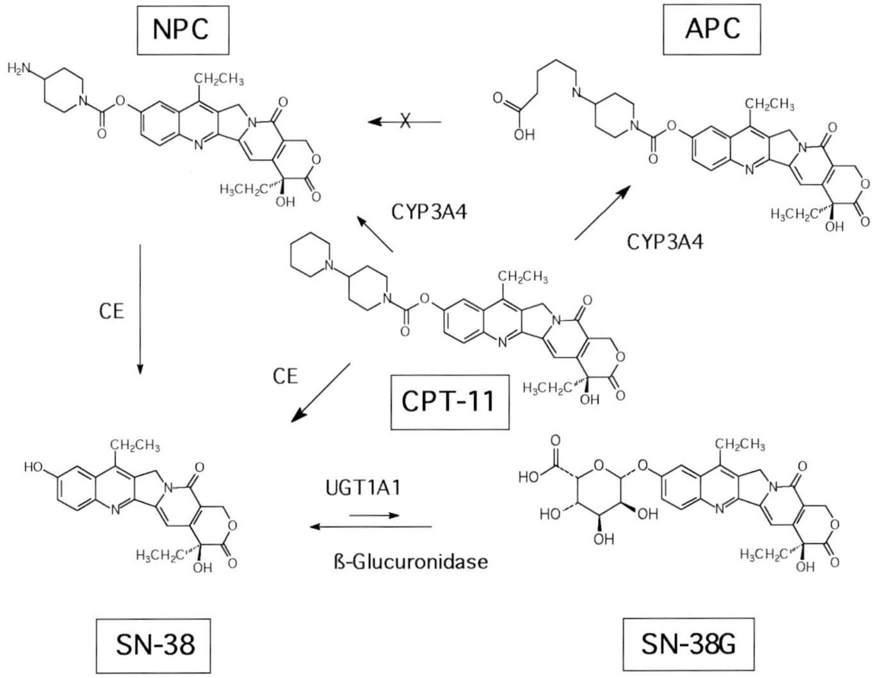

Figure 2. Metabolic pathways of irinotecan (CPT-11) indicating carboxylesterase (CE) mediated conversion to SN-38, cytochrome P450 3A4 (CYP3A4) mediated oxidation to APC and NPC, glucuronidation of SN-38 by uridine diphosphate glucuronosyltransferase isoform 1A1 (UGT1A1) to SN-38 glucuronide (SN-38G), and its deconjugation by bacterial β-glucuronidase.

In single agent regimens, diarrhea is the most important side effect of irinotecan. The early onset diarrhea with abdominal cramping, flushes and transpiration, suggestive of a release of vaso-active compounds, responds well to treatment with atropine. The late onset secretory diarrhea seems to be correlated with the extent of biliary secretion of SN-38 into the intestinal lumen. In addition, a high fecal SN-38 to SN-38G ratio has been found to be related to structural and functional injuries to the intestinal tract by SN-38, leading to diarrhea. In a small clinical study co-administration of the poorly-absorbed aminoglycoside antibiotic neomycin with the second course of irinotecan reduced fecal β-glucuronidase levels to undetectable levels, resulting in a decrease of fecal concentrations of SN-38 without affecting systemic SN-38 levels [85]. The observation that 6 out of 7 patients having diarrhea in the first cycle had less symptoms in the second cycle with neomycin co-administration is currently the subject of a larger confirmatory study.

Recently, several publications have shown a relationship between the occurrence of a TATA-box polymorphism in the promoter of the *UGT1A1* gene (i.e., *UGTA1*28*) and the severity of diarrhea and neutropenia following irinotecan treatment, which is due to a decreased ability to glucuronidate of SN-38 [86]. In addition, genetic polymorphism in the ABCB1 gene, which encodes the efflux transporter P-glycoprotein, was found to be associated with altered exposure to irinotecan [87].

The oral bioavailability of irinotecan was reported to be low, although due to extensive presystemic conversion by enzymes in the small intestine, favorable SN-38 to irinotecan AUC ratios have been observed in both animals and humans.

Clinical development

Colon cancer

The optimal administration schedule for irinotecan in the treatment of colorectal cancer remains unclear. The approved administration schedule in second-line treatment of advanced or metastatic colon cancer in the United States is 125 mg/m^2 weekly for 4 of 6 weeks. In Europe, the most widely used schedule is 350 mg/m^2 every 3 weeks, whereas in Japan 100 mg/m^2 every week or 150 mg/m^2 every other week is being used. A Phase III study comparing the United States *versus* the European schedule in patients with colon cancer showed comparable responses and toxicity profiles [88]. The Food and Drug Administration (FDA) approval for the use of irinotecan in second-line colon cancer in 1998, after accelerated approval in 1996, was based on a response rate of 32% in first-line setting [89] and survival benefit in 5-fluorouracil refractory patients in two Phase III studies [90, 91]. Consequently, two parallel Phase III trials combining irinotecan and 5-fluorouracil in first-line disease were conducted in the US and Europe [92, 93]. Both studies showed significant improvement of response rates (39–49 *versus* 21–31%), progression-free survival (7.0–6.7 *versus* 4.3–4.4 months) and overall survival (14.8–17.4 *versus* 12.6–14.1 months) with the combination as compared to single agent 5-fluouracil.

The development of diarrhea and dehydration in combination with neutropenia needs early recognition and treatment, especially in the bolus regimen [94]. Recent developments with irinotecan in colorectal cancer include combinations with oral 5-fluorouracil analogs [95] and oxaliplatin [96], yielding response rates of 46% and 56% and a median survival of 21 months.

Other malignancies

Apart from the antitumor activity in colorectal carcinoma, encouraging response rates with single agent irinotecan were observed in patients with various tumor types such as mesothelioma, glioblastoma multiforme, (non) small-cell-lung cancer, head- and neck cancer, esophageal cancer, gastric cancer, breast cancer, cervical cancer and ovarian cancer [97]. Combination therapy with, for example, topoisomerase II inhibitors, platinum-derivatives, taxanes and 5-fluorouracil or its analogs seems to be promising in several tumor types as well [92–103]. In Japan, a combination regimen of irinotecan and cisplatin has shown improvement in overall survival over the global standard regimen of etoposide and cisplatin in extensive-stage small-cell-lung cancer [98]. Three randomized trials are in progress to confirm the data [99]. In non-small-cell lung cancer the combination of irinotecan and cisplatin produced superior response rates compared to cisplatin/vindesine [100]. Different doublet and triplet irinotecan combinations are currently being tested against the more commonly used cisplatin combination regimens [101].

Investigational topoisomerase I inhibitors

The development of investigational topoisomerase I inhibitors is based on superior preclinical antitumor activity due to modifications of the camptothecin structure and/or enhanced stability of the chemical active lactone form or due to alternative formulations and vehicles aiming at protracted exposure such as liposomes, microspheres and nanoparticles. Other developments include new dosing strategies in order to increase drug levels at cancer sites with minimal systemic exposure such as intraperitoneal administration and aerosolization, and co-administration of other drugs affecting pharmacology and/or toxicology of the camptothecin derivative. Some of these investigational topoisomerase I inhibitors are summarized in Table 1.

Topoisomerase II inhibitors

Topoisomerase II is the target for many anticancer drugs. The anthracyclines and epipodophyllotoxins have been in widespread use in oncology since the

Table 1. Some investigational topoisomerase I inhibitors and their mechanism of action

Name	Mechanism
9-aminocamptothecin (9-AC)	Synthetic derivate of camptothecin (CPT)
9-nitrocamptothecin (9-NC)	Prodrug of 9-AC
DE-310	Polymer bound exatecan
Diflomotecan (BN80915)	Homocamptothecin (enhanced lactone stability)
Exatecan (DX-8951f)	Hexacyclic CPT-analog
Kareneticin (BNP1350)	Lipophilic, silylated CPT
Liposomal lurtotecan (NX-211)	Liposome encapsulated synthetic CPT-derivate
PEG-camptothecin	CPT-polymer
MAG-camptothecin	CPT-polymer
Silatecan (DB-67)	Lipophilic CPT

1960s, well before their recognition as topoisomerase II poisons. All topoisomerase II-directed agents are able to interfere with at least one step of the catalytic cycle. In this chapter, agents able to stabilize the covalent DNA topoisomerase II complex, traditionally called topoisomerase II poisons, will be discussed. Agents acting on any of the other steps in the catalytic cycle, other than the cleavable complex, called catalytic inhibitors, have been reviewed extensively elsewhere [102, 103].

Topoisomerase II poisons currently in use consist of (i) the anthracyclines epirubicin, doxorubicin, idarubicin and daunorubicin, (ii) the anthracycline related compounds mitoxantrone and amsacrine, and (iii) the epipodophyllotoxins etoposide and teniposide. Since anthracyclines will be discussed separately, this chapter will focus on the epipodophyllotoxins etoposide, etoposide phosphate, teniposide, and on novel topoisomerase II inhibitors that are currently under investigation.

Podophyllotoxins have been used as anticancer medications for over 1,000 years and in 1946, the antimitotic properties were established [104]. Due to toxicity issues new synthetic variants were synthesized with less toxicity in the 1950s. After extensive isolation procedures, the most effective agent was found to be 4'-demethyl-epipodophyllin benzylidene glucoside. Two analogs with increased antineoplastic activity were subsequently synthesized, namely etoposide (VP-16-213) in 1966 and teniposide (VM-26) in 1967 [105].

Pharmacology

To increase solubility, etoposide for iv administration is formulated in polysorbate 80 (Tween 80). Approximately one-third of administered etoposide is excreted in urine, one-third by hepatic metabolism to glucoronide and demethyl metabolites and little of the drug is excreted into the bile [106]. Total etoposide clearance is modestly decreased in patients with renal failure, but not in patients with biliary obstruction [107]. Its half-life in humans is approximately 6.4 h. Etoposide clearance shows limited inter- and intrapatient variability and linear dose-exposure relationships with myelosuppression as the main dose-limiting toxicity, rendering it a popular component of high-dose chemotherapy. In the presence of anticonvulsants, the systemic clearance of etoposide is increased due to induction of cytochrome P-450 hepatic metabolism. Indeed, in patients using anticonvulsants, the systemic clearance was 40% faster in adults [108] and up to 77% faster in children [109]. Co-medication inhibiting P-450 metabolism, such as valspodar [110] or cyclosporin [111] is known to increase etoposide systemic exposure. Etoposide is highly bound to plasma proteins with an average free fraction of 6–8%. Since the free drug is biologically active, one must be aware that physical conditions with reduced serum albumin concentrations (e.g., in cancer patients) can increase free drug concentrations up to 40% and thus lead to exacerbated toxicity.

The oral etoposide preparation, approved by the FDA in 1987, is formulated in a soft gelatin capsule, containing 50 mg of etoposide in a solution of purified water, citric acid, glycerin and polyethylene glycol 400. The bioavailability of oral etoposide ranges from 40–75%. Oral absorption is linear to doses up to 250 mg and decreases with doses greater than 300 mg [112], with intrapatient variability of 16% and interpatient variability of 38%. Inhibition of intestinal drug-transporting proteins like P-glycoprotein may influence the oral absorption of etoposide [113].

Etoposide phosphate is an etoposide prodrug which is rapidly converted to etoposide [114]. It is pharmacokinetically equivalent to etoposide [115]. Due to its improved water solubility, it can be administered more easily. Besides that, the formulation is devoid of Tween 80, held responsible for hypersensitivity reactions with the iv formulation of etoposide [116].

Teniposide is an analog of etoposide with greater *in vitro* anti-cancer activity, probably due to better cellular uptake [117]. Teniposide is even less water soluble than etoposide and allergic reactions are more frequently observed compared with etoposide, possibly as a result of the presence in the clinical formulation of the allergenic excipient, Cremophor EL.

Clinical development

The main toxicity of iv and oral etoposide, etoposide phosphate and teniposide include bone marrow suppression, nausea and vomiting and alopecia. However, induction of secondary malignancies remains a major concern with epipodophyllotoxin-containing treatment modalities. Neutropenia occurs more frequently than thrombocytopenia, and hypersensitivity reactions are more common with iv etoposide and teniposide. Clinical trials with etoposide and teniposide began in 1973 and 1970, leading to their US registration in 1983 for combination therapy of refractory testicular cancer and small cell lung cancer for etoposide and in 1993 for combination therapy for induction in patients with refractory acute lymphoblastic anemia for teniposide. Several preclinical studies have suggested that the duration of exposure of cancer cells to etoposide is important [118, 119], probably related to expression of its target topoisomerase II during mitotic phases of the cell cycle. Therefore, chronic scheduling of this agent may be advantageous, comparable to topoisomerase I inhibitors. This theory has led to the development of a soft gelatin capsule of etoposide for oral administration that was approved in 1987 and which is now widely used. Etoposide phosphate was approved for clinical use by the FDA in 1996 based on pharmacokinetic equivalence and the possibility for shorter infusions compared with etoposide. An oral formulation of etoposide phosphate is currently under development.

New developments with topoisomerase II inhibitors

New directions with topoisomerase II inhibitors include the development of agents blocking the catalytic activity of DNA topoisomerase II without stabilizing the cleavable complex. Also drugs capable of inhibiting both topoisomerase I and II are being developed, which include (i) the DNA-intercalators such as DACA (XR5000), intoplicine (RP60475), TAS-103, XR11576, XR5944 and NSC366140, (ii) hybrid molecules and (iii) miscellaneous dual inhibitors such as taflupozide (F-11782) and BN80927 [120, 121]. Other recent developments include the role of topoisomerase II and p53 status in determining chemosensitivity to topoisomerase II inhibitors [122] and the sequential targeting of topoisomerase I and II [123, 124].

Conclusion

Because of its broad spectrum of antitumor activity, topoisomerase I and II inhibitors are clearly among the most important anticancer drugs developed in the last few decades. Their pharmacokinetic behavior has been explored extensively in recent years, which has been of fundamental importance in our understanding of their clinical effects. In addition, a wealth of information has yielded valuable insight into the mechanism of action, the mechanisms of tumor resistance, toxicities, and considerations of dosage and schedule and route of drug administration. However, only through further investigations that may allow better definition of the biochemistry and pharmacologic profiles of these agents can their rational optimization of therapy be achieved. This need has become even more important in light of the current clinical use of camptothecins and epipodophyllotoxins in combination regimens with other antineoplastic drugs or agents specifically administered to modify toxicity profiles. In this respect, continued investigations into the role of individual levels of expression of enzymes and detection of enzyme and transporter polymorphisms will allow more rational and selective chemotherapy with these drugs.

References

1 Hsiang YH, Liu LF (1988) Identification of mammalian DNA topoisomerase-I as an intracellular target of the anticancer drug camptothecin. *Cancer Res* 48(7): 1722–1726

2 Nelson EM, Tewey KM, Liu LF (1984) Mechanism of antitumor drug-action – poisoning of mammalian DNA topoisomerase-II on DNA by 4'-(9-acridinylamino)-methanesulfon-meta-anisidide. *Proc Natl Acad Sci USA* 81(5): 1361–1365

3 Tewey KM, Chen GL, Nelson EM, Liu LF (1984) Intercalative antitumor drugs interfere with the breakage-reunion reaction of mammalian DNA topoisomerase-II. *J Biol Chem* 259(14): 9182–9187

4 Osheroff N, Zechiedrich EL, Gale KC (1991) Catalytic function of DNA topoisomerase-II. *Bioessays* 13(6): 269–275

5 Wang JC (2002) Cellular roles of DNA topoisomerases: A molecular perspective. *Nat Rev Mol Cell Biol* 3(6): 430–440

6 Walker JV, Nitiss JL (2002) DNA topoisomerase II as a target for cancer chemotherapy. *Cancer Invest* 20(4): 570–589

7 de Jonge MJA, Sparreboom A, Verweij J (1998) The development of combination therapy involving camptothecins: a review of preclinical and early clinical studies. *Cancer Treat Rev* 24(3): 205–220

8 Wall ME, Wani MC, Cook CE, Palmer KH, Mcphail AT, Sim GA (1966) Plant antitumor agents. I. Isolation and structure of camptothecin. A novel alkaloidal leukemia and tumor inhibitor from Camptotheca Acuminata. *J Am Chem Soc* 88(16): 3889–3890

9 Moertel CG, Schutt AJ, Reitemei RJ, Hahn RG (1972) Phase II study of camptothecin (NSC-100880) in treatment of advanced gastrointestinal cancer. *Cancer Chemother Rep* Part 1 56(1): 95–101

10 Creaven PJ, Allen LM (1972) Camptothecin plasma levels during a 5-day course – Correlation with toxicity. *Proc AACR* 13: 46

11 Gottlieb JA, Guarino AM, Oliverio VT, Block JB (1970) Preliminary clinical and pharmacological studies with camptothecin sodium (Cs). *Proc AACR* 11: 31

12 Muggia FM, Cohen MH, Hansen HH, Creaven PJ, Selawry OS (1972) Phase I clinical trial of weekly and daily treatment with camptothecin (NSC-100880) – Correlation with preclinical studies. *Cancer Chemother Rep* Part 1 56(4): 515–521

13 Hsiang YH, Hertzberg R, Hecht S, Liu LF (1985) Camptothecin induces protein-linked DNA breaks via mammalian DNA topoisomerase-I. *J Biol Chem* 260(27): 4873–4878

14 Giovanella BC, Stehlin JS, Wall ME, Wani MC, Nicholas AW, Liu LF et al. (1989) DNA topoisomerase-I targeted chemotherapy of human-colon cancer in xenografts. *Science* 246(4933): 1046–1048

15 van der Zee AGJ, Hollema H, de Jong S, Boonstra H, Gouw A, Willemse PHB et al. (1991) P-glycoprotein expression and DNA topoisomerase-I and topoisomerase-II activity in benign-tumors of the ovary and in malignant-tumors of the ovary, before and after platinum cyclophosphamide chemotherapy. *Cancer Res* 51(21): 5915–5920

16 Burris HA, Hanauske AR, Johnson RK, Marshall MH, Kuhn JG, Hilsenbeck SG et al. (1992) Activity of topotecan, a new topoisomerase-I inhibitor, against human tumor colony-forming-units *in vitro*. *J Natl Cancer Inst* 84(23): 1816–1820

17 Daoud SS, Fetouh MI, Giovanella BC (1995) Antitumor effect of liposome-incorporated camptothecin in human-malignant xenografts. *Anti-Cancer Drugs* 6(1): 83–93

18 Giovanella BC, Hinz HR, Kozielski AJ, Stehlin JS, Silber R, Potmesil M (1991) Complete growth-inhibition of human cancer xenografts in nude-mice by treatment with 20-(S)-camptothecin. *Cancer Res* 51(11): 3052–3055

19 Houghton PJ, Cheshire PJ, Myers L, Stewart CF, Synold TW, Houghton JA (1992) Evaluation of 9-dimethylaminomethyl-10-hydroxycamptothecin against xenografts derived from adult and childhood solid tumors. *Cancer Chemother Pharmacol* 31(3): 229–239

20 Houghton PJ, Cheshire PJ, Hallman JD, Lutz L, Friedman HS, Danks MK et al. (1995) Efficacy of topoisomerase-I inhibitors, topotecan and irinotecan, administered at low-dose levels in protracted schedules to mice bearing xenografts of human tumors. *Cancer Chemother Pharmacol* 36(5): 393–403

21 Pantazis P, Hinz HR, Mendoza JT, Kozielski AJ, Williams LJ, Stehlin JS et al. (1992) Complete inhibition of growth followed by death of human-malignant melanoma-cells *in vitro* and regression of human-melanoma xenografts in immunodeficient mice induced by camptothecins. *Cancer*

Res 52(14): 3980–3987

22 Pantazis P, Kozielski A, Rodriguez R, Petry E, Wani M, Wall M et al. (1994) Therapeutic effica-cy of camptothecin derivatives against human-malignant melanoma xenografts. *Melanoma Res* 4(1): 5–10

23 Hochster H, Liebes L, Speyer J, Sorich J, Taubes B, Oratz R et al. (1994) Phase-I trial of low-dose continuous topotecan infusion in patients with cancer – An active and well-tolerated regimen. *J Clin Oncol* 12(3): 553–559

24 Hochster H, Liebes L, Speyer J, Sorich J, Taubes B, Oratz R et al. (1997) Effect of prolonged topotecan infusion on topoisomerase 1 levels: A phase I and pharmacodynamic study. *Clin Cancer Res* 3(8): 1245–1252

25 Stevenson JP, Scher RM, Kosierowski R, Fox SC, Simmonds M, Yao KS et al. (1998) Phase II trial of topotecan as a 21-day continuous infusion in patients with advanced or metastatic adenocarci-noma of the pancreas. *Eur J Cancer* 34(9): 1358–1362

26 Herben VMM, ten Bokkel Huinink WW, Schot ME, Hudson I, Beijnen JH (1998) Continuous infusion of low-dose topotecan: pharmacokinetics and pharmacodynamics during a phase II study in patients with small cell lung cancer. *Anti-Cancer Drugs* 9(5): 411–418

27 Herben VMM, Schellens JHM, Swart M, Gruia G, Vernillet L, Beijnen JH et al. (1999) Phase I and pharmacokinetic study of irinotecan administered as a low-dose, continuous intravenous infu-sion over 14 days in patients with malignant solid tumors. *J Clin Oncol* 17(6): 1897–1905

28 Kindler HL, Kris MG, Smith IE, Miller VA, Grant SC, Krebs JB et al. (1998) Phase II trial of topotecan administered as a 21-day continuous infusion in previously untreated patients with stage IIIB and IV non-small-cell lung cancer. *Am J Clin Oncol* 21(5): 438–441

29 Creemers GJ, Gerrits CJH, Schellens JHM, Planting AST, van der Burg MEL, van Beurden VM et al. (1996) Phase II and pharmacologic study of topotecan administered as a 21-day continuous infusion to patients with colorectal cancer. *J Clin Oncol* 14(9): 2540–2545

30 Kingsbury WD, Boehm JC, Jakas DR, Holden KG, Hecht SM, Gallagher G et al. (1991) Synthesis of water-soluble (aminoalkyl)camptothecin analogs – inhibition of topoisomerase-I and antitu-mor-activity. *J Med Chem* 34(1): 98–107

31 Schellens JHM, Creemers GJ, Beijnen JH, Rosing H, de Boer-Dennert M, McDonald M et al. (1996) Bioavailability and pharmacokinetics of oral topotecan: A new topoisomerase I inhibitor. *Br J Cancer* 73(10): 1268–1271

32 Kruijtzer CMF, Beijnen JH, Rosing H, ten Bokkel Huinink WW, Schot M, Jewell RC et al. (2002) Increased oral bioavailability of topotecan in combination with the breast cancer resistance pro-tein and P-glycoprotein inhibitor GF120918. *J Clin Oncol* 20(13): 2943–2950

33 O'Reilly S, Rowinsky EK, Slichenmyer W, Donehower RC, Forastiere AA, Ettinger DS et al. (1996) Phase I and pharmacologic study of topotecan in patients with impaired renal function. *J Clin Oncol* 14(12): 3062–3073

34 O'Reilly S, Rowinsky E, Slichenmyer W, Donehower RC, Forastiere A, Ettinger D et al. (1996) Phase I and pharmacologic studies of topotecan in patients with impaired hepatic function. *J Natl Cancer Inst* 88(12): 817–824

35 Gelderblom H, Loos WJ, Verweij J, de Jonge MJA, Sparreboom A (2000) Topotecan lacks third space sequestration. *Clin Cancer Res* 6(4): 1288–1292

36 Baker SD, Heideman RL, Crom WR, Kuttesch JF, Gajjar A, Stewart CF (1996) Cerebrospinal fluid pharmacokinetics and penetration of continuous infusion topotecan in children with central nervous system tumors. *Cancer Chemother Pharmacol* 37(3): 195–202

37 Kim JH, Kim SH, Kolozsvary A, Khil MS (1992) Potentiation of radiation response in human car-cinoma-cells *in vitro* and murine fibrosarcoma *in vivo* by topotecan, an inhibitor of DNA topoiso-merase-I. *Int J Radiat Oncol Biol Phys* 22(3): 515–518

38 Loos WJ, Gelderblom H, Sparreboom A, Verweij J, de Jonge MJA (2000) Inter- and intrapatient variability in oral topotecan pharmacokinetics: Implications for body-surface area dosage regi-mens. *Clin Cancer Res* 6(7): 2685–2689

39 Grochow LB, Rowinsky EK, Johnson R, Ludeman S, Kaufmann SH, McCabe FL et al. (1992) Pharmacokinetics and pharmacodynamics of topotecan in patients with advanced cancer. *Drug Metab Dispos* 20(5): 706–713

40 Rowinsky EK, Adjei A, Donehower RC, Gore SD, Jones RJ, Burke PJ et al. (1994) Phase-I and pharmacodynamic study of the topoisomerase I-inhibitor topotecan in patients with refractory acute-leukemia. *J Clin Oncol* 12(10): 2193–2203

41 van Warmerdam LJC, Verweij J, Schellens JHM, Rosing H, Davies BE, de Boer-Dennert M et al.

(1995) Pharmacokinetics and pharmacodynamics of topotecan administered daily for 5 days every 3 weeks. *Cancer Chemother Pharmacol* 35(3): 237–245

42 Gallo JM, Laub PB, Rowinsky EK, Grochow LB, Baker SD (2000) Population pharmacokinetic model for topotecan derived from phase I clinical trials. *J Clin Oncol* 18(12): 2459–2467

43 Montazeri A, Boucaud M, Lokiec F, Pinguet F, Culine S, Deporte-Fety R et al. (2000) Population pharmacokinetics of topotecan: intraindividual variability in total drug. *Cancer Chemother Pharmacol* 46(5): 375–381

44 Montazeri A, Culine S, Laguerre B, Pinguet F, Lokiec FO, Albin N et al. (2002) Individual adaptive dosing of topotecan in ovarian cancer. *Clin Cancer Res* 8(2): 394–399

45 Leger F, Loos WJ, Fourcade J, Bugat R, Goffinet M, Mathijssen RHJ et al. (2004) Factors affecting pharmacokinetic variability of oral topotecan: a population analysis. *Br J Cancer* 90(2): 343–347

46 ten Bokkel Huinink WW, Gore M, Carmichael J, Gordon A, Malfetano J, Hudson I et al. (1997) Topotecan *versus* paclitaxel for the treatment of recurrent epithelial ovarian cancer. *J Clin Oncol* 15(6): 2183–2193

47 McGuire WP, Blessing JA, Bookman MA, Lentz SS, Dunton CJ (2000) Topotecan has substantial antitumor activity as first-line salvage therapy in platinum-sensitive epithelial ovarian carcinoma: A Gynecologic Oncology Group study. *J Clin Oncol* 18(5): 1062–1067

48 Gordon AN, Fleagle JT, Guthrie D, Parkin DE, Gore ME, Lacave AJ (2001) Recurrent epithelial ovarian carcinoma: A randomized phase III study of pegylated liposomal doxorubicin *versus* topotecan. *J Clin Oncol* 19(14): 3312–3322

49 Gore M, Oza A, Rustin G, Malfetano J, Calvert H, Clarke-Pearson D et al. (2002) A randomised trial of oral *versus* intravenous topotecan in patients with relapsed epithelial ovarian cancer. *Eur J Cancer* 38(1): 57–63

50 Schiller JH, Kim KM, Hutson P, DeVore R, Glick J, Stewart J et al. (1996) Phase II study of topotecan in patients with extensive-stage small-cell carcinoma of the lung: An Eastern Cooperative Oncology Group trial. *J Clin Oncol* 14(8): 2345–2352

51 Ardizzoni A, Hansen H, Dombernowsky P, Gamucci T, Kaplan S, Postmus P et al. (1997) Topotecan, a new active drug in the second-line treatment of small-cell lung cancer: A phase II study in patients with refractory and sensitive disease. *J Clin Oncol* 15(5): 2090–2096

52 von Pawel J, Schiller JH, Shepherd FA, Fields SZ, Kleisbauer JP, Chrysson NG et al. (1999) Topotecan *versus* cyclophosphamide, doxorubicin, and vincristine for the treatment of recurrent small-cell lung cancer. *J Clin Oncol* 17(2): 658–667

53 Schiller JH, Adak S, Cella D, Devore RF, Johnson DH (2001) Topotecan *versus* observation after cisplatin plus etoposide in extensive-stage small-cell lung cancer: E7593 – A phase III trial of the Eastern Cooperative Oncology Group. *J Clin Oncol* 19(8): 2114–2122

54 von Pawel J, Gatzemeier U, Pujol JL, Moreau L, Bildat S, Ranson M et al. (2001) Phase II comparator study of oral *versus* intravenous topotecan in patients with chemosensitive small-cell lung cancer. *J Clin Oncol* 19(6): 1743–1749

55 Beran M, Kantarjian H, OBrien S, Koller C, AlBitar M, Arbuck S et al. (1996) Topotecan, a topoisomerase I inhibitor, is active in the treatment of myelodysplastic syndrome and chronic myelomonocytic leukemia. *Blood* 88(7): 2473–2479

56 Beran M, Kantarjian H, OBrien S, Koller C, AlBitar M, Arbuck S et al. (1996) Topotecan, a topoisomerase I inhibitor, is active in the treatment of myelodysplastic syndrome and chronic myelomonocytic leukemia. *Blood* 88(7): 2473–2479

57 Kantarjian HM, Beran M, Ellis A, Zwelling L, Obrien S, Cazenave L et al. (1993) Phase-I study of topotecan, a new topoisomerase-I inhibitor, in patients with refractory or relapsed acute-leukemia. *Blood* 81(5): 1146–1151

58 Rowinsky EK, Adjei A, Donehower RC, Gore SD, Jones RJ, Burke PJ et al. (1994) Phase-I and pharmacodynamic study of the topoisomerase I-inhibitor topotecan in patients with refractory acute-leukemia. *J Clin Oncol* 12(10): 2193–2203

59 Rowinsky EK, Kaufmann SH, Baker SD, Miller CB, Sartorius SE, Bowling MK et al. (1996) A Phase I and pharmacological study of topotecan infused over 30 minutes for five days in patients with refractory acute leukemia. *Clin Cancer Res* 2(12): 1921–1930

60 Beran M, O'Brien S, Thomas DA, Tran HT, Cortes-Franco JE, Giles F et al. (2003) Phase I study of oral topotecan in hematological malignancies. *Clin Cancer Res* 9(11): 4084–4091

61 Crump M, Couban S, Meyer R, Rudinskas L, Zanke B, Gluck S et al. (2002) Phase II study of sequential topotecan and etoposide in patients with intermediate grade non-Hodgkin's lymphoma:

A National Cancer Institute of Canada Clinical Trials Group study. *Leuk Lymphoma* 43(8): 1581–1587

62 Kraut EH, Balcerzak SP, Young D, Davis MP, Jacobs SA (2002) A phase II study of topotecan in non-Hodgkin's lymphoma: An Ohio State University phase II research consortium study. *Cancer Invest* 20(2): 174–179

63 Bookman MA, Blessing JA, Hanjani P, Herzog TJ, Andersen WA (2000) Topotecan in squamous cell carcinoma of the cervix: A phase II study of the gynecologic oncology group. *Gynecol Oncol* 77(3): 446–449

64 Muderspach LI, Blessing JA, Levenback C, Moore JL (2001) A Phase II study of topotecan in patients with squamous cell carcinoma of the cervix: A gynecologic oncology group study. *Gynecol Oncol* 81(2): 213–215

65 Fiorica J, Holloway R, Ndubisi B, Orr J, Grendys E, Boothby R et al. (2002) Phase II trial of topotecan and cisplatin in persistent or recurrent squamous and nonsquamous carcinomas of the cervix. *Gynecol Oncol* 85(1): 89–94

66 Fiorica JV (2003) The role of topotecan in the treatment of advanced cervical cancer. *Gynecol Oncol* 90(3): S16–S21

67 Rose PG, Gordon NH, Fusco N, Fluellen L, Rodriguez M, Ingalls ST et al. (2000) A phase II and pharmacokinetic study of weekly 72-h topotecan infusion in patients with platinum-resistant and paclitaxel-resistant ovarian carcinoma. *Gynecol Oncol* 78(2): 228–234

68 Wadler S, Levy DE, Lincoln ST, Soori GS, Schink JC, Goldberg G (2003) Topotecan is an active agent in the first-line treatment of metastatic or recurrent endometrial carcinoma: Eastern cooperative oncology group study E3E93. *J Clin Oncol* 21(11): 2110–2114

69 Perez Soler R, Fossella FV, Glisson BS, Lee JS, Murphy WK, Shin DM et al. (1996) Phase II study of topotecan in patients with advanced non-small-cell lung cancer previously untreated with chemotherapy. *J Clin Oncol* 14(2): 503–513

70 Kindler HL, Kris MG, Smith IE, Miller VA, Grant SC, Krebs JB et al. (1998) Phase II trial of topotecan administered as a 21-day continuous infusion in previously untreated patients with stage IIIB and IV non-small-cell lung cancer. *Am J Clin Oncol* 21(5): 438–441

71 Morton CL, Wadkins RM, Danks MK, Potter PM (1999) The anticancer prodrug CPT-11 is a potent inhibitor of acetylcholinesterase but is rapidly catalyzed to SN-38 by butyrylcholinesterase. *Cancer Res* 59(7): 1458–1463

72 Kawato Y, Aonuma M, Hirota Y, Kuga H, Sato K (1991) Intracellular roles of Sn-38, a metabolite of the camptothecin derivative CPT-11, in the antitumor effect of CPT-11. *Cancer Res* 51(16): 4187–4191

73 Tsuji T, Kaneda N, Kado K, Yokokura T, Yoshimoto T, Tsuru D (1991) Cpt-11 converting enzyme from rat serum – purification and some properties. *J Pharmacobio-Dynamics* 14(6): 341–349

74 Zamboni WC, Houghton PJ, Thompson J, Cheshire PJ, Hanna SK, Richmond LB et al. (1998) Altered irinotecan and SN-38 disposition after intravenous and oral administration of irinotecan in mice bearing human neuroblastoma xenografts. *Clin Cancer Res* 4(2): 455–462

75 Atsumi R, Okazaki O, Hakusui H (1995) Metabolism of irinotecan to SN-38 in a tissue-isolated tumor-model. *Biol Pharm Bull* 18(7): 1024–1026

76 Herben VMM, Schellens JHM, Swart M, Gruia G, Vernillet L, Beijnen JH et al. (1999) Phase I and pharmacokinetic study of irinotecan administered as a low-dose, continuous intravenous infusion over 14 days in patients with malignant solid tumors. *J Clin Oncol* 17(6): 1897–1905

77 Mathijssen RHJ, van Alphen RJ, Verweij J, Loos WJ, Nooter K, Stoter G et al. (2001) Clinical pharmacokinetics and metabolism of irinotecan (CPT-11). *Clin Cancer Res* 7(8): 2182–2194

78 Rivory LP, Robert J (1995) Identification and kinetics of a beta-glucuronide metabolite of SN-38 in human plasma after administration of the camptothecin derivative irinotecan. *Cancer Chemother Pharmacol* 36(2): 176–179

79 Rivory LP, Riou JF, Haaz MC, Sable S, Vuilhorgne M, Commercon A et al. (1996) Identification and properties of a major plasma metabolite of irinotecan (CPT-11) isolated from the plasma of patients. *Cancer Res* 56(16): 3689–3694

80 Dodds HM, Haaz MC, Riou JF, Robert J, Rivory LP (1998) Identification of a new metabolite of CPT-11 (irinotecan): Pharmacological properties and activation to SN-38. *J Pharmacol Exp Ther* 286(1): 578–583

81 de Bruijn P, Verweij J, Loos WJ, Nooter K, Stoter G, Sparreboom A (1997) Determination of irinotecan (CPT-11) and its active metabolite SN-38 in human plasma by reversed-phase high-performance liquid chromatography with fluorescence detection. *J Chromatogr B Biomed Appl*

698(1–2): 277–285

82 Sparreboom A, de Bruijn P, de Jonge MJA, Loos WJ, Nooter K, Stoter G et al. (1998) Liquid chromatographic determination of irinotecan (CPT-11) and its major metabolites in human plasma, urine and feces. *Ann Oncol* 9: 58

83 Sparreboom A, de Jonge MJA, de Bruijn P, Brouwer E, Nooter K, Loos WJ et al. (1998) Irinotecan (CPT-11) metabolism and disposition in cancer patients. *Clin Cancer Res* 4(11): 2747–2754

84 Raymond E, Boige V, Faivre S, Sanderink GJ, Rixe O, Vernillet L et al. (2002) Dosage adjustment and pharmacokinetic profile of irinotecan in cancer patients with hepatic dysfunction. *J Clin Oncol* 20(21): 4303–4312

85 Kehrer DFS, Mathijssen RHJ, Verweij J, de Bruijn P, Sparreboom A (2002) Modulation of irinotecan metabolism by ketoconazole. *J Clin Oncol* 20(14): 3122–3129

86 Desai AA, Innocenti F, Ratain MJ (2003) UGT pharmacogenomics: implications for cancer risk and cancer therapeutics. *Pharmacogenetics* 13(8): 517–523

87 Mathijssen RHJ, Marsh S, Karlsson MO, Xie RJ, Baker SD, Verweij J et al. (2003) Irinotecan pathway genotype analysis to predict pharmacokinetics. *Clin Cancer Res* 9(9): 3246–3253

88 Fuchs CS, Moore MR, Harker G, Villa L, Rinaldi D, Hecht JR (2003) Phase III comparison of two irinotecan dosing regimens in second-line therapy of metastatic colorectal cancer. *J Clin Oncol* 21(5): 807–814

89 Conti JA, Kemeny NE, Saltz LB, Huang Y, Tong WP, Chou TC et al. (1996) Irinotecan is an active agent in untreated patients with metastatic colorectal cancer. *J Clin Oncol* 14(3): 709–715

90 Cunningham D, Pyrhonen S, James RD, Punt CJA, Hickish TF, Heikkila R et al. (1998) Randomised trial of irinotecan plus supportive care *versus* supportive care alone after fluorouracil failure for patients with metastatic colorectal cancer. *Lancet* 352(9138): 1413–1418

91 Rougier P, Van Cutsem E, Bajetta E, Niederle N, Possinger K, Labianca R et al. (1998) Randomised trial of irinotecan *versus* fluorouracil by continuous infusion after fluorouracil failure in patients with metastatic colorectal cancer. *Lancet* 352(9138): 1407–1412

92 Saltz LB, Cox JV, Blanke C, Rosen LS, Fehrenbacher L, Moore MJ et al. (2000) Irinotecan plus fluorouracil and leucovorin for metastatic colorectal cancer. *N Engl J Med* 343(13): 905–914

93 Douillard JY, Cunningham D, Roth AD, Navarro M, James RD, Karasek P et al. (2000) Irinotecan combined with fluorouracil compared with fluorouracil alone as first-line treatment for metastatic colorectal cancer: a multicentre randomised trial. *Lancet* 355(9209): 1041–1047

94 Rothenberg ML, Meropol NJ, Poplin EA, Van Cutsem E, Wadler S (2001) Mortality associated with irinotecan plus bolus fluorouracil/leucovorin: Summary findings of an independent panel. *J Clin Oncol* 19(18): 3801–3807

95 Bajetta E, Di Bartolomeo M, Mariani L, Cassata A, Artale S, Frustaci S et al. (2004) Randomized multicenter phase II trial of two different schedules of irinotecan combined with capecitabine as first-line treatment in metastatic colorectal carcinoma. *Cancer* 100(2): 279–287

96 Tournigand C, Andre T, Achille E, Lledo G, Flesh M, Mery-Mignard D et al. (2004) FOLFIRI followed by FOLFOX6 or the reverse sequence in advanced colorectal cancer: A randomized GERCOR study. *J Clin Oncol* 22(2): 229–237

97 Rothenberg ML (2001) Irinotecan (CPT-11): recent developments and future directions – colorectal cancer and beyond. *Oncologist* 6(1): 66–80

98 Noda W, Nishiwaki Y, Kawahara M, Negoro S, Sugiura T, Yokoyama A et al. (2002) Irinotecan plus cisplatin compared with etoposide plus cisplatin for extensive small-cell lung cancer. *N Engl J Med* 346(2): 85–91

99 Saijo N (2003) Progress in treatment of small-cell lung cancer: role of CPT-11. *Br J Cancer* 89(12): 2178–2183

100 Negoro S, Masuda N, Takada Y, Sugiura T, Kudoh S, Katakami N et al. (2003) Randomised phase III trial of irinotecan combined with cisplatin for advanced non-small-cell lung cancer. *Br J Cancer* 88(3): 335–341

101 Socinski MA, Sandler AB, Israel VK, Gillenwater HH, Miller LL, Locker PK et al. (2002) Phase II trial of irinotecan, paclitaxel and carboplatin in patients with previously untreated stage IIIB/IV nonsmall cell lung carcinoma. *Cancer* 95(7): 1520–1527

102 Larsen AK, Eseargueil AE, Skladanowski A (2003) Catalytic topoisomerase II inhibitors in cancer therapy. *Pharmacol Ther* 99(2): 167–181

103 Kellner U, Sehested M, Jensen PB, Gieseler F, Rudolph P (2002) Culprit and victim – DNA topoisomerase II. *Lancet* 3(4): 235–243

104 Slevin ML (1991) The clinical-pharmacology of etoposide. *Cancer* 67(1): 319–329
105 Hande KR (1998) Etoposide: Four decades of development of a topoisomerase II inhibitor. *Eur J Cancer* 34(10): 1514–1521
106 Creaven PJ, Allen LM (1975) EPEG, a new antineoplastic epipodophyllotoxin. *Clin Pharmacol Ther* 18(2): 221–226
107 Hande KR, Wolff SN, Greco FA, Hainsworth JD, Reed G, Johnson DH (1990) Etoposide kinetics in patients with obstructive-jaundice. *J Clin Oncol* 8(6): 1101–1107
108 Mross K, Bewermeier P, Kruger W, Stockschlader M, Zander A, Hossfeld DK (1994) Pharmacokinetics of undiluted or diluted high-dose etoposide with or without busulfan administered to patients with hematologic malignancies. *J Clin Oncol* 12(7): 1468–1474
109 Rodman JH, Murry DJ, Madden T, Santana VM (1994) Altered etoposide pharmacokinetics and time to engraftment in pediatric-patients undergoing autologous bone-marrow transplantation. *J Clin Oncol* 12(11): 2390–2397
110 Advani R, Visani G, Milligan D, Saba H, Tallman M, Rowe JM et al. (1999) Treatment of poor prognosis AML patients using PSC833 (valspodar) plus mitoxantrone, etoposide, and cytarabine (PSC-MEC). *Drug Resistance in Leukemia and Lymphoma Iii* 457: 47–56
111 Lum BL, Kaubisch S, Yahanda AM, Adler KM, Jew L, Ehsan MN et al. (1992) Alteration of etoposide pharmacokinetics and pharmacodynamics by cyclosporine in a phase-I trial to modulate multidrug resistance. *J Clin Oncol* 10(10): 1635–1642
112 Hande KR, Krozely MG, Greco FA, Hainsworth JD, Johnson DH (1993) Bioavailability of low-dose oral etoposide. *J Clin Oncol* 11(2): 374–377
113 Leu BL, Huang JD (1995) Inhibition of intestinal P-glycoprotein and effects on etoposide absorption. *Cancer Chemother Pharmacol* 35(5): 432–436
114 Budman DR, Igwemezie LN, Kaul S, Behr J, Lichtman S, Schulman P et al. (1994) Phase-I evaluation of a water-soluble etoposide prodrug, etoposide phosphate, given as a 5-minute infusion on day-1, day-3, and day-5 in patients with solid tumors. *J Clin Oncol* 12(9): 1902–1909
115 Kaul S, Igwemezie LN, Stewart DJ, Fields SZ, Kosty M, Levithan N et al. (1995) Pharmacokinetics and bioequivalence of etoposide following intravenous administration of etoposide phosphate and etoposide in patients with solid tumors. *J Clin Oncol* 13(11): 2835–2841
116 Gelderblom H, Verweij J, Nooter K, Sparreboom A, Cremophor EL (2001) The drawbacks and advantages of vehicle selection for drug formulation. *Eur J Cancer* 37(13): 1590–1598
117 Hande KR (1998) Etoposide: four decades of development of a topoisomerase II inhibitor. *Eur J Cancer* 34(10): 1514–1521
118 Drewinko B, Barlogie B (1976) Survival and cycle-progression delay of human lymphoma cells *in vitro* exposed to VP-16-213. *Cancer Treat Rep* 60(9): 1295–1306
119 Wolff SN, Grosh WW, Prater K, Hande KR (1987) *In vitro* pharmacodynamic evaluation of VP-16-213 and implications for chemotherapy. *Cancer Chemother Pharmacol* 19(3): 246–249
120 Denny WA, Baguley BC (2003) Dual topoisomerase I/II inhibitors in cancer therapy. *Current Topics in Medicinal Chemistry* 3(3): 339–353
121 Clapp JM, Hande KR (2002) Topoisomerase II inhibitors. *Cancer Chemother Biol Response Modif* 20: 125–149
122 Valkov NI, Sullivan DM (2003) Tumor p53 status and response to topoisomerase II inhibitors. *Drug Resistance Updates* 6(1): 27–39
123 Aisner J, Musanti R, Beers S, Smith S, Locsin S, Rubin EH (2003) Sequencing topotecan and etoposide plus cisplatin to overcome topoisomerase I and II resistance: A pharmacodynamically based phase I trial. *Clin Cancer Res* 9(7): 2504–2509
124 Licitra EJ, Vyas V, Nelson K, Musanti R, Beers S, Thomas C et al. (2003) Phase I evaluation of sequential topoisomerase targeting with irinotecan/cisplatin followed by etoposide in patients with advanced malignancy. *Clin Cancer Res* 9(5): 1673–1679

Drugs Affecting Growth of Tumours
Edited by Herbert M. Pinedo and Carolien H. Smorenburg
© 2006 Birkhäuser Verlag/Switzerland

Tubulin interacting agents

Manon T. Huizing

Antwerp University Hospital, Department of Oncology, Edegem, Belgium

Introduction

Microtubules are important structural elements in all eukaryotic cells, and essential for mitosis, intracellular transport, and maintenance of cell shape, cellular motility and attachment. They play a key role in modulating interactions with cell-surface receptors and the transmembrane signals generated by these interactions. Microtubules are formed through polymerisation of two different proteins, α- and β-tubulin, under the influence of co-factors such as guanosine triphosphate (GTP) and microtubule associated proteins (MAPs) [1].

One end of the microtubule is attached to the centrosome by γ-tubulin. This site is referred to as the minus end which is less dynamically active, while the opposite end (the plus end) is kinetically more dynamic. The plus end is attached at the kinetochore site of the centromere by different proteins such as dyenin and kinesin.

Microtubules display two types of unusual dynamic behaviour, 'dynamic instability' and 'treadmilling', which appear to be important for progression through mitosis and the cell cycle. Dynamic instability is the stochastic switching of microtubule ends between phases of relatively slow growth and rapid shortening [2]. Treadmilling is defined as the addition of tubulin subunits at one end of a microtubule (the plus end) and the balanced net loss from the opposite (minus) end [3].

Microtubule dynamics become extremely rapid when mammalian cells progress from interphase to mitosis, when rapid dynamics are required for the construction of the mitotic spindle and for various chromosome movements [4–7].

Under normal circumstances microtubules are in a state of dynamic equilibrium with the tubulin dimers, which can be disrupted by a broad range of anticancer drugs, most of them derived from natural products.

This chapter reviews the two most important classes of drugs interfering with tubulin function, the vinca alkaloids and the taxanes. Furthermore, several novel antimicrotubular agents and antimitotic drugs in early development will be discussed.

Vinca alkaloids

Both the naturally occurring vinca alkaloids (vincristine and vinblastine), found in small quantities in the periwinkle plant *Catharanthus roseus* G Don (Vinca rosea L), and the semisynthetic derivatives (vindesine, vinorelbine and vinflunine) are antimitotic drugs that are widely and successfully used in the treatment of cancer [8].

Since 1653 the periwinkle plant is known to have medicinal potency in treating haemorrhagic diseases and hyperglycaemia. Research in the late 1950s to unravel its pharmacological properties failed to confirm the hypoglycaemic activity but did demonstrate an antitumour effect of the alkaloid fraction extracted from the leaves of the plant. Isolation and characterisation of the different components in the plant extract revealed a large number of structurally closely related alkaloids, having a dimeric skeleton of two very similar multi-ringed compounds, an indole nucleus 'catharanthine' and a dihydroindole nucleus 'vindoline', linked together by a carbon-carbon bond (Fig. 1). The intriguing effects of small structural differences on the pharmacological

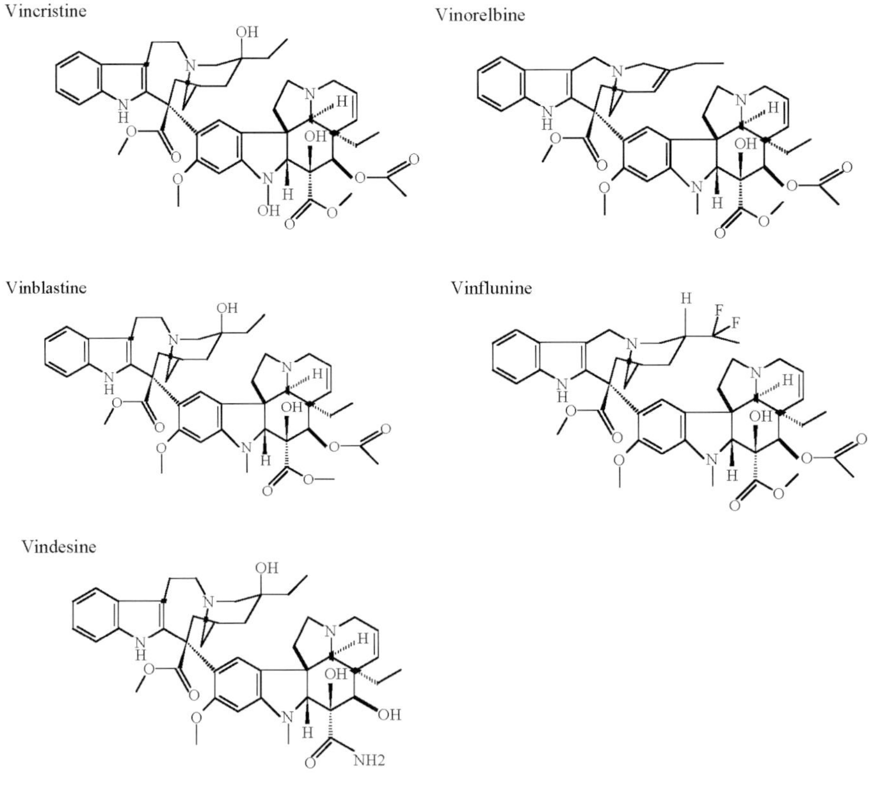

Figure 1. Molecular structures of the vinca alkaloids; vincristine, vinblastine, vindesine, vinorelbine and vinflunine.

activity and severe, disabling neurotoxicity have directed research to the development of semisynthetic derivatives in an effort to design new analogues with improved efficacy and altered antitumour spectrum.

Structurally, vinflunine and vinorelbine differ from vinblastine in the velbanamine moiety. Both drugs were synthesised by a novel method to couple the precursor alkaloids catharanthine and vindoline, which resulted in the formation of an eight-membered rather than a nine-membered ring within the velbanamine portion of the molecule [9–10]. Vinflunine was derived by further modification of vinorelbine, using superacidic chemistry, which specifically introduced two fluorine atoms in the velbanamine moiety [11].

Mechanism of action

Vinca alkaloids inhibit cell proliferation by affecting the dynamics of spindle microtubules. In particular, vinblastine has been shown to bind with high affinity to microtubule ends, strongly suppressing both microtubule dynamic instability and treadmilling. Subsequently, the transition from metaphase to anaphase is blocked, halting mitosis [12–17].

Vinca alkaloids bind to tubulin in intact microtubules with two widely different affinities depending on the localisation of the tubulin binding site either at the microtubule ends or along the surface. The binding sites on the microtubule surface have a low affinity for vinblastine (1–2 sites per molecule of tubulin dimer in microtubules; K_d 0.25–0.3 mM) [18, 19], whereas a high affinity-binding site is located uniquely at one or both microtubule ends (~6 binding sites per microtubule, K_d 1–2 µM) [20]. Low concentrations of a vinca alkaloid will suppress dynamic instability at microtubule plus ends, without reducing the actual microtubule polymer mass [15, 21]. This increased stability at plus ends and decreased stability at minus ends could be an important feature in the powerful blockade of mitosis.

In contrast to the classic vinca alkaloids vincristine and vinblastine, the newer agents vinflunine and vinorelbine alter the dynamic instability by slowing of the microtubule growth rate, an increase in growth duration, and a reduction in shortening duration. Moreover, they neither reduce the rate of shortening nor increase the percentage of time the microtubules spent in an attenuated state. In addition, vinflunine and vinorelbine suppress treadmilling, although less strongly than the other vinca alkaloids. Vinflunine has the capacity to inhibit tubulin assembly, without any stabilising effect on assembled microtubules, at concentrations comparable to those of the other vinca alkaloids tested [22]. However, vinflunine binds relatively weakly to the vinca-binding site, establishing a clear hierarchy of tubulin binding affinities for the different compounds [22].

The diverse actions of these drugs on microtubules are likely to produce different effects on mitotic spindle function, leading to different effects on cell cycle progression and cell killing.

Mechanism of resistance

The development of resistance to chemotherapeutic agents is a significant limiting factor in successful clinical chemotherapy. Until now two important mechanisms of resistance have been described. Firstly, multidrug resistance (MDR), which can be either innate or acquired, is caused by the (over) expression of the MDR1 gene, encoding a membrane-localised glycosylated glycoprotein, called permeability-glycoprotein (P-gp) [23]. It acts as an energy-dependent efflux pump (ATP) for a variety of substrates, including many important cytotoxins (e.g., vinca alkaloids, epipodophyllotoxins, taxanes), thus lowering the intracellular drug levels [24–28].

Secondly, structural and functional alterations in α- or β-tubulin, resulting from either genetic mutations and consequential amino acid substitutions or posttranslational modifications, have been identified in tumour cells with acquired resistance to the vinca alkaloids [29–36].

Clinical pharmacology

In general, vinca alkaloids are administered by short infusions or as continuous infusions. The latter may be advantageous as compared to bolus injection [37–50], preventing high peak plasma concentrations associated with toxic side effects. Furthermore, it was speculated that an increased duration of drug exposure may result in an increased antitumour activity, as these compounds are cell cycle specific agents. However, there is hardly any evidence that supports the theory that prolonged infusion schedules are more effective than bolus schedules.

Radio-immunoassays have been used in most studies to determine the concentrations of the drug levels in biological samples of patients. Although high-pressure liquid chromatography (HPLC) procedures for vinca alkaloids have become available, recently developed analogues are still being studied with less reliable radio-immunoassays.

The plasma pharmacokinetics of all vinca alkaloids following an IV bolus administration can be described by three-compartment kinetics. Peak plasma levels are about 1,000 ng/ml ($\sim 10^{-6}$ M) and fall to ng/ml levels within a few hours. The apparently large volumes of distribution (V_d) indicate an extensive tissue binding. Only the main pharmacokinetic parameters of the clinically investigated vinca alkaloids have been compared. Most studies reported that the intra-patient variabilities are considerable, which has been attributed to differences in protein and tissue binding, hepatic metabolism and biliary clearance [46].

A wide variation in terminal half-lives and total body clearance is reported for vinblastine, vincristine and vindesine. Most studies, including those using HPLC, reported terminal half-lives in the range of 20–35 h for vinblastine and vincristine and 20–25 h for vindesine. A linear correlation exists between the relative toxicities and clearance (Cl) of vincristine, vindesine and vinblastine. Although vincristine appears more toxic due to its extended half-life, other studies have demonstrated that the terminal half-life of vincristine resembles those of vinblastine and vindesine. Analysis of the pharmacokinetic parameters and antitumour activities demonstrated that the increased plasma half-life and lower clearance after continuous infusion were found to coincide with better antitumour activities.

Pharmacokinetic studies using HPLC have shown that vinorelbine has a distinct profile compared to other vinca alkaloids with a longer plasma half-life, an increased clearance and a higher volume of distribution, these parameters being within narrow ranges.

Vincristine is metabolised and excreted primarily by the hepatobiliary system. 72 h after the administration of radiolabelled vincristine, approximately 12% of the radiolabel is excreted in the urine (at least 50% metabolites), and approximately 70–80% is excreted in the faeces (40% metabolites) [37, 40, 43–52].

Like vincristine, vinblastine is principally disposed of through the hepatobiliary system and into the faeces (approximately 95%); however faecal excretion of the parent compound is low, suggesting an extensive hepatic metabolism [51]. Subsequent *in vitro* studies indicate that the cytochrome P-450 CYP3A isoform is primarily responsible for the drug biotransformation [51, 53]. The most important metabolite of vinblastine is 4-deacetyl-vinblastine, or vindesine, which appears to be as active as the parent compound [51, 53]. Renal clearance is negligible, accounting for 1% to 12% of drug disposition [51, 54, 55].

Similarly, for vinorelbine the liver is the principal excretory organ. 33–80% is excreted in the faeces, whereas urinary excretion represents only 16–30% of total drug disposition, the bulk of which is unmetabolised vinorelbine [41, 51, 56–58]. Studies in humans indicate that 4-O-deacetyl-vinorelbine and 3,6-epoxy-vinorelbine are the principal metabolites, and several minor hydroxy-vinorelbine isomer metabolites have been identified [56, 58, 59]. Although most metabolites are inactive, the deacetyl-vinorelbine metabolite may be as active as vinorelbine. Again the cytochrome P-450 CYP3A isoenzyme appears to be principally involved in biotransformation [51, 56, 59].

The clinical use of the oral administration route for vinorelbine has been described. The oral bioavailability of vinorelbine encapsulated in soft gelatine is around 26%.

Drug interactions

The presence of vincristine or vinorelbine enhances the methotrexate accumulation in tumour cells *in vitro*, by a vinca alkaloid-induced blockade of drug efflux. However, the minimal concentrations of vincristine required to achieve this effect occur only transiently *in vivo* [47, 60]. The vinca alkaloids also inhibit the cellular influx of the epipodophyllotoxins *in vitro*, resulting in less cytotoxicity, but the clinical implications of this potential interaction are unknown [61].

L-Asparaginase may reduce the hepatic clearance of the vinca alkaloids, which may result in increased toxicity. To minimise the possibility of this interaction, the vinca alkaloids should be given 12–24 h before L-asparaginase. The combined use of mitomycin C and the vinca alkaloids has been associated with acute dyspnoea and bronchospasm. The onset of these pulmonary toxicities has ranged from within minutes to hours after treatment with the vinca alkaloids or up to 2 weeks after mitomycin C.

Treatment with the vinca alkaloids has precipitated seizures associated with subtherapeutic plasma phenytoin concentrations [60, 61]. Reduced plasma phenytoin levels have been noted from 24 h to 10 days after treatment with vincristine and vinblastine [62].

Because of the importance of the cytochrome P-450 CYP3A isoenzyme in vinca alkaloid metabolism, concurrent administration of erythromycin, H_2-receptor antagonists or other inhibitors of CYP3A may lead to severe toxicity [63]. Conversely, inducers of cytochrome P-450 metabolic processes such as pentobarbital may also influence vincristine clearance [60, 64]. Another potential drug interaction may occur in patients who have Kaposi's sarcoma associated with the acquired immunodeficiency syndrome and are receiving concurrent treatment with 3'-azido-3'-deoxythymidine (AZT), as the vinca alkaloids may impede glucuronidation of AZT to its 5'-O-glucuronide metabolite [65].

Based on a report of a constellation of severe toxicities, including syndrome of inappropriate secretion of antidiuretic hormone (SIADH), bilateral cranial nerve palsies, peripheral neuropathy in upper and lower extremities, cranial nerve palsies, heart failure, and cardiovascular effects after vincristine treatment in children with acute lymphocytic leukaemia who had been receiving treatment with nifedipine and itraconazole, these medications may potentially enhance the neurological and cardiovascular effects of the vinca alkaloids [66].

Clinical activity and toxicity

Vincristine
Vincristine is used in combination with other antitumour agents, as part of potentially curative treatment modalities for lymphomas, leukaemias and testicular cancers.

It is administered intravenously in a bolus dose of 1 to 1.4 mg/m^2 with an absolute dose of 2.0 to 2.5 mg for children and 2.0 mg for adults. It is a potent vesicant and should therefore not be administered intramuscularly, subcutaneously, intravesically or intraperitoneally.

Neuropathy is the most serious, frequent and dose-limiting toxicity, especially in patients above 40 years of age, and it is related to total cumulative dose [67]. Its primary manifestation is a symmetric distal neuropathy affecting both sensory and motor functions. Initial manifestations are usually a loss of the deep tendon reflexes of the lower extremities, followed by paresthesias of the fingers and toes, and ultimately a loss of strength in the dorsiflexors of the lower extremities and in the small musculature of the hand and wrist. 'Footdrop' and 'wristdrop' are observed in patients with advanced vincristine motor neuropathy and are often irreversible or only partially reversible after drug discontinuation.

Cranial motor nerves may also be affected, causing hoarseness, diplopia or facial palsies.

Autonomic neuropathies are unusual and occur primarily as a consequence of high dose vincristine therapy (single dose = 2 mg/m^2) or in patients with altered hepatic function. These patients may develop paralytic ileus, with bloating, abdominal cramps and constipation, as well as urinary retention. Alterations in mental status such as depression, confusion, agitation, insomnia, seizures, coma and visual disturbances have also been described.

Mild and reversible alopecia occurs in approximately 10–20% of the patients treated with vincristine and vinorelbine.

Although acute cardiac ischaemia has been reported, cardio-respiratory symptoms are rare with the use of vinca alkaloids.

Vinblastine

Vinblastine has been administered using a variety of schedules. The most commonly, a bolus of 6 mg/m^2 is injected intravenously once weekly in combination chemotherapy. Although neurotoxicity occurs in a small percentage of patients treated with vinblastine, it is rarely seen at the usual clinical dosage. The dose-limiting toxicity is mainly bone-marrow suppression with thrombocytopenia and leucopenia reaching their nadir 7–10 days after treatment.

Vindesine

Vindesine was the first semisynthetic derivative of vinca alkaloids that turned out to be active and is licensed in various chemotherapeutic regimens. It is a deacetyl derivative of vinblastine.

Vindesine has been given using several schedules in humans, for example 3 mg/m^2 by intravenous bolus, 1.2 mg/m^2/day by 5-day continuous infusion, and 2.0 mg/m^2/day by 2-day continuous infusion.

Primary side effects are a transient leucopenia without thrombocytopenia and vincristine-like neurotoxicity.

Vinorelbine

Vinorelbine is a third generation vinca alkaloid, which has been in clinical development for 15 years. Vinorelbine is usually administered at a dose of 30 mg/m^2 on a weekly or biweekly schedule as a bolus injection, as a short infusion over 20 min, or orally at 80 mg/m^2.

As compared to the other vinca alkaloids it has shown improved efficacy and reduced toxicity. Initially, studies were undertaken to establish the clinical activity of vinorelbine in breast cancer and non-small cell lung cancer. In the treatment of breast cancer significant activity has been seen with the combination of anthracyclines, anthracenediones, antimetabolites and taxanes. The activity of vinorelbine in combination with cisplatin and other agents for the treatment of non-small cell lung cancer is more and more being recognised. Furthermore, it has also demonstrated useful activity in the treatment of a wide variety of other malignancies, such as prostate cancer, multiple myeloma, ovarian cancer, oesophageal cancer, cervical cancer, head and neck cancer and malignant lymphomas.

Taxanes

The history of taxanes started around the turn of the twentieth century, when a British official in the Indian subcontinent noted that parts of the European Yew, *Taxus baccata*, were used in a clarified butter preparation for the treatment of cancer [68, 69]. Several decades later, in 1962, crude bark extracts of the related Pacific (or western) Yew, *Taxus brevifolia*, were provided to the National Cancer Institute (NCI, USA) by the US Forest service, as part of a NCI programme to evaluate US plants for anticancer activity. The crude alcohol extract was shown to be cytotoxic against several murine tumours and was eventually isolated in 1971 when the active component was characterised as paclitaxel [69, 70]. The scarcity of the crude material, initial difficulties in developing a suitable clinical intravenous (IV) pharmaceutical formulation, and the belief that its mechanism of action was identical to that of the vinca alkaloids delayed the development of paclitaxel until the 1970s. However, when its unique mechanism of cytotoxic action was unravelled in 1979 the interest in paclitaxel rekindled.

It was found to act as a promotor of microtubule assembly shifting the physiological equilibrium between tubulins and microtubule toward polymerisation [69, 71]. This mechanism of action is in contrast to the action of other antimicrotubule agents (e.g., vinca alkaloids, colchicine), which induce depolymerisation of microtubules [12, 72, 73].

The search for paclitaxel derivatives from more abundant and renewable resources led to the development of docetaxel, which is synthesised from 10-deacetylbaccatin III, an inactive taxane precursor found in the needles of *Taxus baccata* and esterified with a chemically synthesised side chain (Fig. 2) [74].

Mechanism of action

Paclitaxel and docetaxel are both strong inhibitors of eukaryotic cell replication, blocking cells in the G2 mitotic phase of the cell cycle. They promote microtubule assembly by shifting the dynamic equilibrium towards microtubule assembly and stabilise microtubules, even in the absence of GTP or MAPs, preventing their depolymerisation. The taxanes bind to the interior surface of the microtubule lumen at binding sites that are distinct from the vinca alkaloids.

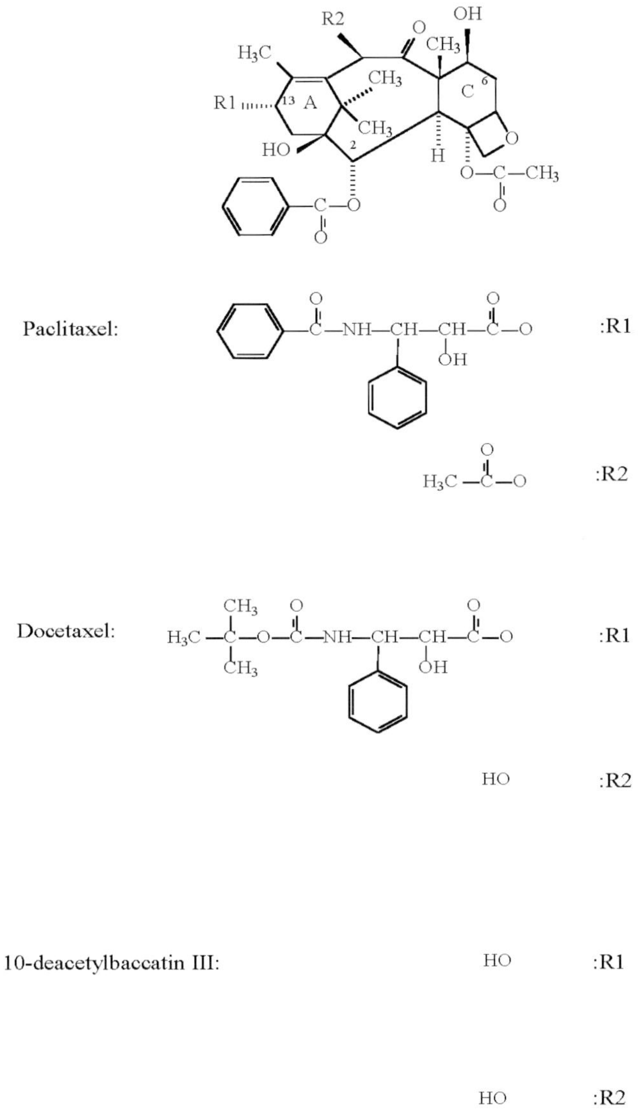

Figure 2. Molecular structures of paclitaxel and docetaxel.

On the basis of photoaffinity labelling and crystallographic analyses, it was demonstrated that both paclitaxel and docetaxel inhibit the function of tubulin by binding to a similar, highly defined region within β-tubulin [75]. Docetaxel, which is slightly more water-soluble than paclitaxel, stabilises microtubules and enhances microtubule polymerisation at stoichiometric concentrations twice as effectively as paclitaxel [70, 76–80].

Tubulin polymers produced by two different tubulin promoters, Tau and MAP2, depolymerise at different rates and efficiencies in the presence of paclitaxel as compared with docetaxel, which suggests that the polymers generated by paclitaxel differ structurally from those generated by docetaxel [76]. Docetaxel does not alter the number of protofilaments in microtubules (13) like its naturally occurring taxane congener paclitaxel (12).

Recent developments indicate that the antineoplastic activity of taxanes may originate in part from induction of genes encoding transcription factors with tumour suppressor effects as well as enzymes governing proliferation, apoptosis, inflammation, and other antiproliferative factors [81–83].

The radiosensitising effects of taxanes have been investigated extensively on the rationale that G_2+M is the most radiosensitive phase of the cell cycle [84, 85]. Most combination studies found a significant radiation potentiating effect of both paclitaxel and docetaxel, with both a block in G_2-M phase and also a high cell killing (~90%). In the clinical setting however, this could result in damage to normal tissue [86, 87]. Docetaxel significantly increases radioresponsiveness *in vitro* by a factor of 2.5- to 3-fold [88, 89]. *In vivo,* a synergistic effect of docetaxel with radiation in murine MCa-K tumours increased the tumour growth delay by a factor of up to 2.64 [90]. Docetaxel produced higher radiosensitivity effects than equimolar concentrations of paclitaxel.

Mechanism of resistance

In tissue culture, resistance to paclitaxel or docetaxel can be attributed to (*a*) overexpression of drug efflux pumps such as P-glycoprotein, (*b*) acquired mutations at the drug binding site of tubulin, (*c*) differential expression of tubulin isoforms, (*d*) alteration in apoptotic mechanisms, (*e*) activation of growth factor pathways, or (*f*) other unknown mechanisms [91, 92]. The contribution of each of these mechanisms to clinical resistance remains uncertain, although correlations have been made with P-glycoprotein expression levels in some tumour types.

Resistance to anti-microtubule agents can also be mediated by altered expression of tubulin and/or MAPs as well as tubulin mutations [91]. In recent years, several paclitaxel- and docetaxel-resistant tumour cell lines harbouring single-point mutations resulting in amino acid substitutions in β-tubulin have been identified.

Clinical pharmacology

Paclitaxel
The pharmacokinetics of paclitaxel appears to be non-linear which may lead to dramatic differences in drug exposure, in terms of area under the curve (AUC), when dosages and/or schedules are changed [93, 94]. This non-linear-

ity probably occurs at the level of saturable hepatic metabolism. Several metabolic products of paclitaxel have been detected in bile of rats and humans, and have been isolated and structurally identified by the use of chromatographic methods, mass spectrometry and nuclear magnetic resonance spectroscopy [95–99]. The identification of the three major metabolites 6α-hydroxypaclitaxel, 3'-p-hydroxypaclitaxel and 6α, 3'-p-dihydroxypaclitaxel of paclitaxel in human bile and in human plasma are supportive for extensive hepatic metabolism. The metabolic breakdown is achieved through the cytochrome P-450 isoforms 3A4 and 2C8 [98].

The hydroxy substituted metabolites were shown to have lost their cytotoxicity in *in vitro* clonogenic assays, using the A2780 human ovarian carcinoma and CC531 rat colon carcinoma tumour cell lines. These metabolites showed reduced myelotoxic effects as compared with paclitaxel in an *in vitro* haemopoietic progenitor toxicity assay [99].

Besides the saturable hepatic metabolism, the pharmaceutical vehicle Cremophor EL is the other determinant in the (pseudo) non-linear pharmacokinetic behaviour of paclitaxel [100].

Pharmacodynamic analysis showed a positive correlation between the $T \geq 0.05$ µmol/L and $T \geq 0.1$ µmol/L and bone marrow suppression according a sigmoidal E_{max} model [93, 94].

In patients with liver metastasis, there is an association between the paclitaxel C_{ss} above 0.07 µmol/L and the clinical toxicity [101]. Higher paclitaxel AUC levels and a prolonged duration of $T \geq 0.1$ µmol/L were associated with liver disease and higher AP levels (liver metastasis or cholelithiasis) [102].

Pharmacokinetic analysis of unbound paclitaxel in elderly patients (>70 years) as compared to their younger counterparts showed a reduced paclitaxel clearance of 50% (124 ± 35.0 *versus* 244 ± 58.8 L/h/m^2) while no enhanced toxicity profile was observed [103]. This observation may be a result from the altered Cremophor EL® clearance in this patient cohort.

There is a clear therapeutic advantage for intraperitoneal administration of paclitaxel (dose 120 mg/m^2) when dissolved in Cremophor EL®. The terminal disposition half-life of paclitaxel was substantially prolonged after intraperitoneal administration (28.7 ± 8.72 h), when compared to the low systemic disposition (17.0 ± 11.3 h) [104]. This is particularly attractive in patients with ovarian cancer and other tumour types confined to the abdominal cavity, such as peritoneal mesothelioma [105].

Docetaxel
In contrast to paclitaxel, the pharmacokinetic behaviour of docetaxel is linear, independent of dose and schedule. The docetaxel plasma profile is typically triphasic, with a terminal half-life ranging from 11–18 h, with a plasma clearance of around 21 litres/h/m^2 and a distribution volume of 72 litres/m^2. Around 75% of the delivered dose is excreted in the faeces and less than 5% of the unchanged drug was excreted renally [106, 107]. The pharmacokinetics of docetaxel shows a considerable interpatient variability. Docetaxel clearance is the

most important parameter in analysing the pharmacokinetic variability and is related to a1-glycoprotein levels, age, body surface and hepatic function [108].

The AUC correlates with the percentage decrease of neutrophils in a sigmoid E_{max} model [106, 109]. No differences in pharmacological behaviour were observed for patients older than 65, but they experienced more profound neutropenia with or without neutropenic fever [110].

A diminished hepatic function contributes to a lower plasma clearance, which leads to severe myelosuppression with life threatening infections, stomatitis and toxic death. A dose reduction to 75 mg/m^2 has been recommended for these patients [107].

Fluid retention may be related to increased AUC levels of the 3'-[3-(5,5-dimethyl-2,4-dioxo-1,3-oxazolidinyl)]-docetaxel metabolite [111].

Studies performed with ^{14}C-labelled docetaxel, carried out in mice, dogs, pigs, mini-pigs and rats, have shown that the drug is extensively metabolised [112–114]. Less than 10% of the dose was excreted unchanged in the faeces, whereas three to four metabolites accounted for about 75% of the dose. In patients given [^{14}C] docetaxel as a 1 h infusion, the major part of the radioactivity was recovered also in the faeces, in the form of both the parent compound and a, by then, unidentified metabolite. *In vitro* metabolism studies with isolated liver microsomes from the mouse, dog, rat and human showed a cytochrome P450 3A-enzyme dependent metabolic profile, similar to that observed in *in vivo* animal studies. The structure of four human metabolites of docetaxel has been established by using HPLC, tandem-mass spectrometry and nuclear magnetic resonance (NMR) spectroscopy [115]. All metabolites originated from oxidation reactions of the tert-butyl moiety in the C13- side chain of the parent compound. Docetaxel is oxidised into the primary alcohol metabolite 3'-(1-OH-2-methyl)-docetaxel, followed by ring closure giving two isomeric hydroxyoxazolidinones metabolites 3'-[3-(5,5-dimethyl-4-OH-2-oxo-1,3-oxazolidinyl)]-docetaxel, via a putative aldehyde. An alternate oxidation pathway of the 3'-(1-OH-2-methyl)-docetaxel metabolite via a putative corresponding acid would give the oxazolidinedione metabolite 3'-[3-(5,5-dimethyl-2,4-dioxo-1,3-oxazolidinyl)]-docetaxel, after cyclisation [115, 116].

The metabolic products were used for evaluating their cytotoxic activities against a human ovarian cancer (A2780) and a rat colon cancer (CC531) cell line, and their myelosuppressive effects in a haematopoietic progenitor toxicity assay. Although distinctions in biological activities between the compounds were evident, all metabolites showed a marked reduction in both cytotoxic and myelotoxic properties [116].

Drug interactions

Extensive biliary excretion and hepatic metabolism of both paclitaxel and docetaxel mediated by the P450 cytochrome C oxidase system may result in drug interaction with drugs using the same metabolic pathway [93, 98, 99, 116].

The anticonvulsants phenytoin and phenobarbital induce accelerated metabolism with subsequent detoxification of both paclitaxel and docetaxel with high tolerance for both drugs.

H_2-receptor blockers, used as pre-medication for hypersensitivity reactions during paclitaxel therapy, have a variable effect on P450 functions that may influence the pharmacological profile, and therefore, toxicity and antitumour effect of the drug. However, no toxicological or pharmacological differences were noted between these agents in randomised clinical trials [117–120]. Both ketoconazole and fluconazole, potent inhibitors of the P450 3A4 system, decreased the formation *in vitro* of one of the two observed paclitaxel metabolites in a preparation of human liver slices and microsomes [117]. Concomitant administration of ketoconazole *in vivo* produced a moderate increase in paclitaxel levels, but a dramatic decrease in biliary metabolite excretion [118, 121].

Several preclinical studies demonstrated a sequence dependent cytotoxicity for the combination cisplatin–paclitaxel [122–124], showing the combination of paclitaxel followed by cisplatin to be most toxic. In human pharmacology an inverse sequence dependency was observed for the 24 h paclitaxel infusion with profound clinical consequences [125].

Although no sequence dependent pharmacokinetic effect of paclitaxel–carboplatin was observed, less thrombocytopenia was observed for the combination as compared to historical carboplatin data [126, 127].

Neutropenia and mucositis are more severe when paclitaxel is administered prior to doxorubicin, as compared to the reverse sequence, which is most likely due to an approximately 32% reduction in the clearance rates of doxorubicin and doxorubicinol [128]. Although neither sequence-dependent pharmacologic nor toxicologic interactions between doxorubicin and paclitaxel (3 h schedule) have been noted, pharmacologic interactions occur with both sequences, and combined treatment with paclitaxel (3 h schedule) and doxorubicin as a bolus infusion is associated with a higher incidence of congestive cardiotoxicity than would have been expected from an equivalent cumulative doxorubicin dose given without paclitaxel [129–131]. Data suggest that docetaxel also enhances the metabolism of doxorubicin to toxic species in the human heart. Similar decrements in the clearance of epirubicin and its metabolites have been noted in studies of paclitaxel combined with epirubicin, but cardiotoxicity does not appear to be enhanced [132]. Competition for the hepatic or biliary P-gp transport of the anthracyclines with paclitaxel or its polyoxyethylated castor oil vehicle (Cremophor EL), or both, may be another explanation [130, 131].

Administration of topotecan on days 1–4 and docetaxel on day 4 resulted in an approximately 50% decrease in docetaxel clearance and was associated with increased neutropenia [133].

Clinical activity and toxicity

Both paclitaxel and docetaxel are widely used in the treatment of a variety of tumours, including breast-, ovarian-, lung-, and prostate cancer.

For both paclitaxel and docetaxel, a high incidence of hypersensitivity reactions (~30%) has been observed during the initial clinical development. Symptoms of hypersensitivity reaction include rash, facial flushing, pruritis, urticaria, fever, and angio edema – sometimes aggravating to hypotension and dyspnoea with or without severe bronchospasm [134]. The introduction of pre-treatment regimens, consisting of corticosteroids with or without H1- and H2 receptor antagonists, led to a substantial decrease in major hypersensitivity reactions (~1.5%) [74, 134, 135]. After cessation of the hypersensitivity symptoms, a reinfusion can be successfully performed in most patients (~80%) for both drugs. Paclitaxel can be administered using a reduced infusion rate and with maximal premedication. In case of mild hypersensitivity reactions occurring after docetaxel administration, reinfusion without premedication can be safely attempted [134–136], while for severe hypersensitivity reactions premedication with prednisolone, cetirizine and ketotifen is warranted [136].

Paclitaxel

Paclitaxel is commonly used as a 3 weekly, 3 h infusion at a dose of 175 mg/m^2 or as 135 mg/m^2 over 24 h infusion period. More recently, a weekly 1 h infusion at a dose of 70–80 mg/m^2 is recommended in taxane resistant tumours or when myelosuppression should be avoided.

Paclitaxel administration into the peritoneal cavities (60 mg/m^2) has a clear clinical advantage due to high local exposure with low system plasma levels in patients with peritoneal seeded ovarian cancer [105]. When paclitaxel and cisplatin are given both intravenously and intraperitoneally, promising 2 year survival rates in women with optimally debulked ovarian cancer were reported and warranted further investigations [137].

Neutropenia and neurotoxicity are the main encountered toxicities in the administration of paclitaxel. Neutropenia is seen more in the high dose and prolonged infusions, while neurotoxicity is seen in the high dose and dose-intense schedules (weekly administration) and in patients with impaired liver enzymes. Neutropenia is commonly of short duration, non-cumulative, and there is a positive correlation between the $T \geq 0.05 - T \geq 0.1$ µmol/L and the severity of bone marrow suppression [93, 94]. Transient myalgia and arthralgia occur 24–48 h after paclitaxel administration. Muscular weakness is frequently reported when patients receive higher doses of paclitaxel or when paclitaxel is combined with either cisplatin or carboplatin [114, 115, 119, 120]. Patients complained about weakness of the upper extremities and difficulty in climbing stairs and rising from a sitting position [138, 139].

Several studies reported a stronger correlation between neuromuscular toxicity and paclitaxel AUC levels than with the administered dose [93, 140]. A significant association between liver metastasis and paclitaxel clearance and a correlation between the paclitaxel steady state concentration (C_{ss}) above 0.07 μmol/L and the clinical toxicity was reported [101]. Peripheral neurotoxicity has frequently been observed during the early development of paclitaxel. The incidence and severity of neurotoxicity is dose-related, cumulative and progressively worsens after multiple courses and was found to be dose limiting in combination with cisplatin [138, 141, 142]. After cessation of therapy symptoms usually improve or resolve within several months after discontinuation of paclitaxel therapy.

There are neurosensory manifestations, including symptoms of numbness and paresthesias in a glove- and stocking distribution. Electrophysiological findings included decreased nerve conduction velocities in sensory nerves, with relative sparing of motor nerves [142, 143], with significant elevations in vibratory and thermal thresholds, supporting both axonal degeneration and demyelinisation as mechanisms for paclitaxel-induced neurotoxicity.

Motor neuropathy is characterised by mild weakness of the extensor hallucis longus and diminished grip strength with reduction in peroneal nerve-evoked amplitude of the extensor digitorum brevis. Paralytic ileus and symptomatic orthostatic hypotension are autonomic neuropathy manifestations of paclitaxel [125, 144, 145].

Paclitaxel can cause asymptomatic atrioventricular conduction abnormalities in association with sinus bradycardia (heart rates range from 30–50 bpm) in patients who received paclitaxel as a single agent or in combination with cisplatin [125, 146].

In animal experiments, the administration of paclitaxel following doxorubicin treatment was shown to cause extensive myocardial necrosis compared with those rats treated with either doxorubicin alone or the reverse sequence of administration. Moreover, rats treated with paclitaxel 24 h after doxorubicin treatment showed exaggeration of the combination-induced cardiotoxicity. In conclusion, paclitaxel might synergistically aggravate doxorubicin-induced cardiotoxicity. The effect might be much more pronounced with those rats treated with paclitaxel 24 h after doxorubicin treatment [147].

Profound cardiotoxicity has been observed for the combination doxorubicin–paclitaxel. Congestive heart failure occurred more frequently for this combination than with other doxorubicin combinations, when the cumulative doxorubicin dose exceeds 360 mg/m the risk of severe cardiac toxicity increases [129, 130, 148].

Cardiac dysfunction grade 3 is observed (8% of the patients) for the combination paclitaxel and trastuzumab, therefore regular cardiac monitoring when using this combination is being advised [149].

Docetaxel
Docetaxel administration is registered as a 3 weekly, 1 h infusion of 60, 75 or 100 mg/m^2. Alternative a weekly schedule can be used with a dose of 30–35 mg/m^2 as 1 h infusion.

Docetaxel as a radiosensitiser is used in a dose of 20 mg/m^2 [150].

Neutropenia is the main and dose-limiting toxicity accompanied with the administration of docetaxel. The incidence and severity of the neutropenia depends on the administered dose, age, the number of prior chemotherapeutic regimens and combination with other cytotoxic drugs [74, 110]. In patients receiving docetaxel 100 mg/m^2 as a 1 h infusion every 3 weeks, grade 4 and febrile neutropenia occur in up to 84% and 11.8% of patients [110, 151]. Anaemia and thrombocytopenia are uncommon in monotherapy docetaxel.

The final pharmacokinetic–pharmacodynamic model might provide a tool for calculation of white blood cell time course, and hence, for prediction of nadir day and duration of leucopenia in breast cancer patients treated with the epirubicin/docetaxel regimen [152].

The toxicity profile of docetaxel is markedly altered when the drug is administered by a weekly schedule. Weekly administration of docetaxel may provide a better tolerance profile [153]. While myelosuppression is mild and uncommon, fatigue and asthenia are the dose limiting toxicities. Other non-haematologic toxicities are rare and include peripheral oedema and neuropathy. The arthralgia/myalgia syndrome was not observed.

Dermatitis developed immediately after docetaxel extravasation but disappears within 24 h. Delayed dermatitis developed with symptoms appears after 5 days and consists of brown discolouration and skin hyperplasia. Topical administration of isotonic saline and dimethyl sulfoxide in combination with local hypothermia is advised. No surgical intervention is needed [154, 155].

Other adverse effects of docetaxel include alopecia, asthenia, neurotoxicity, cutaneous reactions, fluid retention, and stomatitis.

Cutaneous reactions are frequently observed and principally manifested as an erythematous, pruritic maculopapular rash that occasionally progresses to oedema and desquamation of the hand and feet (palmar-plantar erythrodysesthesia). Treatment with hypothermia and pyridoxine seems to be helpful [156–158]. Radiation recall reaction with redness and dermal desquamation of the breast may occur with a latency period of 2 years [159].

The progressive development of peripheral oedema, non-malignant pleural effusions, and ascites was noted in early clinical trials with docetaxel [135, 160–164]. This side effect was observed in more than half of the patients who received a total dose of at least 400 mg/m^2. The fluid retention syndrome resolves very slowly after stopping the docetaxel treatment. Administration of diuretics is usually only moderately successful in dealing with this cumbersome drug toxicity. This phenomenon may be explained by capillary protein leakage [165]. The use of steroids reduces the incidence and delays the onset of the fluid-retention syndrome to a cumulative median dose to 550 mg/m^2 [158].

Skin and nail toxicity is also a unique feature of docetaxel. It consists of a dry itchy skin, maculopapular rashes and desquamation. Up to two-thirds of the patients will experience one or more of these symptoms, although in less than 10% it is severe. Onycholysis, a progressive thickening and discolouration with subsequent loss of the nails, is a disabling toxicity interfering with daily life. It is cumulative, and therefore, like the oedema, more likely to be noted in patients who are receiving prolonged treatment.

Asthenia is a common toxicity in schedules using the 1 h infusion day 1 and 8. Diarrhoea and nausea are generally mild and do not require prophylactic antiemetics [106, 109, 155, 164, 166].

Cardiac conduction disturbances, commonly noted with paclitaxel as asymptomatic bradycardia, have not been documented with docetaxel.

Severe mucositis in combination with neutropenia is associated with prolonged docetaxel infusions (6 and 24 h infusion) and repeated dosing [106, 109, 155, 164, 166].

Neurotoxicity seems less severe during docetaxel administration as compared to paclitaxel. Docetaxel neurotoxicity consisting of mild paresthesias and abolition of the tendon reflexes appeared above doses of 85 mg/m^2 [167]. Moderate or severe reactions occurred more frequently in patients pretreated with platinum compounds or vinca alkaloids.

New developments in taxane compounds

Despite the ability of taxanes and vinca alkaloids to inhibit the progression of some cancers, inherent resistance to antimicrotubule agents is encountered in many tumour types, and acquired resistance usually occurs during multiple cycles of therapy [8, 92]. Beyond this, the side effects are significant and can be attributed to the compound itself and/or the vehicle required for administration. Therefore, there has been great interest in identifying novel antimicrotubule drugs that overcome various modes of resistance and have an improved pharmacological profile. Several cytotoxic drugs have been developed to achieve a higher therapeutic index by evading mechanisms of taxane resistance with less toxicity.

The research into the development of a useful oral formulation of the taxanes has demonstrated that the interpatient variability in the systemic exposure after oral drug administration was of the same order as after intravenous infusion. Co-administration of cyclosporin A strongly enhanced the oral bioavailability of both paclitaxel and docetaxel. Although these findings encouraged the development of oral taxanes with a different pharmacokinetic profile and improved bioavailability, further research is required.

BMS-275183

The modifications of the C-4-methylcarbonate analogue of paclitaxel, which is not clinically useful as an oral agent, into its C-3'-t-butyl-3'-N-t-butyloxycarbonyl analogue has increased the bioavailability and has oral efficacy in preclinical models that is comparable to iv administered paclitaxel. BMS-275183 shares the mechanism of the registered taxanes and when given orally was as effective as intravenous paclitaxel in five tumour models, including murine M109 lung and C3H mammary 16/C, and human A2780 ovarian and HCT/pk colon. In a schedule dependency study, increasing the interval of time between oral administrations resulted in greater cumulative dose tolerance and improved therapeutic outcome. BMS-275183 is currently in Phase I clinical trials at multiple sites.

MAC-321

MAC-321 is a novel analogue of docetaxel. It is a Microtubule/Apoptosis/Cytotoxic: 5β,20-epoxy-1,2α-, 4-,7β-,10β-, 13α-hexahydroxytax-11-en-9-one 4 acetate 2 benzoate 7-propionate 13-ester with (2R,3S)-N-tertbutoxycarbonyl-3-(2-furyl) isoserine, that overcomes P-glycoprotein-mediated resistance to paclitaxel and docetaxel in preclinical model systems. In a highly P-glycoprotein, resistance cell line (KB-V1) MAC-321 was 80-fold resistant compared with that of paclitaxel (1,400-fold) and docetaxel (670-fold). In addition, equivalent or less resistance to MAC-321 compared with paclitaxel or docetaxel was observed in four cell lines that contain distinct point mutations within the taxane-binding site of β-tubulin [168, 169].

Another potential advantage of MAC-321 administration lies in its ability to be formulated in a vehicle that is not expected to induce a hypersensitivity reaction. On the basis of these results, MAC-321 is being evaluated in Phase II clinical trials for the treatment of cancer in humans. Highly efficient taxane-based MDR reversal agents have been developed. Extensive structure–activity relationship studies have led to the development of new generation taxanes that possess 2–3 orders of magnitude higher potencies against human cancer cell lines expressing the MDR phenotype [169]. Second generation taxanes such as SB-T-1213 and SB-T-1 10131 (IDN5109, BAY59-8862), are semisynthetic, orally available taxanes that are up to 400-fold more active than paclitaxel against drug-resistant cancer cell lines as well as human tumour xenografts in mice. SB-T-1213 induces unusual microtubules with attached extra protofilaments or open sheets, and IDN5109 induces large protofilamentous sheets. They target microtubules but alter their polymerisation and structure differently than paclitaxel. These differences may play a role in their enhanced cytotoxicity and efficacy [170]. Both drugs posses an excellent bioavailability profile, and are currently under Phase II clinical trials.

New developments in non-taxane compounds

Hundreds of tubulin inhibitors, naturally occurring, semisynthetic or synthetic, are the subject of active investigation. Several classes of natural occurring antimitotic products include the epothilones, eleutherobins, discodermolides, sarcodictyins, laulimalides and small peptides [171–177]. All these compounds have in common their low-level or no substrate affinity for P-gp and other MDR transporters and retain various degrees of activity against taxane-resistant cells *in vitro*, but the clinical significance of these characteristics is not clear [171, 172, 178].

Epothilones

The epothilones were isolated from the myxobacterium *Sorangium cellulosum*. The epothilones, like the taxanes, induce tubulin polymerisation in the absence of GTP or MAPs, resulting in microtubules that are relatively long, rigid and resistant to destabilisation. However, the epothilones are generally more potent than the taxanes, possessing IC_{50} values in the sub- or low nanomolar range [171, 172, 178–180]. In contrast to taxanes and vinca alkaloids, overexpression of P-gp minimally affects the cytotoxicity of epothilones A and B [171, 172, 178–180]. In addition, various point mutations in α-tubulin, which confer resistance to the taxanes *in vitro*, are not necessarily responsible for resistance to the epothilones, but the significance of β-tubulin isotypes in conferring clinical resistance to tubulin-polymerising agents is not clear. Epothilone B (EPO906) and the epothilone B analogue ixabepilone (BMS-247550) are currently undergoing clinical evaluations [181, 182]. Ixabepilone is metabolised by cytochrome P450 systems, whereas EPO906 is metabolised by carboxyesterases [181, 182]. These differences may be responsible for their different principal toxicities, namely diarrhoea (EPO906), myelosuppression and neurotoxicity (ixapebilone) [181–183]. In early clinical trials, antitumour responses have been noted in patients with breast, lung and ovarian cancers, some of which recurred after or during treatment with the taxanes [181, 182]. Antitumour activity has also been observed with EPO906 in patients with colorectal and renal cancers, which are almost always unresponsive to antimicrotubule agents, but the magnitude of appreciable activity in cancers with primary or acquired taxane resistance is negligible [181, 182].

Epothilone D (desoxyepothilone B; KOS862), which possesses equivalent potency and less toxicity than the taxane and epothilone B analogues in preclinical studies, is also undergoing clinical development [184].

Eleutherobin

Eleutherobin is a novel natural product isolated from a marine soft coral that is extremely potent for inducing tubulin polymerisation *in vitro* and is cytotoxic for cancer cells with an IC_{50} similar to that of paclitaxel. This compound is cross-resistant along with other multidrug-resistant agents against P-glycoprotein-expressing cells and is cross-resistant for structural altered tubulin.

Human colon carcinoma cells exposed to eleutherobin contain multiple micronuclei and microtubule bundles, and they arrest in mitosis, depending on concentration, cell line, and length of exposure. These morphological abnormalities appearing in cultured cells are indistinguishable from those induced by paclitaxel. Thus, eleutherobin has promising potential as a new anticancer agent [174].

Discodermolide

Similar to the epothilones A and B, discodermolide-induced tubulin polymers are very stable to treatment with calcium and composed of short microtubules instead of tubulin spirals [171, 172, 185]. In addition to complete cross-resistance to P-gp-overexpressing cancer cells, paclitaxel and epothilone-resistant human tumour cells that express mutant α-tubulin retain sensitivity to discodermolide [171, 172, 185]. Furthermore, discodermolide and paclitaxel have demonstrated synergistic cytotoxicity *in vitro*, suggesting that their tubulin-binding sites may not be identical [186]. Early clinical evaluations with discodermolide (XAA296) have begun in patients with advanced solid malignancies.

Laulimalide and isolaulimalide

Laulimalide and isolaulimalide are chemically related compounds, with isolaulimalide being a decomposition product of laulimalide. Their mechanism of action showed that these agents are paclitaxel-like stabilisers of microtubules that cause alterations of both interphase and mitotic microtubules.

Laulimalide is a potent inhibitor of cell proliferation and initiates mitotic arrest, micronuclei formation, and ultimately apoptosis. These compounds are superior to paclitaxel in their ability to circumvent P-glycoprotein-mediated drug resistance. The laulimalides represent a new class of paclitaxel-like microtubule-stabilising agents with properties that may provide advantages over the taxanes.

The difference between these two compounds is in the size and attachment points of the oxygen-containing ring within the top portion of the molecules. Laulimalide contains a three-membered epoxide ring involving carbons C-16 and C-17, whereas isolaulimalide contains a five-membered tetrahydrofuran ring linking carbon C-17 with side chain carbon C-20. This slight chemical difference between laulimalide and isolaulimalide results in a difference in potency of greater than two orders of magnitude in their ability to inhibit cell proliferation. Laulimalide initiated short thick bundles of microtubules that were more prevalent in the cell periphery and appeared to form many nucleation centres. In contrast, paclitaxel-induced microtubule bundles were long and thick and aligned in the central areas of the cells surrounding the nucleus, consistent with nucleation from one or two centres.

The mitotic spindles formed in the presence of laulimalide were abnormal and formed unique starburst arrays in contrast to the short thickened tri- and tetra-polar spindles formed in the presence of paclitaxel. Laulimalide-treated mitotic cells exhibited chromatin condensation, loss of the nuclear envelope and abnormal chromatin alignment. The aberrant mitotic spindles were associated with circular chromatin arrays, suggesting that the microtubules were coordinating a specific, but abnormal structuring of the chromatin. Disruption of the mitotic apparatus by laulimalide treatment lead to mitotic arrest, followed by the initiation of apoptosis, as determined by the increase in cells in G_2-M and the activation of the caspase cascade [176, 187]. The initial studies suggest that there are intriguing differences in the mechanisms of action of laulimalide and paclitaxel.

Hemiasterlin

Hemiasterlin is a natural product derived from marine sponges that, like other structurally diverse peptide-like molecules, binds to the vinca peptide site in tubulin, disrupts normal microtubule dynamics, and at stoichiometric amounts depolymerises microtubules. Total synthesis of hemiasterlin and its analogues has been accomplished, and optimal pharmacological features of the series have been explored. HTI-286 inhibited the polymerisation of purified tubulin, disrupted microtubule organisation in cells, and induced mitotic arrest, as well as apoptosis. HTI-286 was a potent inhibitor of proliferation (mean $IC_{50} = 2.5 \pm 2.1$ nM in 18 human tumour cell lines) and had substantially less interaction with multidrug resistance protein (P-glycoprotein) than currently used antimicrotubule agents, including paclitaxel, docetaxel, vinorelbine, or vinblastine. Resistance to HTI-286 was not detected in cells overexpressing the drug transporters MRP1 or MXR. Moreover, HTI-286 inhibited the growth of human tumour xenografts (e.g., HCT-15, DLD-1, MX-1W, and KB-8-5) where paclitaxel and vincristine were ineffective because of inherent or acquired resistance associated with P-glycoprotein. Efficacy was also achieved with oral administration of HTI-286. These data suggest that HTI-286 has excellent preclinical properties that may translate into superior clinical activity, and that it is a useful synthetic reagent to probe the drug contact sites of peptide-like molecules that interact with tubulin [188].

Small peptides

Cryptophycin depsipeptides

Other natural products and semisynthetic antimicrotubule compounds under evaluation interact with tubulin in the vinca alkaloid- or colchicine-binding domains. Among the most potent are the cryptophycin depsipeptides, which are a family of cyanobacterial macrolides that deplete microtubules in intact cells, including cells with the MDR phenotype [171, 172, 189].

The cryptophycins also have impressive activity against a wide array of human tumour xenografts, including those resistant to the vinca alkaloids. However, the clinical development of one semisynthetic analogue, cryptophycin-52, was terminated after an unacceptably low level of antitumour activity and significant toxicity, particularly neurotoxicity.

Dolastatins

The dolastatins constitute a series of oligopeptides isolated from the sea hare, *Dolabela auricularia* [171, 172, 175]. Two of the most potent dolastatins, dolastatin-10 and -15, noncompetitively inhibit the binding of the vinca alkaloids to tubulin, inhibit tubulin polymerisation and tubulin-dependent GTP hydrolysis, stabilise the colchicine-binding activity of tubulin, and possess cytotoxic activity in the picomolar to low nanomolar range. Dolastatin-10 and semisynthetic dolastatin analogues are undergoing clinical evaluations [171, 172].

Phomopsin A, halichondrin B, homohalichondrin B, and spongistatin 1

Phomopsin A, halichondrin B, homohalichondrin B, and spongistatin 1, which competitively inhibit vinca alkaloid binding to tubulin, are also in various stages of development [171, 172, 175, 190].

Halichondrin B, a large polyether macrolide originally isolated from the marine sponge *Halicondrin okadai*, and less complex synthetic marocyclic ketone analogues (ER-076349 and ER-086526) are undergoing clinical development. These compounds bind to tubulin, inhibit tubulin polymerisation, disrupt mitotic spindle formation, induce mitotic arrest, and inhibit the growth of tumours at subnanomolar concentrations.

References

1 Manfredi JJ, Parness J, Horwitz SB (1982) Taxol binds to cellular microtubules. *J Cell Biol* 94: 688–696
2 Mitchison T, Kirschner M (1984) Dynamic instability of microtubule growth. *Nature* 312: 237–242
3 Margolis RL, Wilson L (1998) Microtubule treadmilling: what goes around comes around. *Bioessays* 20: 830–836
4 Hayden JH, Bowser SS, Rieder CL (1990) Kinetochores capture astral microtubules during chromosome attachment to the mitotic spindle: direct visualization in live newt lung cells. *J Cell Biol* 111: 1039–1045
5 Gliksman NR, Skibbens RV, Salmon ED (1993) How the transition frequencies of microtubule dynamic instability (nucleation, catastrophe, and rescue) regulate microtubule dynamics in interphase and mitosis: analysis using a Monte Carlo computer simulation. *Mol Biol Cell* 4: 1035–1050
6 Wordeman L, Mitchison TJ (1995) Identification and partial characterization of mitotic centromere-associated kinesin, a kinesin-related protein that associates with centromeres during

mitosis. *J Cell Biol* 128: 95–104

7 Zhai Y, Kronebusch PJ, Simon PM, Borisy GG (1996) Microtubule dynamics at the G2/M transition: abrupt breakdown of cytoplasmic microtubules at nuclear envelope breakdown and implications for spindle morphogenesis. *J Cell Biol* 135: 201–214

8 De Vita V, Hellman S, Rosenberg S (eds): (2005) *Cancer, principles and practice of oncology review*. Lippincott Williams and Wilkins, Philadelphia, PA

9 Langlois N, Gueritte F, Langlois Y, Potier P (1976) Application of a modification of the Polonovski reaction to the synthesis of vinblastine-type alkaloids. *J Am Chem Soc* 98: 7017–7024

10 Fahy J (2001) Modifications in the 'upper' velbenamine part of the Vinca alkaloids have major implications for tubulin interacting activities. *Curr Pharm Des* 7: 1181–1197

11 Fahy J DA, Ribet JP, Jacquesy JC, Berrier C, Jouannetaud MP, Zunino F (1997) Vinca alkaloids in superacidic media: a method for creating a new family of antitumor derivatives. *J Am Chem Soc* 119: 8576–8577

12 Jordan MA, Thrower D, Wilson L (1991) Mechanism of inhibition of cell proliferation by Vinca alkaloids. *Cancer Res* 51: 2212–2222

13 Panda D, Jordan MA, Chu KC, Wilson L (1996) Differential effects of vinblastine on polymerization and dynamics at opposite microtubule ends. *J Biol Chem* 271: 29807–29812

14 Jordan MA, Himes RH, Wilson L (1985) Comparison of the effects of vinblastine, vincristine, vindesine, and vinepidine on microtubule dynamics and cell proliferation *in vitro*. *Cancer Res* 45: 2741–2747

15 Dhamodharan R, Jordan MA, Thrower D, Wilson L, Wadsworth P (1995) Vinblastine suppresses dynamics of individual microtubules in living interphase cells. *Mol Biol Cell* 6: 1215–1229

16 Jordan MA, Thrower D, Wilson L (1992) Effects of vinblastine, podophyllotoxin and nocodazole on mitotic spindles. Implications for the role of microtubule dynamics in mitosis. *J Cell Sci* 102 (Pt 3): 401–416

17 Toso C, Lindley C (1995) Vinorelbine: a novel vinca alkaloid. *Am J Health Syst Pharm* 52: 1287–1304; quizz 1340–1341

18 Jordan MA, Margolis RL, Himes RH, Wilson L (1986) Identification of a distinct class of vinblastine binding sites on microtubules. *J Mol Biol* 187: 61–73

19 Singer WD, Jordan MA, Wilson L, Himes RH (1989) Binding of vinblastine to stabilized microtubules. *Mol Pharmacol* 36: 366–370

20 Wilson L, Jordan MA, Morse A, Margolis RL (1982) Interaction of vinblastine with steady-state microtubules *in vitro*. *J Mol Biol* 159: 125–149

21 Toso RJ, Jordan MA, Farrell KW, Matsumoto B, Wilson L (1993) Kinetic stabilization of microtubule dynamic instability *in vitro* by vinblastine. *Biochemistry* 32: 1285–1293

22 Etievant C, Barret JM, Kruczynski A, Perrin D, Hill BT (1998) Vinflunine (20',20'-difluoro-3',4'-dihydrovinorelbine), a novel Vinca alkaloid, which participates in P-glycoprotein (Pgp)-mediated multidrug resistance *in vivo* and *in vitro*. *Invest New Drugs* 16: 3–17

23 Gottesman MM, Pastan I (1993) Biochemistry of multidrug resistance mediated by the multidrug transporter. *Annu Rev Biochem* 62: 385–427

24 Dumontet C (2000) Mechanisms of action and resistance to tubulin-binding agents. *Expert Opin Investig Drugs* 9: 779–788

25 Endicott JA, Ling V (1989) The biochemistry of P-glycoprotein-mediated multidrug resistance. *Annu Rev Biochem* 58: 137–171

26 Chen CJ, Chin JE, Ueda K, Clark DP, Pastan I, Gottesman MM, Roninson IB (1986) Internal duplication and homology with bacterial transport proteins in the mdr1 (P-glycoprotein) gene from multidrug-resistant human cells. *Cell* 47: 381–389

27 Ueda K, Cardarelli C, Gottesman MM, Pastan I (1987) Expression of a full-length cDNA for the human 'MDR1' gene confers resistance to colchicine, doxorubicin, and vinblastine. *Proc Natl Acad Sci USA* 84: 3004–3008

28 Lockhart AC, Tirona RG, Kim RB (2003) Pharmacogenetics of ATP-binding cassette transporters in cancer and chemotherapy. *Mol Cancer Ther* 2: 685–698

29 Correia JJ, Lobert S (2001) Physiochemical aspects of tubulin-interacting antimitotic drugs. *Curr Pharm Des* 7: 1213–1228

30 Luduena RF (1998) Multiple forms of tubulin: different gene products and covalent modifications. *Int Rev Cytol* 178: 207–275

31 Khan IA, Luduena RF (2003) Different effects of vinblastine on the polymerization of isotypically purified tubulins from bovine brain. *Invest New Drugs* 21: 3–13

32 Drukman S, Kavallaris M (2002) Microtubule alterations and resistance to tubulin-binding agents (review). *Int J Oncol* 21: 621–628

33 Houghton JA, Houghton PJ, Hazelton BJ, Douglass EC (1985) *In situ* selection of a human rhabdomyosarcoma resistant to vincristine with altered beta-tubulins. *Cancer Res* 45: 2706–2712

34 Cabral F, Barlow SB (1991) Resistance to antimitotic agents as genetic probes of microtubule structure and function. *Pharmacol Ther* 52: 159–171

35 Reichle A, Diddens H, Altmayr F, Rastetter J, Andreesen R (1995) Beta-tubulin and P-glycoprotein: major determinants of vincristine accumulation in B-CLL cells. *Leuk Res* 19: 823–829

36 Hari M, Wang Y, Veeraraghavan S, Cabral F (2003) Mutations in alpha- and beta-tubulin that stabilize microtubules and confer resistance to colcemid and vinblastine. *Mol Cancer Ther* 2: 597–605

37 Gidding CE, Kellie SJ, Kamps WA, de Graaf SS (1999) Vincristine revisited. *Crit Rev Oncol Hematol* 29: 267–287

38 Johnson IS, Armstrong JG, Gorman M, Burnett JJ (1963) The vinca alkaloids: a new class of oncolytic agents. *Cancer Res* 23: 1390–1427

39 Rowinsky EK, Donehower RC (1991) The clinical pharmacology and use of antimicrotubule agents in cancer chemotherapeutics. *Pharmacol Ther* 52: 35–84

40 Joel S (1996) The comparative clinical pharmacology of vincristine and vindesine: does vindesine offer any advantage in clinical use? *Cancer Treat Rev* 21: 513–525

41 Budman DR (1997) Vinorelbine (Navelbine): a third-generation vinca alkaloid. *Cancer Invest* 15: 475–490

42 Workman P, Graham MA (eds) (1993) *Pharmacokinetics and cancer chemotherapy*. Cold Spring Harbor Laboratory Press, Plainview, NY

43 Nelson RL, Dyke RW, Root MA (1980) Comparative pharmacokinetics of vindesine, vincristine and vinblastine in patients with cancer. *Cancer Treat Rev* 7 Suppl 1: 17–24

44 Rahmani R, Bruno R, Iliadis A, Favre R, Just S, Barbet J, Cano JP (1987) Clinical pharmacokinetics of the antitumor drug navelbine (5'-noranhydrovinblastine). *Cancer Res* 47: 5796–5799

45 Jehl F, Quoix E, Leveque D, Pauli G, Breillout F, Krikorian A, Monteil H (1991) Pharmacokinetic and preliminary metabolic fate of navelbine in humans as determined by high performance liquid chromatography. *Cancer Res* 51: 2073–2076

46 Lokich JJ (ed.) (1990) *Cancer chemotherapy by infusion: 2nd edition*. Precept Press, Chicago

47 Bender RA, Castle MC, Margileth DA, Oliverio VT (1977) The pharmacokinetics of [^{3}H]-vincristine in man. *Clin Pharmacol Ther* 22: 430–435

48 Sethi VS, Jackson DV Jr, White DR, Richards F, Stuart JJ, Muss HB, Cooper MR, Spurr CL (1981) Pharmacokinetics of vincristine sulfate in adult cancer patients. *Cancer Res* 41: 3551–3555

49 Zhou XJ, Martin M, Placidi M, Cano JP, Rahmani R (1990) *In vivo* and *in vitro* pharmacokinetics and metabolism of vinca alkaloids in rat. II. Vinblastine and vincristine. *Eur J Drug Metab Pharmacokinet* 15: 323–332

50 Owellen RJ, Root MA, Hains FO (1977) Pharmacokinetics of vindesine and vincristine in humans. *Cancer Res* 37: 2603–2607

51 Rahmani R, Zhou XJ (1993) Pharmacokinetics and metabolism of vinca alkaloids. *Cancer Surv* 17: 269–281

52 Jackson DV Jr, Castle MC, Bender RA (1978) Biliary excretion of vincristine. *Clin Pharmacol Ther* 24: 101–107

53 Zhou-Pan XR, Seree E, Zhou XJ, Placidi M, Maurel P, Barra Y, Rahmani R (1993) Involvement of human liver cytochrome P450 3A in vinblastine metabolism: drug interactions. *Cancer Res* 53: 5121–5126

54 Jordan MA, Wilson L (1990) Kinetic analysis of tubulin exchange at microtubule ends at low vinblastine concentrations. *Biochemistry* 29: 2730–2739

55 Zhou XJ, Zhou-Pan XR, Gauthier T, Placidi M, Maurel P, Rahmani R (1993) Human liver microsomal cytochrome P450 3A isozymes mediated vindesine biotransformation. Metabolic drug interactions. *Biochem Pharmacol* 45: 853–861

56 Johnson SA, Harper P, Hortobagyi GN, Pouillart P (1996) Vinorelbine: an overview. *Cancer Treat Rev* 22: 127–142

57 Leveque D, Merle-Melet M, Bresler L, Didelot JP, Aymard JP, Wihlm J, Jehl F (1993) Biliary elimination and pharmacokinetics of vinorelbine in micropigs. *Cancer Chemother Pharmacol* 32: 487–490

58 Krikorian A, Rahmani R, Bromet M, Bore P, Cano JP (1989) Pharmacokinetics and metabolism of Navelbine. *Semin Oncol* 16: 21–25

59 Himes RH (1991) Interactions of the catharanthus (Vinca) alkaloids with tubulin and microtubules. *Pharmacol Ther* 51: 257–267

60 Chan JD (1998) Pharmacokinetic drug interactions of vinca alkaloids: summary of case reports. *Pharmacotherapy* 18: 1304–1307

61 Yalowich JC (1987) Effects of microtubule inhibitors on etoposide accumulation and DNA damage in human K562 cells *in vitro*. *Cancer Res* 47: 1010–1015

62 Jarosinski PF, Moscow JA, Alexander MS, Lesko LJ, Balis FM, Poplack DG (1988) Altered phenytoin clearance during intensive chemotherapy for acute lymphoblastic leukemia. *J Pediatr* 112: 996–999

63 Tobe SW, Siu LL, Jamal SA, Skorecki KL, Murphy GF, Warner E (1995) Vinblastine and erythromycin: an unrecognized serious drug interaction. *Cancer Chemother Pharmacol* 35: 188–190

64 Crom WR, de Graaf SS, Synold T, Uges DR, Bloemhof H, Rivera G, Christensen ML, Mahmoud H, Evans WE (1994) Pharmacokinetics of vincristine in children and adolescents with acute lymphocytic leukemia. *J Pediatr* 125: 642–649

65 Rajaonarison JF, Lacarelle B, Catalin J, Durand A, Cano JP (1993) Effect of anticancer drugs on the glucuronidation of 3'-azido-3'-deoxythymidine in human liver microsomes. *Drug Metab Dispos* 21: 823–829

66 Sathiapalan RK, El-Solh H (2001) Enhanced vincristine neurotoxicity from drug interactions: case report and review of literature. *Pediatr Hematol Oncol* 18: 543–546

67 Weiss HD, Walker MD, Wiernik PH (1974) Neurotoxicity of commonly used antineoplastic agents (second of two parts). *N Engl J Med* 291: 127–133

68 Song JI, Dumais MR (1991) From yew to us: the curious development of taxol. *JAMA* 266: 1281

69 Huizing MT, Misser VH, Pieters RC, ten Bokkel HuW, Veenhof CH, Vermorken JB, Pinedo HM, Beijnen JH (1995) Taxanes: a new class of antitumor agents. *Cancer Invest* 13: 381–404

70 Wani MC, Taylor HL, Wall ME, Coggon P, McPhail AT (1971) Plant antitumor agents. VI. The isolation and structure of taxol, a novel antileukemic and antitumor agent from Taxus brevifolia. *J Am Chem Soc* 93: 2325–2327

71 Schiff PB, Fant J, Horwitz SB (1979) Promotion of microtubule assembly *in vitro* by taxol. *Nature* 277: 665–667

72 Binet S, Chaineau E, Fellous A, Lataste H, Krikorian A, Couzinier JP, Meininger V (1990) Immunofluorescence study of the action of navelbine, vincristine and vinblastine on mitotic and axonal microtubules. *Int J Cancer* 46: 262–266

73 Kruczynski A, Barret JM, Etievant C, Colpaert F, Fahy J, Hill BT (1998) Antimitotic and tubulin-interacting properties of vinflunine, a novel fluorinated Vinca alkaloid. *Biochem Pharmacol* 55: 635–648

74 Cortes JE, Pazdur R (1995) Docetaxel. *J Clin Oncol* 13: 2643–2655

75 Downing KH, Nogales E (1999) Crystallographic structure of tubulin: implications for dynamics and drug binding. *Cell Struct Funct* 24: 269–275

76 Diaz JF, Andreu JM (1993) Assembly of purified GDP-tubulin into microtubules induced by taxol and taxotere: reversibility, ligand stoichiometry, and competition. *Biochemistry* 32: 2747–2755

77 Schiff PB, Fant J, Horwitz SB (1979) Promotion of microtubule assembly *in vitro* by taxol. *Nature* 277: 665–667

78 Parness J, Horwitz SB (1981) Taxol binds to polymerized tubulin *in vitro*. *J Cell Biol* 91: 479–487

79 Gueritte-Voegelein F, Guenard D, Lavelle F, Le Goff MT, Mangatal L, Potier P (1991) Relationships between the structure of taxol analogues and their antimitotic activity. *J Med Chem* 34: 992–998

80 Ringel I, Horwitz SB (1991) Studies with RP 56976 (taxotere): a semisynthetic analogue of taxol. *J Natl Cancer Inst* 83: 288–291

81 Horwitz SB (1992) Mechanism of action of taxol. *Trends Pharmacol Sci* 13: 134–136

82 Moos PJ, Fitzpatrick FA (1998) Taxane-mediated gene induction is independent of microtubule stabilization: induction of transcription regulators and enzymes that modulate inflammation and apoptosis. *Proc Natl Acad Sci USA* 95: 3896–3901

83 Rodi DJ, Janes RW, Sanganee HJ, Holton RA, Wallace BA, Makowski L (1999) Screening of a library of phage-displayed peptides identifies human bcl-2 as a taxol-binding protein. *J Mol Biol* 285: 197–203

84 Tishler RB, Schiff PB, Geard CR, Hall EJ (1992) Taxol: a novel radiation sensitizer. *Int J Radiat Oncol Biol Phys* 22: 613–617

85 Liebmann J, Cook JA, Fisher J, Teague D, Mitchell JB (1994) Changes in radiation survival curve parameters in human tumor and rodent cells exposed to paclitaxel (Taxol). *Int J Radiat Oncol Biol Phys* 29: 559–564

86 Hei TK, Hall EJ (1993) Taxol, radiation, and oncogenic transformation. *Cancer Res* 53: 1368–1372

87 Liebmann J, Cook JA, Fisher J, Teague D, Mitchell JB (1994) *In vitro* studies of Taxol as a radiation sensitizer in human tumor cells. *J Natl Cancer Inst* 86: 441–446

88 Hennequin C, Giocanti N, Favaudon V (1996) Interaction of ionizing radiation with paclitaxel (Taxol) and docetaxel (Taxotere) in HeLa and SQ20B cells. *Cancer Res* 56: 1842–1850

89 Milas L, Milas MM, Mason KA (1999) Combination of taxanes with radiation: preclinical studies. *Semin Radiat Oncol* 9: 12–26

90 Mason KA, Hunter NR, Milas M, Abbruzzese JL, Milas L (1997) Docetaxel enhances tumor radioresponse *in vivo*. *Clin Cancer Res* 3: 2431–2438

91 Dumontet C, Sikic BI (1999) Mechanisms of action of and resistance to antitubulin agents: microtubule dynamics, drug transport, and cell death. *J Clin Oncol* 17: 1061–1070

92 Gottesman MM, Fojo T, Bates SE (2002) Multidrug resistance in cancer: role of ATP-dependent transporters. *Nat Rev Cancer* 2: 48–58

93 Huizing MT, Keung AC, Rosing H, van der KuV, ten Bokkel HuW, Mandjes IM, Dubbelman AC, Pinedo HM, Beijnen JH (1993) Pharmacokinetics of paclitaxel and metabolites in a randomized comparative study in platinum-pretreated ovarian cancer patients. *J Clin Oncol* 11: 2127–2135

94 Gianni L, Kearns CM, Giani A, Capri G, Vigano L, Lacatelli A, Bonadonna G, Egorin MJ (1995) Nonlinear pharmacokinetics and metabolism of paclitaxel and its pharmacokinetic/pharmacodynamic relationships in humans. *J Clin Oncol* 13: 180–190

95 Huizing MT, Rosing H, Koopman F, Keung AC, Pinedo HM, Beijnen JH (1995) High-performance liquid chromatographic procedures for the quantitative determination of paclitaxel (Taxol) in human urine. *J Chromatogr B Biomed Appl* 664: 373–382

96 Harris JW, Katki A, Anderson LW, Chmurny GN, Paukstelis JV, Collins JM (1994) Isolation, structural determination, and biological activity of 6 alpha-hydroxytaxol, the principal human metabolite of taxol. *J Med Chem* 37: 706–709

97 Monsarrat B, Mariel E, Cros S, Gares M, Guenard D, Gueritte-Voegelein F, Wright M (1990) Taxol metabolism. Isolation and identification of three major metabolites of taxol in rat bile. *Drug Metab Dispos* 18: 895–901

98 Monsarrat B, Alvinerie P, Wright M, Dubois J, Gueritte-Voegelein F, Guenard D, Donehower RC, Rowinsky EK (1993) Hepatic metabolism and biliary excretion of Taxol in rats and humans. *J Natl Cancer Inst Monogr* 39–46

99 Sparreboom A, Huizing MT, Boesen JJ, Nooijen WJ, van Tellingen O, Beijnen JH (1995) Isolation, purification, and biological activity of mono- and dihydroxylated paclitaxel metabolites from human feces. *Cancer Chemother Pharmacol* 36: 299–304

100 Sparreboom A, Verweij J, van der BuM, Loos WJ, Brouwer E, Vigano L, Locatelli A, de Vos AI, Nooter K, Stoter G et al. (1998) Disposition of Cremophor EL in humans limits the potential for modulation of the multidrug resistance phenotype *in vivo*. *Clin Cancer Res* 4: 1937–1942

101 Wilson WH, Berg SL, Bryant G, Wittes RE, Bates S, Fojo A, Steinberg SM, Goldspiel BR, Herdt J, O'Shaughnessy J et al. (1994) Paclitaxel in doxorubicin-refractory or mitoxantrone-refractory breast cancer: a phase I/II trial of 96-hour infusion. *J Clin Oncol* 12: 1621–1629

102 Huizing MT, Vermorken JB, Rosing H, ten Bokkel HuW, Mandjes I, Pinedo HM, Beijnen JH (1995) Pharmacokinetics of paclitaxel and three major metabolites in patients with advanced breast carcinoma refractory to anthracycline therapy treated with a 3-hour paclitaxel infusion: a European Cancer Centre (ECC) trial. *Ann Oncol* 6: 699–704

103 Smorenburg CH, ten Tije AJ, Verweij J, Bontenbal M, Mross K, van Zomeren DM, Seynaeve C, Sparreboom A (2003) Altered clearance of unbound paclitaxel in elderly patients with metastatic breast cancer. *Eur J Cancer* 39: 196–202

104 Gelderblom H, Verweij J, van Zomeren DM, Buijs D, Ouwens L, Nooter K, Stoter G, Sparreboom A (2002) Influence of Cremophor El on the bioavailability of intraperitoneal paclitaxel. *Clin Cancer Res* 8: 1237–1241

105 Markman M, Brady MF, Spirtos NM, Hanjani P, Rubin SC (1998) Phase II trial of intraperitoneal paclitaxel in carcinoma of the ovary, tube, and peritoneum: a Gynecologic Oncology Group

Study. *J Clin Oncol* 16: 2620–2624

106 Extra JM, Rousseau F, Bruno R, Clavel M, Le Bail N, Marty M (1993) Phase I and pharmacokinetic study of Taxotere (RP 56976; NSC 628503) given as a short intravenous infusion. *Cancer Res* 53: 1037–1042

107 Bruno R, Riva A, Hille D, Lebecq A, Thomas L (1997) Pharmacokinetic and pharmacodynamic properties of docetaxel: results of phase I and phase II trials. *Am J Health Syst Pharm* 54: S16–S19

108 Bruno R, Hille D, Riva A, Vivier N, ten Bokkel HuW, van Oosterom AT, Kaye SB, Verweij J, Fossella FV, Valero V et al. (1998) Population pharmacokinetics/pharmacodynamics of docetaxel in phase II studies in patients with cancer. *J Clin Oncol* 16: 187–196

109 Bissett D, Setanoians A, Cassidy J, Graham MA, Chadwick GA, Wilson P, Auzannet V, Le Bail N, Kaye SB, Kerr DJ et al. (1993) Phase I and pharmacokinetic study of taxotere (RP 56976) administered as a 24-hour infusion. *Cancer Res* 53: 523–527

110 ten Tije AJ, Verweij J, Carducci MA, Graveland W, Rogers T, Pronk T, Verbruggen MP, Dawkins F, Baker SD (2005) Prospective evaluation of the pharmacokinetics and toxicity profile of docetaxel in the elderly. *J Clin Oncol* 23: 1070–1077

111 Rosing H, Lustig V, van Warmerdam LJ, Huizing MT, ten Bokkel HuW, Schellens JH, Rodenhuis S, Bult A, Beijnen JH (2000) Pharmacokinetics and metabolism of docetaxel administered as a 1-h intravenous infusion. *Cancer Chemother Pharmacol* 45: 213–218

112 Bissery MC, Nohynek G, Sanderink GJ, Lavelle F (1995) Docetaxel (Taxotere): a review of preclinical and clinical experience. Part I: Preclinical experience. *Anticancer Drugs* 6: 339–355, 363–368

113 Sparreboom A, van Tellingen O, Nooijen WJ, Beijnen JH (1998) Preclinical pharmacokinetics of paclitaxel and docetaxel. *Anticancer Drugs* 9: 1–17

114 Vaclavikova R, Soucek P, Svobodova L, Anzenbacher P, Simek P, Guengerich FP, Gut I (2004) Different *in vitro* metabolism of paclitaxel and docetaxel in humans, rats, pigs, and minipigs. *Drug Metab Dispos* 32: 666–674

115 Monegier B, Gaillard C, Sable S, Vuilhorgne M (1994) Structures of the major human metabolites of docetaxel (RP 56976-Taxotere®). *Tetrahedron Lett* 35: 3715–3718

116 Sparreboom A, Van Tellingen O, Scherrenburg EJ, Boesen JJ, Huizing MT, Nooijen WJ, Versluis C, Beijnen JH (1996) Isolation, purification and biological activity of major docetaxel metabolites from human feces. *Drug Metab Dispos* 24: 655–658

117 Jamis-Dow CA, Klecker RW, Katki AG, Collins JM (1995) Metabolism of taxol by human and rat liver *in vitro*: a screen for drug interactions and interspecies differences. *Cancer Chemother Pharmacol* 36: 107–114

118 Klecker RW, Jamis-Dow CA, Egorin MJ, Erkmen K, Parker RJ, Stevens R, Collins JM (1994) Effect of cimetidine, probenecid, and ketoconazole on the distribution, biliary secretion, and metabolism of [³H]taxol in the Sprague-Dawley rat. *Drug Metab Dispos* 22: 254–258

119 Jamis-Dow CA, Klecker RW, Sarosy G, Reed E, Collins JM (1993) Steady-state plasma concentrations and effects of taxol for a 250 mg/m² dose in combination with granulocyte-colony stimulating factor in patients with ovarian cancer. *Cancer Chemother Pharmacol* 33: 48–52

120 Slichenmyer WJ, Donehower RC, Chen TL, Bowling MK, McGuire WP, Rowinsky EK (1995) Pretreatment H2 receptor antagonists that differ in P450 modulation activity: comparative effects on paclitaxel clearance rates and neutropenia. *Cancer Chemother Pharmacol* 36: 227–232

121 Jamis-Dow CA, Pearl ML, Watkins PB, Blake DS, Klecker RW, Collins JM (1997) Predicting drug interactions *in vivo* from experiments *in vitro*. Human studies with paclitaxel and ketoconazole. *Am J Clin Oncol* 20: 592–599

122 Parker RJ, Dabholkar MD, Lee KB, Bostick-Bruton F, Reed E (1993) Taxol effect on cisplatin sensitivity and cisplatin cellular accumulation in human ovarian cancer cells. *J Natl Cancer Inst Monogr* 83–88

123 Rowinsky EK, Citardi MJ, Noe DA, Donehower RC (1993) Sequence-dependent cytotoxic effects due to combinations of cisplatin and the antimicrotubule agents taxol and vincristine. *J Cancer Res Clin Oncol* 119: 727–733

124 Liebmann JE, Fisher J, Teague D, Cook JA (1994) Sequence dependence of paclitaxel (Taxol) combined with cisplatin or alkylators in human cancer cells. *Oncol Res* 6: 25–31

125 Rowinsky EK, Gilbert MR, McGuire WP, Noe DA, Grochow LB, Forastiere AA, Ettinger DS, Lubejko BG, Clark B, Sartorius SE et al. (1991) Sequences of taxol and cisplatin: a phase I and pharmacologic study. *J Clin Oncol* 9: 1692–1703

126 Huizing MT, Giaccone G, van Warmerdam LJ, Rosing H, Bakker PJ, Vermorken JB, Postmus PE, van Zandwijk N, Koolen MG, ten Bokkel HuW et al. (1997) Pharmacokinetics of paclitaxel and carboplatin in a dose-escalating and dose-sequencing study in patients with non-small-cell lung cancer. The European Cancer Centre. *J Clin Oncol* 15: 317–329

127 Kearns CM, Egorin MJ (1997) Considerations regarding the less-than-expected thrombocytopenia encountered with combination paclitaxel/carboplatin chemotherapy. *Semin Oncol* 24: S291–S296

128 Holmes FA, Madden T, Newman RA, Valero V, Theriault RL, Fraschini G, Walters RS, Booser DJ, Buzdar AU, Willey J et al. (1996) Sequence-dependent alteration of doxorubicin pharmacokinetics by paclitaxel in a phase I study of paclitaxel and doxorubicin in patients with metastatic breast cancer. *J Clin Oncol* 14: 2713–2721

129 Gianni L, Munzone E, Capri G, Fulfaro F, Tarenzi E, Villani F, Spreafico C, Laffranchi A, Caraceni A, Martini C et al. (1995) Paclitaxel by 3-hour infusion in combination with bolus doxorubicin in women with untreated metastatic breast cancer: high antitumor efficacy and cardiac effects in a dose-finding and sequence-finding study. *J Clin Oncol* 13: 2688–2699

130 Gianni L, Vigano L, Locatelli A, Capri G, Giani A, Tarenzi E, Bonadonna G (1997) Human pharmacokinetic characterization and *in vitro* study of the interaction between doxorubicin and paclitaxel in patients with breast cancer. *J Clin Oncol* 15: 1906–1915

131 Vigano L, Locatelli A, Grasselli G, Gianni L (2001) Drug interactions of paclitaxel and docetaxel and their relevance for the design of combination therapy. *Invest New Drugs* 19: 179–196

132 Gennari A, Salvadori B, Donati S, Bengala C, Orlandini C, Danesi R, Del Tacca M, Bruzzi P, Conte PF (1999) Cardiotoxicity of epirubicin/paclitaxel-containing regimens: role of cardiac risk factors. *J Clin Oncol* 17: 3596–3602

133 Zamboni WC, Egorin MJ, Van Echo DA, Day RS, Meisenberg BR, Brooks SE, Doyle LA, Nemieboka NN, Dobson JM, Tait NS et al. (2000) Pharmacokinetic and pharmacodynamic study of the combination of docetaxel and topotecan in patients with solid tumors. *J Clin Oncol* 18: 3288–3294

134 Nannan Panday VR, Huizing MT, WW tB Vermorken JB, Beijnen JH (1997) Hypersensitivity reactions to the taxanes paclitaxel and docetaxel. *Clin Drug Investig* 14: 418–427

135 Schrijvers D, Wanders J, Dirix L, Prove A, Vonck I, van Oosterom A, Kaye S (1993) Coping with toxicities of docetaxel (Taxotere). *Ann Oncol* 4: 610–611

136 Peereboom DM, Donehower RC, Eisenhauer EA, McGuire WP, Onetto N, Hubbard JL, Piccart M, Gianni L, Rowinsky EK (1993) Successful re-treatment with taxol after major hypersensitivity reactions. *J Clin Oncol* 11: 885–890

137 Rothenberg ML, Liu PY, Braly PS, Wilczynski SP, Hannigan EV, Wadler S, Stuart G, Jiang C, Markman M, Alberts DS et al. (2003) Combined intraperitoneal and intravenous chemotherapy for women with optimally debulked ovarian cancer: results from an intergroup phase II trial. *J Clin Oncol* 21: 1313–1319

138 Rowinsky EK, Eisenhauer EA, Chaudhry V, Arbuck SG, Donehower RC (1993) Clinical toxicities encountered with paclitaxel (Taxol). *Semin Oncol* 20: 1–15

139 Garrison JA, McCune JS, Livingston RB, Linden HM, Gralow JR, Ellis GK, West HL (2003) Myalgias and arthralgias associated with paclitaxel. *Oncology (Williston Park)* 17: 271–277; discussion 281–282

140 Sonnichsen DS, Hurwitz CA, Pratt CB, Shuster JJ, Relling MV (1994) Saturable pharmacokinetics and paclitaxel pharmacodynamics in children with solid tumors. *J Clin Oncol* 12: 532–538

141 Rowinsky EK, Chaudhry V, Cornblath DR, Donehower RC (1993) Neurotoxicity of Taxol. *J Natl Cancer Inst Monogr* 107–115

142 Postma TJ, Vermorken JB, Liefting AJ, Pinedo HM, Heimans JJ (1995) Paclitaxel-induced neuropathy. *Ann Oncol* 6: 489–494

143 Chaudhry V, Rowinsky EK, Sartorius SE, Donehower RC, Cornblath DR (1994) Peripheral neuropathy from taxol and cisplatin combination chemotherapy: clinical and electrophysiological studies. *Ann Neurol* 35: 304–311

144 Wiernik PH, Schwartz EL, Einzig A, Strauman JJ, Lipton RB, Dutcher JP (1987) Phase I trial of taxol given as a 24-hour infusion every 21 days: responses observed in metastatic melanoma. *J Clin Oncol* 5: 1232–1239

145 Wiernik PH, Schwartz EL, Strauman JJ, Dutcher JP, Lipton RB, Paietta E (1987) Phase I clinical and pharmacokinetic study of taxol. *Cancer Res* 47: 2486–2493

146 Arbuck SG, Strauss H, Rowinsky E, Christian M, Suffness M, Adams J, Oakes M, McGuire W,

Reed E, Gibbs H et al. (1993) A reassessment of cardiac toxicity associated with Taxol. *J Natl Cancer Inst Monogr* 117–130

147 Saad SY, Najjar TA, Alashari M (2004) Cardiotoxicity of doxorubicin/paclitaxel combination in rats: effect of sequence and timing of administration. *J Biochem Mol Toxicol* 18: 78–86

148 Biganzoli L, Cufer T, Bruning P, Coleman RE, Duchateau L, Rapoport B, Nooij M, Delhaye F, Miles D, Sulkes A et al. (2003) Doxorubicin-paclitaxel: a safe regimen in terms of cardiac toxicity in metastatic breast carcinoma patients. Results from a European Organization for Research and Treatment of Cancer multicenter trial. *Cancer* 97: 40–45

149 Gori S, Colozza M, Mosconi AM, Franceschi E, Basurto C, Cherubini R, Sidoni A, Rulli A, Bisacci C, De Angelis V et al. (2004) Phase II study of weekly paclitaxel and trastuzumab in anthracycline- and taxane-pretreated patients with HER2-overexpressing metastatic breast cancer. *Br J Cancer* 90: 36–40

150 Mauer AM, Masters GA, Haraf DJ, Hoffman PC, Watson SM, Golomb HM, Vokes EE (1998) Phase I study of docetaxel with concomitant thoracic radiation therapy. *J Clin Oncol* 16: 159–164

151 Minami H, Ohe Y, Niho S, Goto K, Ohmatsu H, Kubota K, Kakinuma R, Nishiwaki Y, Nokihara H, Sekine I et al. (2004) Comparison of pharmacokinetics and pharmacodynamics of docetaxel and Cisplatin in elderly and non-elderly patients: why is toxicity increased in elderly patients? *J Clin Oncol* 22: 2901–2908

152 Sandstrom M, Lindman H, Nygren P, Lidbrink E, Bergh J, Karlsson MO (2005) Model describing the relationship between pharmacokinetics and hematologic toxicity of the epirubicin-docetaxel regimen in breast cancer patients. *J Clin Oncol* 23: 413–421

153 Hainsworth JD, Burris HA 3r, Greco FA (1999) Weekly administration of docetaxel (Taxotere): summary of clinical data. *Semin Oncol* 26: 19–24

154 Berghammer P, Pohnl R, Baur M, Dittrich C (2001) Docetaxel extravasation. *Support Care Cancer* 9: 131–134

155 Pazdur R, Newman RA, Newman BM, Fuentes A, Benvenuto J, Bready B, Moore D Jr, Jaiyesimi I, Vreeland F, Bayssas MM et al. (1992) Phase I trial of Taxotere: five-day schedule. *J Natl Cancer Inst* 84: 1781–1788

156 Vukelja SJ, Baker WJ, Burris HA 3r, Keeling JH, Von Hoff D (1993) Pyridoxine therapy for palmar-plantar erythrodysesthesia associated with taxotere. *J Natl Cancer Inst* 85: 1432–1433

157 Zimmerman GC, Keeling JH, Lowry M, Medina J, Von Hoff DD, Burris HA (1994) Prevention of docetaxel-induced erythrodysesthesia with local hypothermia. *J Natl Cancer Inst* 86: 557–558

158 Piccart MJ, Klijn J, Paridaens R, Nooij M, Mauriac L, Coleman R, Bontenbal M, Awada A, Selleslags J, Van Vreckem A et al. (1997) Corticosteroids significantly delay the onset of docetaxel-induced fluid retention: final results of a randomized study of the European Organization for Research and Treatment of Cancer Investigational Drug Branch for Breast Cancer. *J Clin Oncol* 15: 3149–3155

159 Morkas M, Fleming D, Hahl M (2002) Challenges in oncology. Case 2 Radiation recall associated with docetaxel. *J Clin Oncol* 20: 867–869

160 Gelmon K (1994) The taxoids: paclitaxel and docetaxel. *Lancet* 344: 1267–1272

161 Chevallier B, Fumoleau P, Kerbrat P, Dieras V, Roche H, Krakowski I, Azli N, Bayssas M, Lentz MA, Van Glabbeke M et al. (1995) Docetaxel is a major cytotoxic drug for the treatment of advanced breast cancer: a phase II trial of the Clinical Screening Cooperative Group of the European Organization for Research and Treatment of Cancer. *J Clin Oncol* 13: 314–322

162 Fossella FV, Lee JS, Berille J, Hong WK (1995) Summary of phase II data of docetaxel (Taxotere), an active agent in the first- and second-line treatment of advanced non-small cell lung cancer. *Semin Oncol* 22: 22–29

163 Pazdur R, Kudelka AP, Kavanagh JJ, Cohen PR, Raber MN (1993) The taxoids: paclitaxel (Taxol) and docetaxel (Taxotere). *Cancer Treat Rev* 19: 351–386

164 Tomiak E, Piccart MJ, Kerger J, Lips S, Awada A, de Valeriola D, Ravoet C, Lossignol D, Sculier JP, Auzannet V et al. (1994) Phase I study of docetaxel administered as a 1-hour intravenous infusion on a weekly basis. *J Clin Oncol* 12: 1458–1467

165 Semb KA, Aamdal S, Oian P (1998) Capillary protein leak syndrome appears to explain fluid retention in cancer patients who receive docetaxel treatment. *J Clin Oncol* 16: 3426–3432

166 Burris H, Irvin R, Kuhn J, Kalter S, Smith L, Shaffer D, Fields S, Weiss G, Eckardt J, Rodriguez G et al. (1993) Phase I clinical trial of taxotere administered as either a 2-hour or 6-hour intravenous infusion. *J Clin Oncol* 11: 950–958

167 New PZ, Jackson CE, Rinaldi D, Burris H, Barohn RJ (1996) Peripheral neuropathy secondary

to docetaxel (Taxotere). *Neurology* 46: 108–111

168 Giannakakou P, Gussio R, Nogales E, Downing KH, Zaharevitz D, Bollbuck B, Poy G, Sackett D, Nicolaou KC, Fojo T et al. (2000) A common pharmacophore for epothilone and taxanes: molecular basis for drug resistance conferred by tubulin mutations in human cancer cells. *Proc Natl Acad Sci USA* 97: 2904–2909

169 Sampath D, Discafani CM, Loganzo F, Beyer C, Liu H, Tan X, Musto S, Annable T, Gallagher P, Rios C et al. (2003) MAC-321, a novel taxane with greater efficacy than paclitaxel and docetaxel *in vitro* and *in vivo*. *Mol Cancer Ther* 2: 873–884

170 Jordan MA, Ojima I, Rosas F, Distefano M, Wilson L, Scambia G, Ferlini C (2002) Effects of novel taxanes SB-T-1213 and IDN5109 on tubulin polymerization and mitosis. *Chem Biol* 9: 93–101

171 Jordan MA (2002) Mechanism of action of antitumor drugs that interact with microtubules and tubulin. *Curr Med Chem Anti-Canc Agents* 2: 1–17

172 Kavallaris M, Verrills NM, Hill BT (2001) Anticancer therapy with novel tubulin-interacting drugs. *Drug Resist Updat* 4: 392–401

173 Kowalski RJ, Giannakakou P, Gunasekera SP, Longley RE, Day BW, Hamel E (1997) The microtubule-stabilizing agent discodermolide competitively inhibits the binding of paclitaxel (Taxol) to tubulin polymers, enhances tubulin nucleation reactions more potently than paclitaxel, and inhibits the growth of paclitaxel-resistant cells. *Mol Pharmacol* 52: 613–622

174 Long BH, Carboni JM, Wasserman AJ, Cornell LA, Casazza AM, Jensen PR, Lindel T, Fenical W, Fairchild CR (1998) Eleutherobin, a novel cytotoxic agent that induces tubulin polymerization, is similar to paclitaxel (Taxol). *Cancer Res* 58: 1111–1115

175 Hamel E, Sackett DL, Vourloumis D, Nicolaou KC (1999) The coral-derived natural products eleutherobin and sarcodictyins A and B: effects on the assembly of purified tubulin with and without microtubule-associated proteins and binding at the polymer taxoid site. *Biochemistry* 38: 5490–5498

176 Mooberry SL, Tien G, Hernandez AH, Plubrukarn A, Davidson BS (1999) Laulimalide and isolaulimalide, new paclitaxel-like microtubule-stabilizing agents. *Cancer Res* 59: 653–660

177 Altmann KH, Wartmann M, O'Reilly T (2000) Epothilones and related structures – a new class of microtubule inhibitors with potent *in vivo* antitumor activity. *Biochim Biophys Acta* 1470: M79–M91

178 Wartmann M, Altmann KH (2002) The biology and medicinal chemistry of epothilones. *Curr Med Chem Anti-Canc Agents* 2: 123–148

179 Kowalski RJ, Giannakakou P, Hamel E (1997) Activities of the microtubule-stabilizing agents epothilones A and B with purified tubulin and in cells resistant to paclitaxel (Taxol(R)). *J Biol Chem* 272: 2534–2541

180 Kamath K, Jordan MA (2003) Suppression of microtubule dynamics by epothilone B is associated with mitotic arrest. *Cancer Res* 63: 6026–6031

181 Rothermel J, Wartmann M, Chen T, Hohneker J (2003) EPO906 (epothilone B): a promising novel microtubule stabilizer. *Semin Oncol* 30: 51–55

182 Lee FY, Borzilleri R, Fairchild CR, Kim SH, Long BH, Reventos-Suarez C, Vite GD, Rose WC, Kramer RA (2001) BMS-247550: a novel epothilone analog with a mode of action similar to paclitaxel but possessing superior antitumor efficacy. *Clin Cancer Res* 7: 1429–1437

183 Low JA, Wedam SB, Lee JJ, Berman AW, Brufsky A, Yang SX, Poruchynsky MS, Steinberg SM, Mannan N, Fojo T et al. (2005) Phase II clinical trial of ixabepilone (BMS-247550), an epothilone B analog, in metastatic and locally advanced breast cancer. *J Clin Oncol* 23: 2726–2734

184 Chou TC, Zhang XG, Harris CR, Kuduk SD, Balog A, Savin KA, Bertino JR, Danishefsky SJ (1998) Desoxyepothilone B is curative against human tumor xenografts that are refractory to paclitaxel. *Proc Natl Acad Sci USA* 95: 15798–15802

185 ter Haar E, Kowalski RJ, Hamel E, Lin CM, Longley RE, Gunasekera SP, Rosenkranz HS, Day BW (1996) Discodermolide, a cytotoxic marine agent that stabilizes microtubules more potently than taxol. *Biochemistry* 35: 243–250

186 Martello LA, McDaid HM, Regl DL, Yang CP, Meng D, Pettus TR, Kaufman MD, Arimoto H, Danishefsky SJ, Smith AB 3r et al. (2000) Taxol and discodermolide represent a synergistic drug combination in human carcinoma cell lines. *Clin Cancer Res* 6: 1978–1987

187 Pryor DE, O'Brate A, Bilcer G, Diaz JF, Wang Y, Wang Y, Kabaki M, Jung MK, Andreu JM, Ghosh AK et al. (2002) The microtubule stabilizing agent laulimalide does not bind in the taxoid

site, kills cells resistant to paclitaxel and epothilones, and may not require its epoxide moiety for activity. *Biochemistry* 41: 9109–9115

188 Loganzo F, Discafani CM, Annable T, Beyer C, Musto S, Hari M, Tan X, Hardy C, Hernandez R, Baxter M et al. (2003) HTI-286, a synthetic analogue of the tripeptide hemiasterlin, is a potent antimicrotubule agent that circumvents P-glycoprotein-mediated resistance *in vitro* and *in vivo*. *Cancer Res* 63: 1838–1845

189 Panda D, DeLuca K, Williams D, Jordan MA, Wilson L (1998) Antiproliferative mechanism of action of cryptophycin-52: kinetic stabilization of microtubule dynamics by high-affinity binding to microtubule ends. *Proc Natl Acad Sci USA* 95: 9313–9318

190 Towle MJ, Salvato KA, Budrow J, Wels BF, Kuznetsov G, Aalfs KK, Welsh S, Zheng W, Seletsk BM, Palme MH et al. (2001) *In vitro* and *in vivo* anticancer activities of synthetic macrocyclic ketone analogues of halichondrin B. *Cancer Res* 61: 1013–1021

Drugs Affecting Growth of Tumours
Edited by Herbert M. Pinedo and Carolien H. Smorenburg
© 2006 Birkhäuser Verlag/Switzerland

Vaccination therapies in solid tumors

Alfonsus J.M. van den Eertwegh

Division of Immunotherapy, Department of Medical Oncology, Vrije Universiteit Medical Center, De Boelelaan 1117, 1081 HV Amsterdam, The Netherlands

Introduction

Over the last two decades there has been a great deal of interest in specific immunotherapies. Particularly in the field of passive immunotherapy, using tumor-specific antibodies, some interesting successes have been reported. The humanized monoclonal antibody, Herceptin, directed to the Her-2-neu antigen is now an established standard modality in the treatment of breast cancer patients, whose tumor is overexpressing the Her-2-neu antigen [1]. Cetuximab, a monoclonal antibody specific for another epidermal growth factor receptor, is about to be registered for the treatment of metastatic colon cancer [2]. The treatment with anti-CD20 monoclonal antibodies improves the prognosis of lymphoma patients and is now considered as a standard immunotherapy for B cell lymphomas [3]. All together it took more than 30 years before monoclonal antibodies have evolved to a standard treatment in cancer. It is important to realize that it was no more than 10 years ago that the perspectives of this type of passive immunotherapy were not so promising. The humanization of monoclonal antibodies was a real breakthrough and opened the way for this type of treatment.

Active specific immunotherapies are now facing similar problems as the tumor-specific antibodies a decade ago. These vaccination treatments are logistically demanding, expensive and only small studies have shown its value in the treatment of cancer. However, so far no real survival advantage has been reported and consequently vaccination is not yet accepted as a standard treatment for cancer patients. As several studies have demonstrated, the requirement is now to fine tune this treatment and unequivocally demonstrate its efficacy in the treatment of cancer. Hopefully we can then also add vaccination therapies to the armament of the oncologist.

Vaccination therapies

Vaccination differs from nonspecific immune-based therapies in that the goal is not general but rather specific activation of the immune system to eliminate tumor cells without affecting surrounding normal tissue [4]. It is generally assumed that specific vaccination should result in activation of the two main arms of the immune system, namely the humoral (antibody producing B cells) and the cellular immune response (T cells) [5, 6]. B cells recognize the tumor antigens in their native protein state at the cell surface, whereas T lymphocytes recognize proteins as peptide fragments, presented in the context of major histocompatibility complex (MHC) antigens on the surface of the tumor cells. There are two types of T cells, CD4 and CD8, which recognize antigens through a specific T cell receptor. These antigens are presented by a group of specialized cells called antigen-presenting cells (APC). A variety of cells are capable in processing and presenting antigens including B cells, monocytes, macrophages, and dendritic cells (DCs). DCs are the most efficient APC, expressing co-stimulatory molecules and high levels of MHC Class I and Class II molecules required for the activation of CD8 and CD4 positive T cells, respectively. CD4 positive T cells, also called helper T cells, secrete cytokines that regulate B cells, cytotoxic cells and other immune cells, but can also have a cytotoxic activity. CD8 positive cytotoxic T cells (CTL) are at this moment considered to be the most potent cells to eradicate specifically tumor cells. The purpose of most vaccination strategies is to activate this specific subset of T cells. DCs are essential for the specific activation of T cells and these cells are found in the lymphoid organs, blood and skin. There are several ways DCs can be used to induce a specific immune response. Antigens can be injected in the skin where they are taken up by dermal DCs and these professional APC migrate to the lymph node to meet specific T cells. Another possibility is to collect DC precursors, culture these cells into DCs and load them with tumor antigens. These professional APC can be injected in the skin, lymph nodes or intravenously, and it is expected that these cells migrate to the lymphoid organs to encounter and activate tumor specific T cells. Specific elements of the vaccine and vaccination are very critical for generating a successful anti-tumor immune response [7]. A specific tumor antigen (or antigens) must be present in the vaccine. Once a tumor antigen is identified, a platform is required that can induce the immune response. Current platforms include tumor cell-based vaccines, peptides/proteins, DCs, and recombinant viral vectors. Different types of platforms may be decisive in the type of immune response that will be induced. Finally, recent studies suggest that the application of antibodies against CTLA-4 or T regulatory cells are very potent strategies to break tolerance or to overcome immune escape mechanisms of tumor cells, thereby enhancing the efficacy of cancer vaccines. All these aspects of vaccination will be addressed in this review.

Peptide and protein based vaccines

On the basis of their tissue distribution, T cell specific tumor-associated antigens (TAA) are classified in 5 groups [8, 9]; 1) differentiation antigens, which are expressed in a lineage-related manner and are also detected in normal tissue (e.g., MAGE, BAGE, GAGE, NY-ESO-1, SSX); 2) tumor-restricted antigens, which are expressed only on cancer cells (e.g., Melan A/MART-1, tyrosinase, gp100, CEA, NY-BR-1, rab 38); 3) unique tumor restricted antigens, including point mutations of normal tumor antigens (e.g., β-catenin, MUM-1, CDK-4, p53, ras); 4) overexpressed antigens of normal tissue (e.g., HER-2/neu, p53, MUC-1); and 5) viral antigens (e.g., human papillomavirus, hepatitis B virus, Epstein-Barr virus).

The majority of known TAA peptides are presented in association with Class I MHC molecules and are recognized by tumor-specific $CD8^+$ T cells, whereas a small number of TAA is recognized by $CD4^+$ T cells in the context of MHC Class II. Most of the known TAA peptides are expressed by melanoma, while a few TAA epitopes have been characterized in other tumors [8, 9].

Melanoma peptides were the first to be tested in patients with metastatic melanoma. In general, clinical responses were observed in 0–30% of the treated patients. Cormier et al. [10] vaccinated melanoma patients with Melan-A/MART-1 with incomplete Freund's adjuvant and 15 out of 16 patients developed a specific CTL response in their blood, but no clinical responses were observed. In contrast, Rosenburg vaccinated patients with modified gp100 peptide and a high dose of IL-2 and demonstrated in 42% of the patients a clinical response [11]. Whether these responses could be attributed to the vaccination or the systemic treatment with IL-2 is hard to determine. Jager et al. treated three patients with metastatic melanoma with a vaccine consisting of a mixture of Melan-A/MART-1/gp100/tyrosinase peptides and the adjuvant GM-CSF. Specific immunity and tumor regression were observed in all three patients [12]. In a larger study of 51 patients, this group observed 11 clinical responses [9]. Slingluff et al. vaccinated metastatic melanoma patients with a mixture of four gp100 and tyrosinase peptides, plus a tetanus helper peptide either in an emulsion with GM-CSF and montanide ISA-51 adjuvant or pulsed on immature monocyte-derived DCs [13]. They observed that peptide vaccination generated in higher percentage of the patient T cell responses in draining lymph nodes as compared to DC vaccination (80% *versus* 13%), suggesting that *in vivo* vaccination is at least as effective as DC-based approaches. In 15% of the patients a clinical response was observed after peptide vaccination.

Peptides have also been used to immunize patients with other solid tumors. About 90% of pancreatic cancer cells have a specific mutation in the K-RAS oncogene. Gjertsen et al. vaccinated patients with pancreatic cancer (10 surgical resected and 38 patients with advanced disease) with K-RAS peptides and GM-CSF and found in more than 50% of the patients a T cell response [14]. Moreover, the patient group with such a T cell response survived longer (median survival 148 *versus* 61 days) than those without an immune response, suggesting a potential clinical benefit [14].

In patients with Her-2/neu-positive breast, ovarian cancer vaccination and non-small-cell lung cancers, vaccination with Class II HLA-restricted Her-2-

neu peptides plus GM-CSF was investigated. In 92% patients peptide-specific T cell responses were detected and 24/27 had a positive DTH response against the peptide [15]. In addition, epitope spreading was observed in 84% of the patients, which means that patients developed immunity against another epitope of HER-2-neu than was present in the vaccine. At 1 year follow-up, immunity to the HER-2-neu antigen persisted in 38% of patients. Whether the development of an HER-2-neu immune response results in clinical benefit is currently unknown.

CEA is a 180 kD oncofetal glycoprotein present predominantly in fetal gut and is also expressed by endodermally derived neoplasms of gastrointestinal, respiratory tract, etc. [16]. It has also been identified in small amounts in normal adult mucosa of colon. CEA is considered a self-antigen by the immune system and patients with CEA-positive tumors are immunologically tolerant to CEA.

Samanci et al. cloned the CEA gene from human colon adenocarcinoma cells and introduced it into a baculovirus which was used for the production of recombinant CEA. This protein was used for the vaccination of colorectal cancer patients without macroscopic disease [17]. One group was vaccinated with GM-CSF and the control without. All patients in the GM-CSF group developed a strong rhCEA-specific proliferative T cell response, whereas patients vaccinated without GM-CSF showed a weak response. A cellular response against native human CEA could be found in 8/9 patients in the GM-CSF group, although at a significantly lower level than against recombinant CEA and warrants further studies in man to optimize vaccination strategies with CEA antigen.

Arlen et al. investigated different CEA-targeted vaccination strategies in patients with solid tumors expressing CEA [18]. They first vaccinated patients with a CEA peptide in an adjuvant and used an ELISPOT assay for immunomonitoring. Hardly any CEA-specific T cell responses could be demonstrated after vaccination with the peptide. Interestingly, patients who were vaccinated with vaccinia CEA followed by avipox-CEA, or avipox-CEA alone showed significant increases in CEA-specific T cell responses and antibody responses. Although this study was not randomized the results suggested that pox-virus recombinant-based vaccines are more potent in the induction of tumor-specific immune responses than vaccines using peptides.

The great advantage of peptide vaccination is that tumor antigens are well defined, giving the possibility to use patient-specific vaccines and making the specific immunological evaluation of immunotherapy more easy. The possibility to produce relatively easy in large quantities for a relatively low price are important advantages of this approach. A possible drawback is the fact that the vaccines contain only a limited number of T cell epitopes, which increases the chance for immune selection of tumors with genetic variations that no longer express the peptide epitope. Jager et al. showed this in patients who initially responded to a peptide vaccine, but who relapsed despite the presence peptide-specific T cells [19]. They showed after repeated tumor biopsies during the course of disease a gradual loss of antigen expression. A way to prevent this form of antigen loss might be a vaccination strategy that uses a cocktail of peptides.

In conclusion, it has been clearly demonstrated that peptide-based vaccines have antitumor activity. However, the selection of the best peptides and most optimal vaccination schedule is still not defined. Moreover, best results of specific immunotherapies are most likely in low residual disease, requiring large randomized trials.

Anti-idiotype antibody vaccines

The murine monoclonal antibody CEA Vac mimics a highly restricted CEA epitope that has no cross-reactivity with CEA expressed by normal human tissues. This antibody acts as a surrogate tumor antigen, inducing anti-CEA antibody responses and specific T cell responses, and was demonstrated to have a major antitumor effect in a murine tumor model [20]. In a study in 23 patients with advanced colorectal cancer, 17 generated anti-anti-idiotype responses, and 13 of these were proven to be true anti-CEA responses [21]. However, none of the patients had objective clinical responses and toxicity was limited to local swelling and minimal pain at vaccination site. CEA Vac has also been evaluated in the setting of adjuvant therapy of high risk colorectal cancer [22]. 32 patients were included in this study, 4 stage II, 11 stage III, 11 completed resected stage IV and nine stage IV patients with minimal residual disease. 15 patients received 5-FU-based chemotherapy, simultaneously with the CEA Vac. All patients had high-titer polyclonal anti-CEA responses which were not negatively affected by chemotherapy. Although no responses were observed, there appeared to be a biological effect since in a number of patients a prolonged period of stable disease was observed. A Phase III trial is planned by the American College of Surgeons Oncology Group; stage III patients' will be randomized to 5-FU/leucovorin *versus* 5-FU/leucovorin and CEA Vac.

Recombinant vaccines expressing tumor associated antigens

The immunogenic nature of CEA in humans is unclear, and the induction of T cell responses with protein vaccination is weak. Therefore, co-presentation of CEA with a strong immunogen such as a virus might increase its immunogenicity and induce strong anti-CEA immune responses [23]. Vaccinia viruses are highly immunogenic and stimulate both humoral and cellular mediated immune responses. In a Phase I trial immunization with a CEA-encoding recombinant vaccinia (rV-CEA) was investigated over a limited dose range [24]. Toxicity was limited to modest local inflammation at the inoculation site as well as low grade fever and fatigue. Unfortunately, there was no evidence of CEA-specific T cell proliferation, antibody responses or DTH responses.

The fact that patients were treated with only two vaccinations may explain the absence of specific immune activation. Because vaccinia virus proteins are highly immunogenic, vaccinia recombinants can only be administered once or twice due to the induction of neutralizing antivaccinia immune responses. By using a different immunization strategy Marshall et al. were able to demonstrate CEA-specific T cell activation [25]. They used another anti-CEA vaccine, the canary pox ALVAC-CEA (avipox-CEA). Unlike vaccinia, which is highly immunogenic and cannot be used for serial use, ALVAC can only replicate in avian species. In mammals, ALVAC infects cells, expresses its transgene products for 14–21 days, and is unable to infect other cells. Another advantage of ALVAC virus is that most humans have not been exposed to this virus. In their first clinical trial using ALVAC-CEA in 20 advanced CEA-positive cancer patients, Marshall et al. showed that treatment was well tolerated at all dose levels [25]. Mild skin reaction and injection site soreness were occasionally reported. In addition, they showed that ALVAC-CEA was able to induce CEA-specific CTL responses. However, besides one CEA-normalization no objective tumor-responses were reported.

In another trial of this group, patients with stage IV disease but without radiographic evidence of disease were randomized to receive either rV-CEA followed by three ALVAC-CEA vaccinations, or three times ALVAC-CEA followed by rV-CEA [26]. The first schedule was superior to the second in the generation of CEA-specific T cell responses, measured by an ELISPOT assay. When GM-CSF was given with subsequent vaccinations a further increase in CEA-specific T-cell precursors was observed. Survival was unrelated to pre-treatment T cell levels, while higher post vaccination T cell levels were associated with better survival. When the vaccination schedule is optimized, randomized trials are needed to investigate whether this type of vaccination is effective in the adjuvant treatment of colon carcinoma.

Improvement of vaccination could be achieved by increasing the antigen-presenting capacity of DC. When a DC presents an antigen two signals are required to activate a naive T cell. The first signal is provided by MHC Class I or Class II antigens presenting fragments of peptides to T cells [5, 6]. The second co-stimulatory signal can be given by B7.1 or B7.2, also known as CD80 and CD86, respectively. Without the second signal, the T cell develops an anergic response to the antigen. Thus, vaccine strategies that result in the coordinated presentation of antigen with a co-stimulatory molecule may result in improved immunity. The group of Schlom prepared a canary pox vector encoding the gene for CEA and for B7.1, called ALVAC-CEA B7.1 [27, 28]. In a pilot study they vaccinated patients with CEA-expressing metastatic colorectal cancer who had failed standard therapy. The therapy was well tolerated and after four vaccinations it was possible to demonstrate increases in CEA-specific T cell precursor frequencies. Except from skin reactions at injection site, flu-like symptoms, and mild gastrointestinal problems no toxicity was observed. No tumor responses were observed in patients, although 6 out of 17 patients with elevated CEA levels experienced a decline of CEA levels after

ALVAC-CEA vaccinations. The number of prior chemotherapy regimens was inversely correlated with the ability to generate a T cell response, suggesting that the real clinical impact of vaccination strategies can only be determined in a patient population without immune compromise.

The tricom vaccine consists of a triad of co-stimulatory molecules: B7.1, ICAM-1, LFA-3 (rV-CEA-TRICOM). Preclinical studies indicated that continued boosting with vaccine was required to maintain CEA-specific T cell responses and that co-administration of GM-CSF and/or IL-2 enhanced the antitumor activity. Therefore, Marshall et al. conducted a Phase I clinical trial to investigate the immunogenicity of rV-CEA-TRICOM and the most optimal vaccination schedule in advanced cancer patients with CEA-positive tumors [29]. The treatment was well tolerated and all HLA-A2 patients developed a CEA-specific immune response. In 1 out of 30 patients a clinical response was observed. The most optimal vaccination schedule was not defined, but they showed that the immune response tended to decrease when vaccines were administered every 3 months, suggesting that a vaccination schedule on a monthly basis is probably more beneficial.

Autologous tumor cells based vaccines

One of the main advantages of autologous tumor cell vaccination is that all potential tumor antigens are presented to the immune system. In the 1980s Hanna et al. established a guinea pig hepatocarcinoma model for the study of active specific immunotherapy as adjuvant treatment [30]. They demonstrated the value of a vaccine prepared from viable metabolically active tumor cells mixed with Bacillus Calmette-Guérin (BCG). A correct ratio of BCG organisms to tumor cells and an optimal vaccination schedule enabled them to control hematogenous and lymphatic metastases from surgically excised primary tumors.

On the basis of these preclinical studies Hoover et al. conducted a trial using irradiated autologous tumor cells and BCG in patients with stage II and stage III colorectal cancer [31]. After surgical resection of their primary tumors, patients were randomized to vaccination or observation and stratified by both disease type and stage. 3–4 weeks after surgery patients were vaccinated with two weekly vaccinations with tumor cells and BCG. One week later a third vaccine was administered, not containing BCG. An intention-to-treat analysis showed no significant clinical benefit, but a subgroup analysis of overall and disease-free survival in colon cancer patients showed a significant trend for ASI being superior to surgery alone. Immunized patients showed delayed type hypersensitivity reactions to autologous tumor cells that were stronger than background responses to autologous mucosal cells, suggesting the presence of tumor-specific immunity. The absence of a survival benefit in the rectal cancer group was thought to be caused by the radiotherapy that was given close to the draining lymph nodes of the vaccination site. However, it is important to realize that because of the low number of vaccinated rectal cancer patients, this low powered study does not allow a reliable analysis about the efficacy of ASI in rectal cancer. Nowadays, rectal cancer patients are being irradiated before removal of the carcinoma, which precludes ASI in its current form. Side effects were minimal and the most prominent were ulcerations at the site of the first two vaccinations and were caused by BCG.

These promising results were the reason to perform a large Phase III study with stage II and stage III colon cancer patients under the auspices of the Eastern Cooperative Oncology Group (ECOG) [32]. This study differed from the Hoover study in that, due to large number and wide geographic distribution of sites involved, each site performed its own vaccine manufacturing. It is possible that because of the fact that these centers were not making vaccines on a daily basis, the quality of vaccines was not always according to the required standards. In an intent-to-treat analysis of all randomized patients, there were no significant differences between the two treatment arms in time to recurrence or overall survival. In the ECOG study, 12% of all vaccines failed to meet quality control specifications (cell number/viability), and 15% of the vaccinated patients failed to have adequate DTH reactions. It was hypothesized that the poor quality of a part of the vaccinations could have caused the disappointing results of this study. Therefore, an explorative survival analysis was performed on patients who were treated with vaccines that met standardized criteria and developed antitumor immunity (DTH response to third vaccine >5 mm) and compared to control patients. In this subgroup analysis a significant improvement in overall survival was demonstrated in patients treated with ASI, suggesting that optimal immunization strategies are essential for a successful adjuvant treatment of colon cancer patients. This hypothesis was supported by the observation that the size of DTH response to autologous tumor cells correlated with survival, which has also been described in metastatic melanoma [33].

A third Phase III study was conducted in the Netherlands involving 254 patients with stage II and stage III colon cancer [34]. This pivotal study differed from the previous clinical trials in that treated patients received a booster with irradiated tumor cells alone, administered 6 months after surgical resection. In contrast to the previous study a centralized manufacturing laboratory supported the 12 participating hospitals, which prepared 98% quality approved vaccines. We showed that 97% of the vaccinated patients had DTH responses greater than 5 mm, suggesting that the centralized method of vaccine manufacturing is very important for vaccine quality. In an intent-to-treat analysis, ASI significantly reduced the rate of disease recurrence by 44% in patients with stage II and stage III colon cancer, but the overall survival was not significantly better. The major impact was seen in stage II disease in which there was 61% risk reduction for recurrences and a trend toward improved overall survival. The absence of a significant survival benefit has probably the same explanation as is mentioned for the adjuvant chemotherapy trials in stage II colon cancer. The relatively high non-colon cancer related mortality in this aged patient group together with the relatively good overall survival rate of stage II colon cancer patients requires a very large (more than 1,000 patients) randomized study to detect a survival benefit for any adjuvant treatment.

Therefore, a meta-analysis was performed which included the above-mentioned three randomized trials [35]. In the intent-to-treat meta-analysis of all 723 patients who received either a three- or a four-vaccine regimen, recurrence-free survival was significantly improved by ASI. In the meta-analysis of patients who met quality control specifications and protocol eligibility, recurrence free survival was significantly improved and disease-specific survival approached significance when compared with controls. In general, patients with a distant recurrence will eventually die from colon cancer. However, despite the fact that recurrences were significantly reduced by ASI, no significant survival benefit could be demonstrated in the intent-to-treat meta-analysis, indicating that a large adequate powered randomized trial is required. In conclusion, these studies showed that ASI has minimal side effects and that the most pronounced clinical benefit can be seen in stage II colon cancer.

In stage III colon cancer patients ASI did not result in a significant clinical benefit, which could be explained by the lack of statistical power of these studies. Furthermore, the residual tumor load in stage III patients is definitely larger than in stage II patients, which could be relevant since it is known that ASI is more effective in a minimal residual disease setting [30].

In preclinical models ASI and chemotherapy were shown to have a synergistic antitumor effect [36]. Apart from the capacity to directly destroy micrometastases, ASI has been demonstrated to disrupt the characteristically compact structure of metastatic foci, enabling chemotherapy to reach deeper into the cancer tissue. Furthermore, chemotherapy reduces the tumor burden, thereby increasing the possibility of ASI to eliminate the residual malignant cells. In preparation for a large Phase III trial, we performed a feasibility study on the combination of ASI and chemotherapy in stage III colon cancer. We showed that the combination ASI and 5-FU/leucovorin did not result in more toxicity and that the ASI-induced antitumor immunity (DTH response) was hardly impaired by consecutive chemotherapy [37]. A randomized trial should prove that these two modalities have indeed a synergistic antitumor effect.

Another way to increase immunogenicity of autologous vaccines is to transfect these tumor cells with genes of cytokines or chemokines. Dranoff et al. tested more than 30 different potent immunological substances in a preclinical B16 melanoma tumor model and showed very convincingly that GM-CSF producing tumor cells generated the best antitumor immunity [38].

Subsequently, Soiffer et al. [39] conducted a Phase I study in patients with metastatic melanoma, investigating the toxicity and immunogenicity of tumor cells infected with retroviral viruses expressing GM-CSF and irradiated with 15,000 cGy. In all patients DTH responses to injections of non-transduced autologous tumor cells were demonstrated. The most convincing evidence that vaccination enhanced anti-melanoma immunity was revealed by pathological examination of the host response to tumor cells. Whereas metastatic lesions resected before vaccination were minimally infiltrated with immune cells, metastatic lesions resected after vaccination were densely infiltrated with T and B cells, and a part of these cells were specific for the melanoma cells. In addition, extensive tumor destruction was observed in tumors of 11 of 16 patients. One patient showed a partial remission, while in three other patients minor responses were observed. Because retroviral vectors have major logistical problems to be used in large clinical trials, this group has replaced the retroviral vector by an adenoviral vector expressing the GM-CSF gene and had about the same clinical results [40]. Although the tumor-specific immune responses induced were impressive, most patients eventually died because of disease progression. To increase its efficacy, it is interesting to investigate GM-CSF producing vaccines in combination with chemotherapy, to reduce tumor load, IL-2 to promote CTL activity or anti-CTLA-4 antibodies to prevent the diminished effector function of activated CTLs, as will be discussed later.

Heat shock proteins based vaccines

Heat shock proteins (HSPs) are the most abundant and ubiquitous soluble intracellular proteins. They perform a multitude of housekeeping functions that are essential for cellular survival and their ability to interact with a wide range of proteins and peptides has made them suitable to participate in innate and adaptive immune responses [41, 42]. Heat shock proteins are present in cells under perfectly normal conditions. They act like 'chaperones', making sure that the cell's proteins are in the right shape and in the right place at the right time. Because of the normal functions of heat shock proteins inside the cell HSPs end up binding virtually every protein made within the cell. This means that at any given time, HSPs can be found inside the cell bound to a wide array of peptides that represent a 'library' of all the proteins inside the cell. This library contains normal peptides that are found in all cells as well as abnormal peptides that are found in cancer cells. Thus, using HSPs from tumor cells for vaccination enables us to immunize patients with the whole repertoire of peptides in the tumor cell. These HSP-peptide complexes are stable and very immunogenic, inducing tumor-specific CD4$^+$ and CD8$^+$ T cell responses. This approach has two advantages. First, vaccination with HSP-peptide complexes does not require identification of immunogenic epitopes and second, the immunization with these complexes is against the entire antigenic repertoire of the tumor, reducing the possibility that the tumors of patients escape variants. Because the HSP approach makes use of autologous tumor cells, it is applicable to any type of cancer (renal, melanoma, colon and pancreatic cancer), provided that enough tumor material (a few grams) is available.

Phase I/II studies have been performed in melanoma patients with detectable tumor and in colorectal cancer patients rendered disease-free by complete resection of liver metastasis [43–45]. In these studies, each patient was vaccinated with gp96-peptide complexes isolated from his or her own tumor. In either studies *de novo* induction, or the augmentation of antitumor-specific T-cell response, was achieved in a large proportion of gp96-vaccinated patients. For colorectal cancer, 17 out 29 patients (59%) displayed a statistically significant increase in postvaccination frequency of peripheral blood mononuclear cells that released INF-γ in response to either autologous or allogeneic HLA-matched colon carcinoma cells. A similar frequency of immunological responder patients was detected in the melanoma vaccination study with 11 out of 23 patients (47.8%) showing an increased number of tumor-specific T cells after gp96 vaccination as evaluated by IFN-γ ELISSPOT assay [44]. The antitumor response induced by *in vivo* gp96 vaccination included T cells specific for shared tumor antigens gp100 and Melan-A/MART-1 for melanoma, CEA and EpCam for colorectal cancer, while the presence of T cells directed against individual antigens could not be demonstrated due to the poor viability of fresh tumor cell suspensions. In both studies a real clinical benefit has been observed in a limited number of treated patients. For the melanoma study, complete responses involving regression of both cutaneous and visceral metastasis were observed in 2 out of 28 tumor-bearing patients [43].

The colorectal trial, although the number of patients was small, showed a disease-free and overall survival comparable to historical controls. However, patients with an immunological response had statistically significant survival advantage at 24 months on OS (100% *versus* 50% of non-responding patients) [45]. These results suggest a benefit for the adjuvant treatment of solid tumors after surgery. The results of Phase III trials in high risk renal cell carcinoma and metastatic melanoma are awaited with great eagerness and hopefully will show that adjuvant vaccination therapy can improve survival of cancer patients.

Allogeneic tumor cells based vaccines

Autologous tumor cells are used in various strategies of active specific immunotherapy. A real disadvantage of this approach is that it is logistically very demanding and that a relatively large amount of tumor tissue is required for the preparation of vaccines. Because of this, some groups were investigating the application of allogeneic tumor cell vaccines. One such vaccines is Melacine (Corixa-Montana) consisting of lyophilized lysates of two melanoma cell lines, MSM-M-1 and MSM-M-2, admixed with the immunological adjuvant Detox-PC (Corixa-Montana) immediately before use. A Phase I study in 19 patients with metastatic melanoma proved the feasibility of immunization against melanoma-associated antigens in 50% of the patients and elicited a 29% objective clinical response [46]. The follow up Phase II trial in 25 patients with metastatic melanoma demonstrated a response in 5 patients and immunomonitoring revealed that an increase in precursors of cytolytic T cells against melanoma cells was correlated with clinical benefit [47]. In one study in metastatic melanoma patients who were refractory to Melacine, treatment with IFN-α resulted in a response rate of 44%, including visceral metastatic sites as well as soft tissue and lung [48]. On the basis of these encouraging results, a Phase II trial of Melacine and IFN-α in combination was performed in metastatic melanoma. In this study 47 patients were enrolled, and were treated with cyclophosphamide 3 days before vaccination, followed by Melacine. Melacine was administered at a dose of 2×10^7 tumor cell equivalents per dose admixed with 0.25 ml of Detox-PC s.c. once a week on weeks 1–4 and week 6. Melacine maintenance was then given monthly from week 8, until progression or intolerable toxicity. IFN-α was started in the evening after the fourth dose of Melacine at a dose of 5 MIU units/m^2 3 times a week, and continued until progression. The treatment was well tolerated and the overall objective response rate was 10.2%, but 64% of patients had stabilization of their disease for at least 16 weeks [49].

Morton et al. developed a polyvalent whole tumor cell vaccine consisting of three different allogeneic melanoma cell lines, chosen for their content of highly immunogenic tumor associated antigens [50]. During a 12-week induction phase patients were treated with bi-weekly intradermal vaccinations, the first two of which were given BCG. The patients then received monthly vaccinations during the first year, followed by 3-monthly vaccination until progression of disease. Recently, they performed an analysis on their prospective database of 11,000 patients between 1/1/86 and 1/1/2001 [51, 52]. Although complete regression of evaluable disease was rare among vaccine patients, disease stabilization was common. Survival from initial stage IV diagnosis was prolonged almost two-fold for the entire group of vaccine patients and for all vaccine-treated patients with metastatic disease (p = 0.001). A large multicenter randomized Phase III trial is currently underway and the first results are expected in 2006.

Dendritic cells based vaccines

DCs play a central role in the regulation of immune responses as they can mediate tolerance as well as immunity, depending on their maturation status. Immature DCs are particularly good at antigen ingestion and processing, while mature DCs are potent antigen presenting cells, which can prime naive T cells and induce strong T cell memory responses. The first clinical study was reported in 1996 [53]. Since then a variety of DC-based trials have been started and published, showing occasionally impressive tumor responses in patients with extensive disease. In all these trials different protocols were used with regard to source of DCs, maturation stimuli, number of DCs, site of injection, frequency of vaccination, immunomonitoring, type of cancer and extensiveness of disease. In the first published vaccination study Hsu et al. vaccinated four patients with follicular B-cell lymphoma with DCs loaded *ex vivo* with specific recombinant idiotype protein. Idiotype-specific immune responses were found and three patients experienced a clinical benefit [53]. In a later report the group of patients was extended and the data of the first study was confirmed [54]. Quite a number of groups have investigated DC-based therapies in patients with metastatic melanoma, using peptide, tumor lysates or proteins [55–58]. In all these studies clinical effects were limited to occasional regressions (5–20%) and a correlation between tumor regressions and tumor-specific immune responses was often reported. In metastatic renal cell carcinoma Holtl et al. treated patients with tumor lysates pulsed DCs and observed responses in 10 out of 17 patients, including 2 complete responses [59, 60]. In prostate cancer some small successes were observed using different approaches. Small et al. loaded DCs with a GM-CSF/prostate acid phosphatase fusion protein and demonstrated a significant decrease in serum prostate-specific antigen (PSA), indicating a reduction of tumor load, in some patients [61]. Heiser et al. showed, using prostate-specific RNA loaded DCs, that PSA specific T cell responses were induced as well as some clinical responses [62].

So far, we can conclude that DC based immunotherapies are safe, non-toxic, feasible and effective in some patients. It has made a remarkably quick transformation from fundamental research to clinical application. However, many issues such as route, schedule, antigen and dosage have to be optimized before large randomized studies can prove its value in the treatment of cancer patients.

Antibodies to improve efficacy of vaccines

Activation of naive T cells is dependent on the delivery of at least two signals by the antigen-presenting cells, an antigen-specific signal via the T cell receptor and a second signal via co-stimulatory molecules [5, 6]. CD28, expressed on the cell surface of resting and activated T cells, and its counter-receptors B7-1 and B7-2 expressed on antigen-presenting cells are a major source of co-stimulatory signals to T cells. CTLA-4 is a second high-affinity receptor for the B7 family members that is expressed on activated, but not resting, T cells. However, unlike CD28, CTLA-4 engagement delivers a negative signal, attenuating T-cell responses by raising the threshold of signals needed for T cell activation [5, 6]. Blocking of CTLA-4 signaling in vitro with antibodies leads to enhanced T cell receptor and CD28-dependent proliferation of T cells [63, 64].

In murine studies, blockade of CTLA-4 function in vivo enhanced antitumor T cell-dependent immunity. Treatment of mice with anti-CTLA-4 led to the rejection of immunogenic transplanted tumors but had little or no effect on weakly or non-immunogenic tumors [65, 66]. Rejection of non-immunogenic tumors, including pre-established tumors, was achieved if CTLA-4 blockade was used in combination with an immunization protocol [67–69] or with low-dose chemotherapy [70] under the conditions that neither treatment alone was effective.

Phan et al. treated 14 patients with metastatic melanoma by using serial iv administration of a fully human anti-CTLA-4 antibody in conjunction with subcutaneous vaccination with two modified HLA-A*0201-restricted peptides from the gp100 melanoma-associated antigen [71]. After this treatment grade III/IV autoimmune manifestations in six patients (43%), including dermatitis, enterocolitis, hepatitis, and hypophysitis was observed, which recovered completely after stopping anti-CTLA-4 treatment and starting steroid therapy. Interestingly, objective cancer regression was observed in three patients (21%; two complete and one partial response). This study can be considered as an important breakthrough, because it establishes CTLA-4 as an important molecule regulating tolerance to 'self' antigens in humans and suggests a role for CTLA-4 blockade in breaking tolerance to human cancer antigens for cancer immunotherapy. This new and powerful modality is thus able to enhance the efficacy of vaccination, and opens perspectives for all active specific immunotherapies.

Using affinity-based *in vitro* selection methods, Santulli-Marotto isolated short oligonucleotide aptamers that could bind and block murine CTLA-4 with high affinity and specificity and interfere with its function *in vitro* and *in vivo* [72]. However, compared with the anti-CTLA-4 antibodies markedly more aptamers were required to elicit an effect *in vivo*, meaning that huge doses are required to treat patients, limiting its potential use in the clinic. Aptamers are synthetic chemicals and their manufacturing is easier and less expensive compared with protein-based clinical reagents. However, the *in vivo* bioactivity of aptamers should first be enhanced before it can be used in the clinic.

The recent discovery of human $CD4^+CD25^+$ regulatory T cells (Tregs) has made it feasible to develop strategies that modulate the immunosuppressive effects of Tregs in a vaccination setting [73]. Tregs have been shown to play an important role in the repression of T cell responses to both self and foreign antigens, and the loss of Tregs leads to the development of autoimmunity. Experimental tumor models in mice have revealed that Tregs can inhibit anti-tumor immune responses as well; depletion of such Tregs using anti-CD25 antibody permits the development of an antitumor immune response and tumor rejection [74]. Recent evidence indicates that Tregs may exist in high proportion in human cancer patients, possibly inducing or maintaining tolerance to tumors [75]. Vieweg et al. hypothesized that *in vivo* elimination of Tregs using the fusion protein denileukin diftitox (ONTAK) can enhance the efficacy of tumor RNA-transfected DC vaccines to stimulate a tumor-specific T cell response. They conducted a Phase I clinical trial in which patients received a single dose of ONTAK, followed by vaccination with total tumor RNA transfected DC. The RNA loaded DC were administered intradermally using three cycles of 10^7 cells applied in weekly intervals [76]. They demonstrated that that $CD4^+/CD25^{high}$ Tregs can be effectively eliminated in a dose-dependent manner and showed, as hypothesized, that Treg depletion followed by vaccination reproducibly led to improved stimulation of tumor-specific T-cells when compared to vaccination alone. Phase II/III trials are now required to show that this approach can indeed increase tumor responses in vaccinated cancer patients.

Conclusion

Recent advances in tumor immunology together with the clinical results obtained by vaccination therapies for cancer patients indicate that the immunotherapeutic approach could be an attractive option for complementary treatment after surgery and/or chemotherapy. The limited toxicity of vaccination is a major advantage of this modality as compared to chemotherapy. Nevertheless, the heterogeneity of protocols, the conflicting results of some trials and the relatively small number of patients enrolled in these studies, make it impossible to draw any definitive conclusions about its potential clinical relevance. It is important to realize that most studies have been performed in patients with extensive metastatic disease, while preclinical studies clearly show that immunotherapies are most effective in low residual disease setting. However, studies in the adjuvant setting, requiring high numbers of patients, are very laborious and expensive. Therefore, it is better to first optimize vaccination strategies in small Phase II studies before we move to larger Phase III studies. In this respect the antibodies against CLA-4 and/or Tregs are very promising modalities to enhance the efficacy of vaccination therapies. A number of very interesting new cancer vaccine strategies have entered clinical trials, and we eagerly await their findings.

References

1 Slamon DJ, Leyland-Jones B, Shak S, Fuchs H, Paton V, Bajamonde A, Fleming T, Eiermann W, Wolter J, Pegram M et al. (2001) Use of chemotherapy plus a monoclonal antibody against HER2 for metastatic breast cancer that overexpresses HER2. *N Engl J Med* 344: 783–792

2 Saltz LB, Meropol NJ, Loehrer PJ Sr, Needle MN, Kopit J, Mayer RJ (2004) Phase II trial of cetuximab in patients with refractory colorectal cancer that expresses the epidermal growth factor receptor. *J Clin Oncol* 22: 1201–1208

3 Coiffier B, Lepage E, Briere J, Herbrecht R, Tilly H, Bouabdallah R, Morel P, Van Den Neste E, Salles G, Gaulard P et al. (2002) CHOP chemotherapy plus rituximab compared with CHOP alone in elderly patients with diffuse large-B-cell lymphoma. *N Engl J Med* 346: 235–242

4 Pardoll DM (1998) Cancer vaccines. *Nat Med* 4: 525–531

5 Delves PJ, Roitt IM (2000) The immune system. First of two parts. *N Engl J Med* 343: 37–49

6 Delves PJ, Roitt IM (2000) The immune system. Second of two parts. *N Engl J Med* 343: 108–117

7 Gilboa E (2004) The promise of cancer vaccines. *Nat Rev Cancer* 4: 401–411

8 Parmiani G, Castelli C, Dalerba P, Mortarini R, Rivoltini L, Marincola FM, Anichini A (2002) Cancer immunotherapy with peptide-based vaccines: what have we achieved? Where are we going? *J Natl Cancer Inst* 94: 805–818

9 Jager E, Jager D, Knuth A (2002) Clinical cancer vaccine trials. *Curr Opin Immunol* 14: 178–182

10 Cormier JN, Salgaller ML, Prevette T, Barracchini KC, Rivoltini L, Restifo NP, Rosenberg SA, Marincola FM (1997) Enhancement of cellular immunity in melanoma patients immunized with a peptide from MART-1/Melan A. *Cancer J Sci Am* 3: 37–44

11 Rosenberg SA, Yang JC, Schwartzentruber DJ, Hwu P, Marincola FM, Topalian SL, Restifo NP, Sznol M, Schwarz SL, Spiess PJ et al. (1999) Impact of cytokine administration on the generation of antitumor reactivity in patients with metastatic melanoma receiving a peptide vaccine. *J Immunol* 163: 1690–1695

12 Jager E, Ringhoffer M, Dienes HP, Arand M, Karbach J, Jager D, Ilsemann C, Hagedorn M, Oesch F, Knuth A (1996) Granulocyte-macrophage-colony-stimulating factor enhances immune responses to melanoma-associated peptides *in vivo*. *Int J Cancer* 67: 54–62

13 Slingluff CL Jr, Petroni GR, Yamshchikov GV, Barnd DL, Eastham S, Galavotti H, Patterson JW, Deacon DH, Hibbitts S, Teates D et al. (2003) Clinical and immunologic results of a randomized phase II trial of vaccination using four melanoma peptides either administered in granulocyte-macrophage colony-stimulating factor in adjuvant or pulsed on dendritic cells. *J Clin Oncol* 21: 4016–4026

14 Gjertsen MK, Buanes T, Rosseland AR, Bakka A, Gladhaug I, Soreide O, Eriksen JA, Moller M, Baksaas I, Lothe RA et al. (2001) Intradermal ras peptide vaccination with granulocyte-macrophage colony-stimulating factor as adjuvant: clinical and immunological responses in patients with pancreatic adenocarcinoma. *Int J Cancer* 92: 441–450

15 Disis ML, Gooley TA, Rinn K, Davis D, Piepkorn M, Cheever MA, Knutson KL, Schiffman K (2002) Generation of T-cell immunity to the HER-2/neu protein after active immunization with HER-2/neu peptide-based vaccines. *J Clin Oncol* 20: 2624–2632

16 Benchimol S, Fuks A, Jothy S, Beauchemin N, Shirota K, Stanners CP (1989) Carcinoembryonic antigen, a human tumor marker, functions as an intercellular adhesion molecule. *Cell* 57: 327–334

17 Samanci A, Yi Q, Fagerberg J, Strigard K, Smith G, Ruden U, Wahren B, Mellstedt H (1998) Pharmacological administration of granulocyte/macrophage-colony- stimulating factor is of significant importance for the induction of a strong humoral and cellular response in patients immunized with recombinant carcinoembryonic antigen. *Cancer Immunol Immunother* 47: 131–142

18 Arlen P, Tsang KY, Marshall JL, Chen A, Steinberg SM, Poole D, Hand PH, Schlom J, Hamilton JM (2000) The use of a rapid ELISPOT assay to analyze peptide-specific immune responses in carcinoma patients to peptide *versus* recombinant poxvirus vaccines. *Cancer Immunol Immunother* 49: 517–529

19 Jager E, Ringhoffer M, Altmannsberger M, Arand M, Karbach J, Jager D, Oesch F, Knuth A (1997) Immunoselection *in vivo*: independent loss of MHC class I and melanocyte differentiation antigen expression in metastatic melanoma. *Int J Cancer* 71: 142–147

20 Pervin S, Chakraborty M, Bhattacharya-Chatterjee M, Zeytin H, Foon KA, Chatterjee SK (1997) Induction of antitumor immunity by an anti-idiotype antibody mimicking carcinoembryonic antigen. *Cancer Res* 57: 728–734

21 Foon KA, Chakraborty M, John WJ, Sherratt A, Kohler H, and Bhattacharya-Chatterjee M (1995) Immune response to the carcinoembryonic antigen in patients treated with an anti-idiotype antibody vaccine. *J Clin Invest* 96: 334–342

22 Foon KA, John WJ, Chakraborty M, Das R, Teitelbaum A, Garrison J, Kashala O, Chatterjee SK, Bhattacharya-Chatterjee M (1999) Clinical and immune responses in resected colon cancer patients treated with anti-idiotype monoclonal antibody vaccine that mimics the carcinoembryonic antigen. *J Clin Oncol* 17: 2889–2895

23 Schlom J, Tsang KY, Kantor JA, Abrams SI, Zaremba S, Greiner J, Hodge JW (1999) Strategies in the development of recombinant vaccines for colon cancer. *Semin Oncol* 26: 672–682

24 Conry RM, Khazaeli MB, Saleh MN, Allen KO, Barlow DL, Moore SE, Craig D, Arani RB, Schlom J, LoBuglio AF (1999) Phase I trial of a recombinant vaccinia virus encoding carcinoembryonic antigen in metastatic adenocarcinoma: comparison of intradermal *versus* subcutaneous administration. *Clin Cancer Res* 5: 2330–2337

25 Marshall JL, Hawkins MJ, Tsang KY, Richmond E, Pedicano JE, Zhu MZ, Schlom J (1999) Phase I study in cancer patients of a replication-defective avipox recombinant vaccine that expresses human carcinoembryonic antigen. *J Clin Oncol* 17: 332–337

26 Marshall JL, Hoyer RJ, Toomey MA, Faraguna K, Chang P, Richmond E, Pedicano JE, Gehan E, Peck RA, Arlen P et al. (2000) Phase I study in advanced cancer patients of a diversified prime-and- boost vaccination protocol using recombinant vaccinia virus and recombinant nonreplicating avipox virus to elicit anti-carcinoembryonic antigen immune responses. *J Clin Oncol* 18: 3964–3973

27 Horig H, Lee DS, Conkright W, Divito J, Hasson H, LaMare M, Rivera A, Park D, Tine J, Guito K et al. (2000) Phase I clinical trial of a recombinant canarypoxvirus (ALVAC) vaccine expressing human carcinoembryonic antigen and the B7.1 co-stimulatory molecule. *Cancer Immunol Immunother* 49: 504–514

28 Von Mehren M, Arlen P, Tsang KY, Rogatko A, Meropol N, Cooper HS, Davey M, McLaughlin S, Schlom J, Weiner LM (2000) Pilot study of a dual gene recombinant avipox vaccine containing both carcinoembryonic antigen (CEA) and B7.1 transgenes in patients with recurrent CEA-expressing adenocarcinomas. *Clin Cancer Res* 6: 2219–2228

29 Marshall J (2003) Carcinoembryonic antigen-based vaccines. *Semin Oncol* 30 (Suppl 8): 30–36

30 Hanna MG Jr, Brandhorst JS, Peters LC (1979) Active specific immunotherapy of residual micrometastasis: an evaluation of sources, doses and ratios of BCG with tumor cells. *Cancer Immunother* 7: 165–173

31 Hoover HC Jr, Brandhorst JS, Peters LC, Surdyke MG, Takeshita Y, Madariaga J, Muenz LR, Hanna MG Jr, (1993) Adjuvant active specific immunotherapy for human colorectal cancer: 6.5-year median follow-up of a phase III prospectively randomized trial. *J Clin Oncol* 11: 390–399

32 Harris JE, Ryan L, Hoover HC Jr, Stuart RK, Oken MM, Benson AB III, Mansour E, Haller DG, Manola J, Hanna MG Jr, (2000) Adjuvant active specific immunotherapy for stage II and III colon cancer with an autologous tumor cell vaccine: Eastern Cooperative Oncology Group Study E5283. *J Clin Oncol* 18: 148–157

33 Baars A, Claessen AM, van den Eertwegh AJ, Gall HE, Stam AG, Meijer S, Giaccone G, Meijer CJ, Scheper RJ, Wagstaff J et al. (2000) Skin tests predict survival after autologous tumor cell vaccination in metastatic melanoma: experience in 81 patients. *Ann Oncol* 11: 965–970

34 Vermorken JB, Claessen AM, van Tinteren H, Gall HE, Ezinga R, Meijer S, Scheper RJ, Meijer CJ, Bloemena E, Ransom JH et al. (1999) Active specific immunotherapy for stage II and stage III human colon cancer: a randomised trial. *Lancet* 353: 345–350

35 Hanna MG, Hoover HC, Vermorken JB, Harris JE, Pinedo HM (2001) Adjuvant active specific immunotherapy of stage II and stage III colon cancer with an autologous tumor cell vaccine: first randomized phase III trials show promise. *Vaccine* 19: 2576–2582

36 Hanna MG Jr, Key ME (1982) Immunotherapy of metastases enhances subsequent chemotherapy. *Science* 217: 367–369

37 Baars A, Claessen AME, Gall HE, Zegers I, Scheper RJ, Giaccone G, Meijer S, Meijer CJLM, Wagstaff J, Vermorken JB et al. (2002) A phase II study of Active Specific Immunotherapy in a combination with 5-FU/LV for the adjuvant treatment of stage III colon carcinoma. *Br J Cancer* 86: 1230–1234

38 Dranoff G, Jaffee E, Lazenby A, Golumbek P, Levitsky H, Brose K, Jackson V, Hamada H, Pardoll D, Mulligan RC (1993) Vaccination with irradiated tumor cells engineered to secrete murine granulocyte-macrophage colony-stimulating factor stimulates potent, specific, and long-lasting anti-tumor immunity. *Proc Natl Acad Sci USA* 90: 3539–3543

39 Soiffer R, Lynch T, Mihm M, Jung K, Rhuda C, Schmollinger JC, Hodi FS, Liebster L, Lam P, Mentzer S et al. (1998) Vaccination with irradiated autologous melanoma cells engineered to secrete human granulocyte-macrophage colony-stimulating factor generates potent antitumor immunity in patients with metastatic melanoma. *Proc Natl Acad Sci USA* 95: 13141–13146

40 Soiffer R, Hodi FS, Haluska F, Jung K, Gillessen S, Singer S, Tanabe K, Duda R, Mentzer S, Jaklitsch M et al. (2003) Vaccination with irradiated, autologous melanoma cells engineered to secrete granulocyte-macrophage colony-stimulating factor by adenoviral-mediated gene transfer augments antitumor immunity in patients with metastatic melanoma. *J Clin Oncol* 21: 343–3350

41 Castelli C, Rivoltini L, Rini F, Belli F, Testori A, Maio M, Mazzaferro V, Coppa J, Srivastava PK, Parmiani G (2004) Heat shock proteins: biological functions and clinical application as personalized vaccines for human cancer. *Cancer Immunol Immunother* 53: 227–233

42 Srivastava P (2002) Roles of heat-shock proteins in innate and adaptive immunity. *Nat Rev Immunol* 2: 185–194

43 Belli F, Testori A, Rivoltini L, Maio M, Andreola G, Sertoli MR, Gallino G, Piris A, Cattelan A, Lazzari I et al. (2002) Vaccination of metastatic melanoma patients with autologous tumor-derived heat shock protein gp96-peptide complexes: clinical and immunologic findings. *J Clin Oncol* 20: 4169–4180

44 Rivoltini L, Castelli C, Carabba M, Mazzaferro V, Pilla L, Huber V, Coppa J, Gallino F, Scheibenbogen C, Squarcina P et al. (2003) Human tumor-derived heat shock protein 96 mediates *in vitro* activation and *in vivo* expansion of melanoma- and colon carcinoma-specific T cells. *J Immunol* 171: 3467–3474

45 Mazzaferro V, Coppa J, Carabba M, Rivoltini L, Schiavo M, Regalia E, Mariani L, Camerini T, Marchiano A, Andreola S et al. (2003) Vaccination with autologous tumor derived heat-shock protein gp96 after liver resection for metastatic colorectal cancer. *Clin Cancer Res* 9: 3235–3245

46 Mitchell MS, Kan-Mitchell J, Kempf RA, Harel W, Shau H, Lind S (1988) Active specific immunotherapy for melanoma: Phase I trial of allogeneic lysates and a novel adjuvant. *Cancer Res* 48: 5883–5893

47 Mitchell MS, Harel W, Kempf RA, Hu E, Kan-Mitchell L, Boswell WD, Dean G, Stevenson L

(1990) Active specific immunotherapy for melanoma. *J Clin Oncol* 8: 856–869

48 Mitchell MS, Jakowatz J, Harel W, Dean G, Stevenson L, Boswell WD, Groshen S (1994) Increased effectiveness of interferon α-2b following active specific immunotherapy for melanoma. *J Clin Oncol* 12: 402–411

49 Vaishampayan U, Abrams J, Darrah D, Jones V, Mitchell MS (2002) Active immunotherapy of metastatic melanoma with allogeneic melanoma lysates and interferon alpha. *Clin Cancer Res* 8: 3696–3701

50 Morton DL, Barth A (1996) Vaccine therapy for malignant melanoma. *CA Cancer J Clin* 46: 225–244

51 Morton DL, Hsueh EC, Essner R, Foshag LJ, O'Day SJ, Bilchik A, Gupta RK, Hoon DS, Ravindranath M, Nizze JA et al. (2002) Prolonged survival of patients receiving active immunotherapy with Canvaxin therapeutic polyvalent vaccine after complete resection of melanoma metastatic to regional lymph nodes. *Ann Surg* 236: 438–448

52 Hsueh EC, Gupta RK, Gammon G, Foshag LJ, Essner R, Qi K, Morton DL (2004) Correlation of specific immune responses with survival in AJCC stage IV melanoma patients receiving polyvalent melanoma cell vaccine. *ASCO 2004*, abstract 1675

53 Hsu FJ, Benike C, Fagnoni F, Liles TM, Czerwinski D, Taidi B, Engleman EG, Levy R (1996) Vaccination of patients with B-cell lymphoma using autologous antigen-pulsed dendritic cells. *Nat Med* 2: 52–58

54 Timmerman JM, Czerwinski DK, Davis TA, Hsu FJ, Benike C, Hao ZM, Taidi B, Rajapaksa R, Caspar CB, Okada CY et al. (2002) Idiotype-pulsed dendritic cell vaccination for B-cell lymphoma: clinical and immune responses in 35 patients. *Blood* 99: 1517–1526

55 Nestle FO, Alijagic S, Gilliet M, Sun Y, Grabbe S, Dummer R, Burg G, Schadendorf D (1998) Vaccination of melanoma patients with peptide- or tumor lysate-pulsed dendritic cells. *Nat Med* 4: 328–332

56 Schuler-Thurner B, Schultz ES, Berger TG, Weinlich G, Ebner S, Woerl P, Bender A, Feuerstein B, Fritsch PO, Romani N et al. (2002) Rapid induction of tumor-specific type 1 T helper cells in metastatic melanoma patients by vaccination with mature, cryopreserved, peptide-loaded monocyte-derived dendritic cells. *J Exp Med* 195: 1279–1288

57 Thurner B, Haendle I, Roder C, Dieckmann D, Keikavoussi P, Jonuleit H, Bender A, Maczek C, Schreiner D, von den Driesch P et al. (1999) Vaccination with mage-3A1 peptide-pulsed mature, monocyte-derived dendritic cells expands specific cytotoxic T cells and induces regression of some metastases in advanced stage IV melanoma. *J Exp Med* 190: 1669–1678

58 Banchereau J, Palucka AK, Dhodapkar M, Burkeholder S, Taquet N, Rolland A, Taquet S, Coquery S, Wittkowski KM, Bhardwaj N et al. (2001) Immune and clinical responses in patients with metastatic melanoma to CD34(+) progenitor-derived dendritic cell vaccine. *Cancer Res* 61: 6451–6458

59 Höltl L, Rieser C, Papesh C, Ramoner R, Bartsch G, Thurnher M (1998) CD83+ blood dendritic cells as a vaccine for immunotherapy of metastatic renal-cell cancer. *Lancet* 352: 1358

60 Höltl L, Zelle-Rieser C, Gander H, Papesh C, Ramoner R, Bartsch G, Rogatsch H, Barsoum AL, Coggin JH Jr, Thurnher M (2002) Immunotherapy of metastatic renal cell carcinoma with tumor lysate-pulsed autologous dendritic cells. *Clin Cancer Res* 8: 3369–3376

61 Small EJ, Fratesi P, Reese DM, Strang G, Laus R, Peshwa MV, Valone FH (2000) Immunotherapy of hormone-refractory prostate cancer with antigen-loaded dendritic cells. *J Clin Oncol* 18: 3894–3903

62 Heiser A, Coleman D, Dannull J, Yancey D, Maurice MA, Lallas CD, Dahm P, Niedzwiecki D, Gilboa E, Vieweg J (2002) Autologous dendritic cells transfected with prostate-specific antigen RNA stimulate CTL responses against metastatic prostate tumors. *J Clin Invest* 109: 409–417

63 Tivol EA, Borriello F, Schweitzer AN, Lynch WP, Bluestone JA, Sharpe AH (1995) Loss of CTLA-4 leads to massive lymphoproliferation and fatal multiorgan tissue destruction, revealing a critical negative regulatory role of CTLA-4. *Immunity* 3: 541–547

64 Walunas TL, Lenschow DJ, Bakker CY, Linsley PS, Freeman GJ, Green JM, Thompson CB, Bluestone JA (1994) CTLA-4 can function as a negative regulator of T cell activation. *Immunity* 1: 405–413

65 Leach DR, Krummel MF, Allison JP (1996) Enhancement of antitumor immunity by CTLA-4 blockade. *Science* 271: 1734–1736

66 Yang YF, Zou JP, Mu J, Wijesuriya R, Ono S, Walunas T, Bluestone J, Fujiwara H, Hamaoka T (1997) Enhanced induction of antitumor T-cell responses by cytotoxic T lymphocyte-associated

molecule-4 blockade: the effect is manifested only at the restricted tumor-bearing stages. *Cancer Res* 57: 4036–4041

67 Van Elsas A, Hurwitz AA, Allison JP (1999) Combination immunotherapy of B16 melanoma using anti-cytotoxic T lymphocyte-associated antigen 4 (CTLA-4) and granulocyte/macrophage colony-stimulating factor (GM-CSF)-producing vaccines induces rejection of subcutaneous and metastatic tumors accompanied by autoimmune depigmentation. *J Exp Med* 190: 355–366

68 Hurwitz AA, Foster BA, Kwon ED, Truong T, Choi EM, Greenberg NM, Burg MB, Allison JP (2000) Combination immunotherapy of primary prostate cancer in a transgenic mouse model using CTLA-4 blockade. *Cancer Res* 60: 2444–2448

69 Hurwitz AA, Yu TF, Leach DR, Allison JP (1998) CTLA-4 blockade synergizes with tumor-derived granulocyte-macrophage colony-stimulating factor for treatment of an experimental mammary carcinoma. *Proc Natl Acad Sci USA* 95: 10067–10071

70 Mokyr MB, Kalinichenko T, Gorelik L, Bluestone JA (1998) Realization of the therapeutic potential of CTLA-4 blockade in low-dose chemotherapy-treated tumor-bearing mice. *Cancer Res* 58: 5301–5304

71 Phan GQ, Yang JC, Sherry RM, Hwu P, Topalian SL, Schwartzentruber DJ, Restifo NP, Haworth LR, Seipp CA, Freezer LJ et al. (2003) Cancer regression and autoimmunity induced by cytotoxic T lymphocyte-associated antigen 4 blockade in patients with metastatic melanoma. *Proc Natl Acad Sci USA* 100: 8372–8377

72 Santulli-Marotto S, Nair SK, Rusconi C, Sullenger B, Gilboa E (2003) Multivalent RNA Aptamers That Inhibit CTLA-4 and Enhance Tumor Immunity. *Cancer Res* 63: 7483–7489

73 Shevach EM (2002) CD4+ CD25+ suppressor T cells: more questions than answers. *Nat Rev Immunol* 2: 398–400

74 Jones E, Dahm-Vicker M, Simon AK, Green A, Powrie F, Cerundolo V, Gallimore A (2002) Depletion of CD25+ regulatory cells results in suppression of melanoma growth and induction of autoreactivity in mice. *Cancer Immun* 2:1

75 Liyanage UK, Moore TT, Joo HG, Tanaka Y, Herrmann V, Doherty G, Drebin JA, Strasberg SM, Eberlein TJ, Goedegebuure PS et al. (2002) Prevalence of regulatory T cells is increased in peripheral blood and tumor microenvironment of patients with pancreas or breast adenocarcinoma. *J Immunol* 169: 2756–2761

76 Vieweg J, Su Z, Dannull J (2004) Enhancement of antitumor immunity following depletion of CD4+CD25+ regulatory T cells. *ASCO 2004*, abstract 2506

Drugs Affecting Growth of Tumours
Edited by Herbert M. Pinedo and Carolien H. Smorenburg
© 2006 Birkhäuser Verlag/Switzerland

Oral anticancer agents

Carolien H. Smorenburg[1] and Alex Sparreboom[2]

[1] *Department of Medical Oncology, 10B Vrije Universiteit Medical Center, PO Box 7057, 1007 MB
  Amsterdam, The Netherland*
[2] *Clinical Pharmacology Research Core, Medical Oncology Clinical Research Unit, National
  Cancer Institute, Bethesda, Maryland, USA*

Introduction

In general medicine, oral ingestion is the most common way of drug administration, being convenient, safe and effective for most agents. In contrast, in
oncology most anticancer agents are delivered by intravenous (iv) injection.
This is probably due to the narrow therapeutic index of many antineoplastic
drugs and the pharmacologic observation that oral administration often
results in a large intra- and intersubject variability in drug exposure. However,
the burden of iv administration is evident: every iv injection carries, although
small, a risk of bleeding, extravasation, infection and thrombosis and requires
medically qualified personnel at a hospital setting. Moreover, especially in
cancer patients, repeated iv injections are hampered by the fact that a patient's
accessible vein may disappear during chemotherapy due to flebitis or thrombosis.

Only during recent years, attention has focussed on the development of
oral chemotherapy in oncology [1–3]. In an extensive review on this topic,
DeMario and Ratain address the pharmacokinetic limitations of oral
chemotherapy and discuss novel oral cytotoxics [1]. For reasons of patient
convenience, oral chemotherapy seems a valuable addition to standard iv use.
Besides, the outpatient administration of oral agents may potentially reduce
total healthcare system costs [4]. In addition, oral formulations are of benefit
in therapies that require prolonged exposure by means of a protracted treatment course. This also fits in the concept of 'metronomic scheduling' of frequent administration of chemotherapy at low dose to increase anti-angiogenic
activity [5]. However, apart from pharmacokinetic limitations, oral
chemotherapy has other potential drawbacks, as listed in Table 1. This chapter discusses specific issues regarding patient preferences and compliance,
pharmacokinetics of absorption and bioavailability, barriers of intestinal
absorption, and subsequently summarises available and novel oral anticancer
agents.

Table 1. Advantages and disadvantages of oral anticancer agents

Issue	Advantages	Potential disadvantages
Patient	Convenient	Non-compliance
	Patient active participation in treatment	More extensive patient education needed
Costs	Economical	Increased drug costs
Drug delivery	No injection → no pain, extravasation, bleeding, thrombosis, infection	Patient must be able to swallow, patient should not vomit
	Schedule flexibility	No direct drug availability
		Limited and variable absorption

Patient preference and compliance

For the majority of metastatic solid tumours, chemotherapy offers at most an improvement in symptom relief and only a modest gain in actual survival. In these palliative treatment regimens, the aspect of quality of life is increasingly being recognised. Surprisingly the patient preference for oral *versus* intravenous chemotherapy was only recently examined by Liu et al. [6]. Of 103 patients with metastatic cancer, 92 preferred oral chemotherapy, provided that both oral and iv chemotherapy had a similar efficacy. Major reasons for preferring oral chemotherapy were patient convenience, problems with iv access or needles and control of the environment in which patients would receive treatment. The patient preference was not associated with sex, age or prior chemotherapy experiences. Grober et al. did a survey about patient attitudes towards oral and iv therapy in cancer patients with a history of only oral treatment (n = 109) and only iv treatment (n = 242) [7]. Indeed, orally treated patients regarded their therapy as being more convenient, comfortable and safe (p < .001) than the iv group viewed their iv treatment. Of interest, intravenously treated patients regarded oral therapy as less safe and effective than orally treated patients did (p < .001). These data confirm the patient preference for oral chemotherapy, but also show patients' misconception of oral chemotherapy being less effective and safe.

It is common knowledge for clinicians that many patients fail to comply with prescribed therapy. This probably also holds for oral anticancer agents, although few studies on oral chemotherapy have investigated patient compliance. The lack of data of patient compliance is even more striking and worrisome in dose-finding studies on new oral anticancer agents, as these studies recommend a certain dose level for further use. Partridge et al. reviewed various aspects of compliance on oral chemotherapy [8]. Their search of literature revealed only few patient studies with data on actual medication intake. In 52 breast cancer patients treated with oral cyclophosphamide, Lebovits et al. reported a compliance rate of only 43% [9]. In another study, only 17% of patients with a haematologic malignancy complied with a regimen of oral allopurinol and prednisone [10]. An extensive review on compliance with oral chemotherapy in childhood leukaemia was given by Davies and Lilleyman [11]. Based on published studies, they estimated that at least 10–40% of these children did not comply with oral (and potentially curative!) maintenance therapy. Many factors may be associated with a higher rate of non-compliance: lower socioeconomic status, age, side effects and the complexity, duration and frequency of the treatment regimen. A oral drug that should be taken more often than twice a day is less likely to be taken as compared to drugs taken once or twice daily [12]. For oral chemotherapy, however, Richardson et al. could not detect any correlation between side effects and patient compliance [13]. Methods for detecting non-compliance consist of self-reporting by the patient, and objective measures such as pill counts, electronic devices that detect bottle-opening systems or drug assays of urine or blood samples. Every method has its drawbacks and none of them is completely reliable. To obtain good compliance, it is important to give clear information on the drug and its schedule at the start of therapy, to repeat this throughout the therapy and to give additional written information as well. Frequent patient visits and a medication diary card may also be of help. Unfortunately, despite the growing use of oral chemotherapeutic agents there are hardly any data on improving compliance.

Pharmacology

Intravenous infusion makes, by definition, 100% of the drug available in the blood. Bioavailability is a term used to indicate the fraction of a given drug that actually reaches the systemic circulation following extravascular administration. A drug given by mouth has to overcome various barriers before it finally reaches the central blood compartment. Firstly, there are many mechanical barriers. In oncology, certain malignancies of the head and neck and oesophagus may prevent proper swallowing, while nausea due to various causes frequently is encountered in cancer patients. The extent and rate of absorption from the stomach is usually much smaller than that of the intestines because of its acidic environment, the small surface area and a thick mucus layer on the stomach wall. Any irregularity in propulsion, due to fibrosis, tumour or in the case of an ileus, will prevent a proper absorption. Furthermore, oral drugs may be destructed by low gastric pH, digestives enzymes or intestinal flora. Last but not least, the intestinal epithelium and its intestinal drug transporters and metabolising enzymes may hamper absorption, hereby diminishing the final bioavailability of the drug (Fig. 1). Hellriegel et al. noticed a significant correlation between decreasing bioavailability and increasing interpatient variability in bioavailability [14]. As most anticancer agents have a narrow therapeutic index, a high variability in bioavailability and subsequently in exposure to the active agent may predispose to either toxic or ineffective dosing. For novel oral compounds of available iv anticancer agents not only the bioavailability but

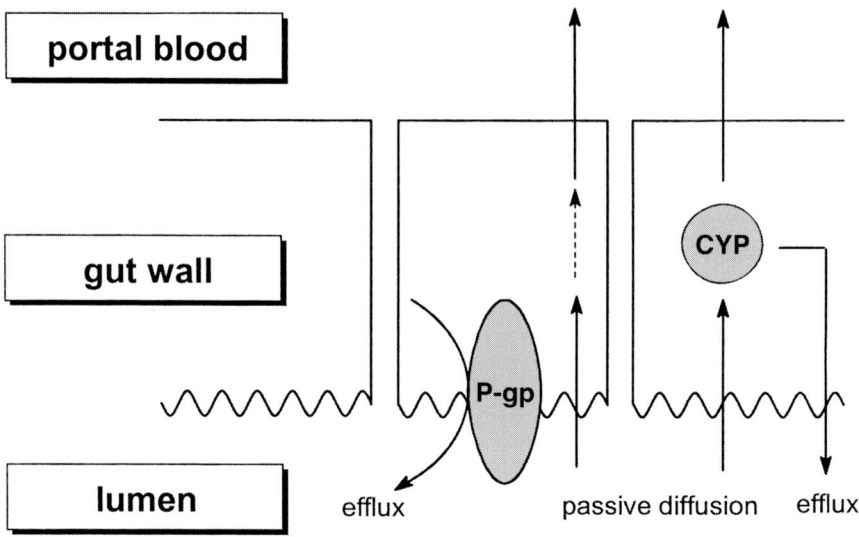

Figure 1. Schematic overview of physiologic barriers in the intestinal epithelium that may diminish the oral bioavailability of anti-cancer drugs. *Abbreviations:* P-gp, P-glycoprotein; CYP, cytochrome P450.

also the intra- and interpatient variability in drug exposure of the oral form should therefore be evaluated and be compared with that as observed in the iv route. However, the prejudice of constant and predictable exposure after iv administration does not hold true for many cytotoxic agents. Data on bioavailability of some anticancer agents are depicted in Table 2.

Intestinal drug transporters

P-glycoprotein (PGP, ABCB1) is a member of the family of ATP-binding cassette drug transporters that is abundantly expressed in the intestinal mucosa [15]. Other members of this family are multidrug-resistance protein (MRP 1, ABCC1 and MRP2, ABCC2) and breast cancer-resistance protein (BCRP, ABCG2) [16]. These drug transporters act as an outward-directed drug efflux pump and have a function in detoxification and protection against toxic compounds. Oral anticancer agents that are PGP substrates include epipodophyllotoxins, anthracyclines, camptothecin analogues, and taxanes. In mdr Ia(–/–)mice, which lack PGP, the bioavailability of oral paclitaxel indeed significantly increased, with a six-fold increase in the area under the curve (AUC) [17]. Due to the drug efflux pump PGP and the interaction with the formulation vehicle Cremophor EL, the bioavailability of oral paclitaxel is less than 10%. In order to improve its absorption, studies have combined oral taxanes with various inhibitors of PGP activity. Meerum Terwogt et al. treated 14 patients with oral paclitaxel together with

Table 2. Bioavailability and indication of approved oral anticancer agents

CLASS Agent	Oral bioavailability (%)	Indication	Comments	Ref.
ALKYLATING AGENTS				
Cyclophosphamide	85	breast and ovarian cancer, leukaemia, lymphoma	Activated by the hepatic CYP2B P450 isoenzyme In breast cancer part of oral CMF schedule	39
Melphalan	58–85	myeloma	Lower bioavailability when taken with food	32
Chloorambucil	70–80	CLL		40
Busulphan	100	CML		40
Temozolamide	100	anaplastic astrocytoma	No clinical significant effect of food on oral bioavailability	41
Lomustine (=CCNU)	100	glioma	Complete conversion during first pass to *cis*- and *trans*-4-hydroxy CCNU Compound of PCV schedule	42
ANTI-METABOLITES				
Methotrexate	5–97	childhood ALL	Highly variable oral bioavailability Sequestration in third space (e.g., in pleural fluid)	43
5-fluorouracil	0–80		Due to unpredictable oral bioavailability not clinically used as oral agent	44
5-fluorouracil and eniluracil	100	colorectal cancer	Eniluracil is a DPD inactivator DLT = diarrhoea Not equivalent efficacy as iv 5FU/LV ⇒ not approved by FDA	25, 26, 45
UFT (tegafur + uracil)	100	colorectal and breast cancer	5-fluorouracil prodrug with a DPD inhibitor DLT = diarrhoea Not approved in the United States	46, 47
Capecitabine	70–95	colorectal and breast cancer	Converted to 5-fluorouracil by three enzymes DLT = diarrhoea and hand-foot syndrome Increased exposure in renal impairment Decreased exposure when taken with food	35, 48, 49

(Continued on next page)

Table 2. (Continued)

CLASS Agent	Oral bioavailability (%)	Indication	Comments	Ref.
ANTI-METABOLITES (continued)				
Mercaptopurine	5–37	ALL	First pass metabolic inactivation by hepatic xanthine oxidase Large interpatient variability in exposure	40
Thioguanine	low	AML	Large interpatient variability in exposure	40
Hydroxurea	80–100	Leukemia, polycythemia	40–80% excreted in urine within 24 h	40
NATURAL COMPOUNDS				
Etoposide	50–67	Childhood leukaemia	Comparable exposure with once or twice daily dosing	50
		Lung cancer	Similar exposure in elderly results in more severe myelotoxicity	51
		Breast cancer	Response rate 0–35% in metastatic disease	52
Idarubicin	10–30	leukaemia, breast cancer	First oral anthracycline Extensive (80%) first pass conversion to idarubicinol	53
OTHER				
Imatinib (=Glivec)	98	CML	EGFR tyrosine kinase inhibitor	54, 55
		Gastrointestinal stromal tumour	No clinical significant effect of food on oral bioavailability	56, 57
Gefitimib (=Iressa)	57	Lung cancer	EGFR tyrosine kinase inhibitor Slowly absorbed ⇒ once daily dosing DLT = diarrhoea	58
Hormones		Breast and prostate cancer	Various compounds	

ALL = Acute lymphoblastic leukaemia
CLL = Chronic lymphoblastic leukaemia
CML = Chronic myeloid leukaemia
DLT = Dose-limiting toxicity
EGFR = epidermal growth factor receptor

or without oral cyclosporin A and observed an eight-fold higher bioavailability of paclitaxel [18]. Likewise, the addition of cyclosporin A to oral docetaxel improved its bioavailability of only 8% to 90% [19]. In search for a better PGP inhibitor than cyclosporin A, which is also an immunosuppressive agent, the PGP (and BCRP) blocker GF120918 has been tested with oral paclitaxel in six patients [20]. The addition of GF120918 resulted in an increase of the paclitaxel AUC which was comparable with that achieved by cyclosporin A, and enhanced paclitaxel bioavailability to 30%. Kruijtzer et al. treated eight patients with oral topotecan with or without GF120918 [21]. The apparent bioavailability of topotecan rose significantly from 40% to 97% (p = .008) in patients treated with only one single dose of GF120918. As topotecan is a substrate for both PGP and BCRP with a lower affinity for PGP, the increase in bioavailability by GF120918 is most likely mediated by inhibition of intestinal BCRP activity. Of note, modulation of intestinal drug transporters may not only increase oral bioavailability but may also reduce interindividual variability in drug exposure and hence variation in toxicity.

Intestinal enzymes

Enzymes involved in drug metabolism are not only present in the liver, but in the intestinal wall as well [22]. Several enzymes located in the enterocyte, like CYP3A4, one of the major subclasses of cytochrome P450 expressed in the intestines, are involved in the presystemic metabolism of many cytotoxic agents, like cyclophosphamide, docetaxel, etoposide, and vinorelbine, thereby limiting the oral absorption of these drugs (Fig. 1). The bioavailability of these drugs might be substantially enhanced by pharmacological modulation of enteric CYP3A4 activity. Several investigators confirmed recently that by inhibiting CYP3A4 activity by co-administration of specific inhibitors such as erythromycin, quinidine, ketoconazole, and cyclosporin A the oral bioavailability of various anticancer agents (e.g., etoposide) could be improved, thereby also diminishing the variability in absorption [23].

One of the best-studied examples of pharmacokinetic biomodulation is the co-administration of eniluracil, an inactivator of dihydropyrimidine dehydrogenase (DPD), with 5-fluorouracil. DPD is the initial and rate-limiting enzyme involved in the degradation of the pyrimidines uracil and thymine and of 5-fluorouracil by a reduction pathway. The high variation in the population in DPD activity accounts for much of the variability observed with the therapeutic use of 5-fluorouracil, including variable drug levels, variable bioavailability, and inconsistent toxicity and activity profiles. Eniluracil has been shown to improve the efficacy of 5-fluorouracil in preclinical models through the selective, irreversible inhibition of DPD-mediated metabolism [24]. In subsequent clinical studies, eniluracil enabled the oral administration of 5-fluorouracil by inhibiting intestinal DPD activity, thereby increasing the oral availability and diminishing the variability in absorption [25, 26].

In addition to CYP3A4 and DPD, several other Phase I and Phase II classes of enzymes are expressed in intestinal epithelium and are known to be involved in anticancer drug metabolism. The best characterised of these include carboxylesterases (CE) and uridine diphosphate glucuronosyltransferases (UGT). The expression of CE in the human gastrointestinal tract has particular relevance to the camptothecin analogue irinotecan, which requires CE-mediated metabolic conversion to its active form SN-38. Therefore it may be possible to effectively administer irinotecan orally with the knowledge that substantial presystemic metabolism could take place. The clinical utility of this concept is currently under further investigation.

Even though intentional pharmacokinetic biomodulation can be beneficial in cancer patients, pharmacokinetic and pharmacodynamic studies have to be conducted in humans to determine at which of these two levels (i.e., kinetic or dynamic) the interaction takes place and whether biomodulation ultimately improves the therapeutic index of anticancer agents.

Effects of concomitant therapy

There is considerable motivation for understanding adverse drug interactions with anticancer agents, particularly when administered orally, because of their narrow therapeutic index, and the numerous concomitant medications that are administered routinely or intermittently to patients. Although progress has been made recently towards a proper understanding of drug interactions resulting in increased chemotherapy-induced toxicity, much less is known about concomitant medications resulting in metabolic inactivation of anticancer drugs given orally as a result of induction of drug transporters or enzymes. Recent studies have shown, for example, that induction of intestinal PGP and MRP2 by rifampin appears to be the underlying mechanism of decreased plasma concentrations of substrates, including digoxin and drug conjugates, with concomitant rifampin therapy [27]. This suggests an underrated new type of steady-state drug interaction affecting compounds, likely including several anticancer drugs, which are subject to transport rather than metabolism. Similarly, induction of several enzymes, including CYP3A4, by some medications has been described and may have a serious impact on anticancer therapy. For example, use of St. John's Wort extracts has been shown to result in increased expression of CYP3A4 and significantly increased clearance or decreased bioavailability of frequently prescribed drugs, leading to complete loss of therapeutic effects [28]. It is expected that induction of CYP3A4 expression by St. John's Wort extracts will result in altered drug clearance and affect toxicity profiles and possibly antitumour activity of numerous anticancer drugs. With these kinds of potential implementations for oral administration of anticancer agents, particularly in an outpatient setting, these intriguing results clearly deserve further investigation in the field of anticancer drug pharmacology.

Effects of food

The intake of drugs together with a daily routine, such as the intake of a meal, may improve drug compliance. However, food can modulate drug absorption in a variety of ways: it may increase, decrease or delay it, depending on the solubility, permeability and dissolution parameters of the drug [29]. Certain food components may also affect drug metabolism by inducing or inhibiting drug transporters or metabolising enzymes [30, 31]. Li et al. summarised some of these. For example, components of grapefruit juice inhibit both PGP and CYP3A. It is increasingly being recognised that the effect of food on the pharmacokinetics of an oral drug should be investigated early in drug development to optimise further trial design [31]. Reece et al. treated eight patients with oral melphalan in a crossover design with and without a standardised breakfast [32]. The median AUC and bioavailability were reduced when taking food (122 ng.h/ml and 58%, respectively) as compared with fasting (179 ng.h/ml and 85%, respectively) (p < .01 and p < .025). A competing amino acid transport was mentioned as a possible explanation for the reduced absorption of melphalan. For oral vinorelbine a similar crossover study was done in 13 patients [33]. Apart from a shorter median time to peak concentration (T_{max}) in fasted patients of 1.63 h *versus* 2.48 h in fed patients (p > .05), the peak concentration (C_{max}) and AUC were similar. In contrast, Herben et al. found an increase in median T_{max} for oral topotecan in fed patients (3.1 h) as compared to fasting patients (2.0 h) (p = .013), but likewise no significant differences in AUC and C_{max} were observed [34]. Reigner et al. showed that concomitant intake of food significantly decreased AUC and C_{max} of capecitabine, and that this effect diminished for its metabolites 5-DFCR, 5-DFUR and 5-fluorouracil [35]. For the main active metabolites, 5-DFUR and 5-fluorouracil, the AUC was not significantly lower. As these results were not related to any clinical outcome, the authors suggest that capecitabine should be given after food, as was done in previous dose-finding and other clinical studies with capecitabine.

These studies on the effect of food on pharmacokinetics used a standardised high-fat breakfast, which is recommended by the United States Food and Drug Administration (FDA). The effect of daily-life variation in diet on the variability in bioavailability of oral anticancer agents has, however, not been examined.

Effects of formulation vehicles

Many poorly water-soluble drugs are manufactured using pharmaceutical vehicles to improve drug formulation. For most vehicles no interaction with drug exposure is intended, but some vehicles such as liposomes may be used intentionally to alter pharmacologic parameters. For example, stealth liposomal doxorubicin (Caelyx, for iv use) was developed to provide a slow release of doxorubicin and to alter its biodistribution to prevent cardiotoxicity. The surfactants Cremophor EL and polysorbate 80 (Tween 80) are used as drug formulation vehicles for a variety of anticancer agents, such as teniposide and paclitaxel (Cremophor EL), and docetaxel and etoposide (Tween 80). The biological and pharmacological effects of these vehicles have recently been reviewed by Ten Tije et al. [36]. Based on *in vitro*, animal and human studies using various oral agents, Tween 80 appears to enhance oral absorption by increasing biomembrane permeability. Malingre et al. investigated the effect of both vehicles on the absorption of oral paclitaxel in a cross-over study in six cancer patients [37]. As compared to Tween 80, the oral formulation with Cremophor EL resulted in a significant reduced plasma paclitaxel concentration (p = .046). Other studies using various oral agents administered orally together with Cremophor EL have confirmed a reduced bioavailability compared with other vehicles. To overcome the current problems associated with oral delivery of paclitaxel, alternative pharmaceutical formulations are being developed with more favourable characteristics that may allow oral administration [38].

Approved and novel oral chemotherapeutic agents

Table 2 lists data on bioavailability and indications of approved oral anticancer agents. Of notice, many of these agents are used in the treatment of haematological malignancies while only a few compounds (e.g., cyclophosphamide

Table 3. Some oral anticancer agents in development

Agent	Comment	Ref.
Vinorelbine	Oral bioavailability 36–40%	59–61
	Similar interindividual variability in exposure between oral and iv route	
	Various ongoing clinical combination studies	
Topotecan	Oral bioavailability 30–40%	62–64
	Slightly less active in ovarian and lung cancer compared with iv route and less toxic	
Paclitaxel or docetaxel + cyclosporine	Cyclosporin inhibits PGP and CYP3A4, proteins that hamper oral absorption of substrate	18, 19
BMS 275183 and other new oral taxanes		
Trofosfamide	Nearly 100% oral bioavailability	
	Metabolised to ifosfamide	
Erlotinib (Tarceva)	EGFR tyrosine kinase inhibitor	
	Phase III completed	
Irinotecan	Phase I studies ongoing	

CYP3A4 = cytochrome P450 3A
EGFR = epidermal growth factor receptor
PGP = P-glycoprotein

and etoposide) have been used for many years in solid tumours. However, during the last decade, the use of new oral formulations of the existing drug 5-fluorouracil for common tumour types as colorectal and breast cancer has rapidly increased.

Table 3 lists some novel oral anticancer agents, which in part are oral formulations of existing cytotoxic agents and in part new targeted molecules. Of notice, the value of thorough investigation of pharmacokinetic and pharmacodynamic aspects, together with research on dose scheduling and food interactions is fortunately increasingly appreciated in the development of oral anticancer agents.

References

1 Demario M, Ratain MJ (1998) Oral chemotherapy: rationale and future directions. *J Clin Oncol* 16: 2557–2567
2 Sparreboom A, de Jonge MJA, Verweij J (2002) The use of oral cytotoxic and cytostatic drugs in cancer treatment. *Eur J Cancer* 38: 18–22
3 O'Neill VJ, Twelves CJ (2002) Oral cancer treatment: developments in chemotherapy and beyond. *B J Cancer* 87: 933–937
4 Navarro RP, Morrow T, Baran R (2002) Pharmacoeconomic and clinical outcomes in oncology using oral chemotherapy. *Manag Care Interface* 15: 55–62
5 Gasparini G (2001) Metronomic scheduling: the future of chemotherapy? *Lancet Oncol* 2: 733–740
6 Liu G, Franssen E, Fitch MI, Warner E (1997) Patient preferences for oral *versus* intravenous palliative chemotherapy. *J Clin Oncol* 15: 110–115
7 Grober SE, Carpenter RC, Glassman M, Blum D (2003) A comparison of patients' perceptions of oral cancer treatments and intravenous cancer treatments: what the health care team needs to know. *Proc Am Soc Clin Oncol* 22: 746 (abstract 3000)
8 Partridge AH, Avorn J, Wang PS, Winer EP (2002) Adherence to therapy with oral antineoplastic agents. *J Natl Cancer Inst* 94: 652–661
9 Lebovits AH, Strain JJ, Steven J, Schleifer SJ, Tanaka JS, Bhardwaj S, Messe MR (1990) Patient noncompliance with self-administered chemotherapy. *Cancer* 65: 17–22
10 Levine AM, Richardson JL, Marks G, Chan K, Graham J, Selser JN, Kishbaugh C, Shelton DR, Johnson CA (1987) Compliance with oral drug therapy in patients with haematologic malignancy. *J Clin Oncol* 5: 1469–1476
11 Davies HA, Lilleyman JS (1995) Compliance with oral chemotherapy in childhood lymphoblastic leukemia. *Cancer Treat Rev* 21: 93–103
12 Greenberg RN (1984) Overview of patient compliance with medication dosing: a literature review. *Clin Ther* 6: 592–599
13 Richardson JL, Marks G, Levine A (1988) The influence of symptoms of disease and side effects of treatment on compliance with cancer therapy. *J Clin Oncol* 6: 1746–1752
14 Hellriegel ET, Bjornsson TD, Hauck WW (1996) Interpatient variability in bioavailability is related to the extent of absorption: Implications for bioavailability and bioequivalence studies. *Clin Pharmacol Ther* 60: 601–607
15 Thiebaut F, Truruo T, Hamada H, Gottesman MM, Pastan I, Willingham MC (1987) Cellular localisation of the multidrug resistance gene product P-glycoprotein in normal human tissue. *Proc Natl Acad Sci USA* 84: 7735–7738
16 Borst P, Evers R, Kool M, Wijnholds J (2000) A family of drug transporters: the multidrug resistance-associated proteins. *J Natl Cancer Inst* 92: 1295–1302
17 Sparreboom A, van Asperen J, Mayer U, Schinkel AH, Smits JW, Meijer DKF, Borst P, Nooijen WJ, Beijnen JH, van Tellingen O (1997) Limited oral bioavailability and active epithelial excretion of paclitaxel caused by P-glycoprotein in the intestine. *Proc Natl Acad Sci USA* 94: 2031–2035

18 Meerum Terwogt JM, Malingré MM, Beijen JH, ten Bokkel Huinink WW, Rosing H, Koopman FJ, van Tellingen O, Swart M, Schellens JHM (1999) Coadministration of oral cyclosporin A enables oral therapy with paclitaxel. *Clin Cancer Res* 5: 3379–3384
19 Malingré MM, Richel DJ, Beijnen JH, Rosing H, Koopman FJ, ten Bokkel Huinink WW, Schot ME, Schellens JHM (2001) Coadministration of cyclosporine strongly enhances the oral bioavailability of docetaxel. *J Clin Oncol* 19: 1160–1166
20 Malingré MM, Beijen JH, Rosing H, Koopman FJ, Jewell RC, Paul EM, ten Bokkel Huinink WW, Schellens JHM (2001) Co-administration of GF120918 significantly increases the systemic exposure to oral paclitaxel in cancer patients. *B J Cancer* 84: 42–47
21 Kruijtzer CMF, Beijen JH, Rosing H, ten Bokkel Huinink WW, Schot M, Jewell RC, Paul EM, Schellens JHM (2002) Increased oral bioavailability of topotecan in combination with the Breast Cancer Resistance Protein (BCRP) and P-glycoprotein inhibitor GF120918. *J Clin Oncol* 20: 2943–2950
22 Kolars JC, Schmiedlin-Ren P, Schuetz JD, Fang C, Watkins PB (1992) Identification of rifampicin-inducible P450IIIA4 (CYP3A4) in human small bowel enterocytes. *J Clin Invest* 90: 1871–1878
23 Sparreboom A, Loow WJ, de Jonge MJA, Verweij J (2002) Clinical trial design: incorporation of pharmacokinetic, pharmacodynamic and pharmacogenetic principles. In: BC Baguley and DJ Kerr (eds): *Anticancer Drug Development*. Academic Press, Chapter 18: 329–351
24 Cao S, Rustum YM, Spector T (1994) 5-Ethynyluracil (776C85): modulation of 5-fluorouracil efficacy and therapeutic index in rats bearing advanced colorectal carcinoma. *Cancer Res* 15: 1507–1510
25 Baker SD, Diasio RB, O'Reilly S, Lucas VS, Khor SP, Sartorius SE, Donehower RC, Grochow LB, Spector T, Hohneker JA et al. (2000) Phase I and pharmacologic study of oral fluorouracil on a chronic daily schedule in combination with the dihydropyrimidine dehydrogenase inactivator eniluracil. *J Clin Oncol* 18: 915–926
26 Adjei AA, Reid JM, Diasio RB, Sloan JA, Smith DA, Rubin J, Pitot HC, Alberts SR, Goldberg RM, Hanson LJ et al. (2002) Comparative pharmacokinetic study of continuous venous infusion fluorouracil and oral fluorouracil with eniluracil in patients with advanced solid tumors. *J Clin Oncol* 20: 1683–1691
27 Niemi M, Backman JT, Fromm MF, Neuvonen PJ, Kivisto KT (2003) Pharmacokinetic interactions with rifampicin: clinical relevance. *Clin Pharmacokinet* 42: 819–850
28 Ioannidis C (2002) Pharmacokinetic interactions between herbal remedies and medicinal drugs. *Xenobiotica* 32: 451–478
29 Singh BN (1999) Effects of food on clinical pharmacokinetics. *Clin Pharmacokinet* 37: 213–255
30 Deferme S, Augustijns P (2003) The effect of food components on the absorption of P-gp substrates: a review. *J Pharm Pharmacol* 55: 153–162
31 Li Z, Vachharajani NN, Krishna R (2002) On the assessment of effects of food on the pharmacokinetics of drugs in early development. *Biopharm Drug Disp* 23: 165–171
32 Reece PA, Kotasek D, Morris RG, Dale BM, Sage RE (1986) The effect of food on oral melphalan absorption. *Cancer Chemother Pharmacol* 16: 194–197
33 Bugat R, Variol P, Roché H, Fumoleau P, Robinet G, Senac I (2002) The effects of food on the pharmacokinetic profile of oral vinorelbine. *Cancer Chemother Pharmacol* 50: 285–290
34 Herben VMM, Rosing H, ten Bokkel Huinink WW, van Zomeren DM, Batchelor D, Doyle E, Beusenberg FD, Beijnen JH, Schellens JHM (1999) Oral topotecan: bioavailability and effect of food co-administration. *B J Cancer* 80: 1380–1386
35 Reigner B, Verweij J, Dirix L, Cassidy J, Twelves C, Allman D, Weidekamm E, Roos B, Banken L, Utoh M et al. (1998) Effect of food on the pharmacokinetics of capecitabine and its metabolites following oral administration in cancer patients. *Clin Cancer Res* 4: 941–948
36 ten Tije AJ, Verweij J, Loos WJ, Sparreboom A (2003) Pharmacological effects of formulation vehicles: Implications for cancer chemotherapy. *Clin Pharmacokinet* 42: 665–685
37 Malingré MM, Schellens JHM, van Tellingen O, Ouwehand M, Bardelmeijer HA, Rosing H, Koopman FJ, Jansen SE, Schot ME, ten Bokkel Huinink WW et al. (2001) The co-solvent Cremophor EL limits absorption of orally administered paclitaxel in cancer patients. *B J Cancer* 85: 1472–1477
38 Nuijen B, Bouma M, Schellens JH, Beijnen JH (2001) Progress in the development of alternative pharmaceutical formulations of taxanes. *Invest New Drugs* 19: 143–153
39 Gheuens E, Slee PH, de Bruijn EA (1990) Bioavailability of cyclophosphamide in the CMF regi-

men. *Onkologie* 13: 203–206

40 Chabner BA (2001) Antineoplastic agents. In: JG Hardman, LE Limbird, A Goodman Gilman (eds): *The pharmacological basis of therapeutics*. Medical Publishing Division, New York, 1389–1459

41 Brada M, Judson I, Beale P, Moore S, Reidenberg P, Statkevich P, Dugan M, Batra V, Cutler D (1999) Phase I dose-escalation and pharmacokinetic study of temozolomide (SCH 52365) for refractory or relapsing malignancies. *Br J Cancer* 81: 1022–1030

42 Lee FYF, Workman P, Roberts JT, Bleehen NM (1985) Clinical pharmacokinetics of oral CCNU (Lomustine). *Cancer Chemother Pharmacol* 14: 125–131

43 Balis FM, Holcenberg JS, Poplack DG, Ge J, Sather HN, Murphy RF, Ames MM, Waskerwitz MJ, Tubergen DG, Zimm S et al. (1998) Pharmacokinetics and pharmacodynamics of oral methotrexate and mercaptopurine in children with lower risk acute lymphoblastic leukemia: a Joints Children's Cancer Group and Pediatric Oncology Branch Study. *Blood* 92: 3569–3577

44 Diasio RB, Harris BE (1989) Clinical Pharmacology of 5-fluouracil. *Clin Pharmacokinet* 16: 215–237

45 Schilsky RL, Levin J, West WH, Wong A, Colwell B, Thirlwell MP, Ansari RH, Bell WN, White RL, Yates BB et al. (2002) Randomized, open-label, phase III study of a 28-day oral regimen of eniluracil plus fluouracil *versus* intravenous fluouracil plus leucovorin as first-line therapy in patients with metastatic/advanced colorectal cancer. *J Clin Oncol* 20: 1519–1526

46 Douillard JY, Hoff PM, Skillings JR, Eisenberg P, Davidson N, Harper P, Vincent MD, Lembersky BC, Thompson S, Maniero A et al. (2000) Multicenter phase III study of uracil/tegafur and oral leucovorin *versus* fluouracil and leucovorin in patients with previously untreated metastatic colorectal cancer. *J Clin Oncol* 20: 3605–3616

47 Carmichael J, Popiela T, Radstone D, Falk S, Borner M, Oza A, Skovsgaard T, Munier S, Martin C (2002) Randomized comparative study of tegafur/uracil and oral leucovorin *versus* parenteral fluouracil and leucovorin in patients with previously untreated metastatic colorectal cancer. *J Clin Oncol* 20: 3617–3627

48 Judson IR, Beale PJ, Trigo JM, Aherne W, Crompton T, Jones J, Bush E, Reigner B (1999) A human capecitabine excretion balance and pharmacokinetic study after administration of a single oral dose of ^{14}C-labelled drug. *Invest New Drugs* 17: 49–56

49 Poole C, Gardiner J, Twelves C, Johnston P, Harper P, Cassidy J, Monkhouse J, Banken L, Weidekamm E, Reigner B (2002) Effect of renal impairment on the pharmacokinetics and tolerability of capecitabine in cancer patients. *Cancer Chemother Pharmacol* 49: 225–234

50 Edick MJ, Gajjar A, Mahmoud HH, van de Poll MEC, Harrison PL, Panetta JC, Rivera GK, Ribeiro RC, Sandlund JT, Boyett JM et al. (2003) Pharmacokinetics and pharmacodynamics of oral etoposide in children with relapsed or refractory acute lymphoblastic leukemia. *J Clin Oncol* 21: 1340–1346

51 Ando M, Minami H, Ando Y, Sakai S, Shimono Y, Sugiura S, Saka H, Shimokata K, Hasewaga Y (1999) Pharmacological analysis of etoposide in elderly patients with lung cancer. *Clin Cancer Res* 5: 1690–1695

52 Bontenbal M, Planting ASTh, Verweij J, de Wit R, Kruit WHJ, Stoter G, Klijn JGM (1995) Second-line chemotherapy with long-term low-dose oral etoposide in patients with advanced breast cancer. *Breast Cancer Res Treat* 34: 185–189

53 Toffoli G, Sorio R, Aita P, Crivellari D, Corona G, Bearz A, Robieux I, Colussi AM, Stocco F, Boiocchi M (2000) Dose-finding and pharmacologic study of chronic oral idarubicin therapy in metastatic breast cancer patients. *Clin Cancer Res* 6: 2279–2287

54 van Oosterom AT, Judson I, Verweij J, Stroobants S, Donato di Paola E, Dimitrijevic S, Martens M, Webb A, Sciot R, van Glabbeke M et al. (2001) Safety and efficacy of imatinib (STI571) in metastatic gastrointestinal stromal tumours: a phase I study. *Lancet* 358: 1421–1423

55 Peng B, Hayes M, Racine-Poon A, Druker BJ, Talpaz M, Sawyers CL, Resta D, Ford JM, Lloyd P, Capdeville R (2001) Clinical investigation of the pharmacokinetic and pharmacodynamic relationship for Glivec (STI571): a novel inhibitor of signal transduction. *Proc Am Soc Clin Oncol* 20: abstract 280

56 Demetri GD, von Mehren M, Blanke CD, van den Abbeele AD, Eisenberg B, Roberts PJ, Heinrich MC, Tuveson DA, Singer S, Janicek M et al. (2002) Efficacy and safety of imatinib mesylate in advanced gastrointestinal stromal tumors. *NEJM* 347: 472–480

57 Reckman AH, Fischer T, Peng B, Hayes M, Mehring G, Reese SF, Resta D, Ben-Am M, Gschaidmeier H, Huber Ch et al. (2001) Effect of food on STI571 Glivec pharmacokinetics and

bioavailability. *Proc Am Soc Clin Oncol* 20: abstract 1223

58 Ranson M, Hammond LA, Ferry D, Kris M, Tullo A, Murray PI, Miller V, Averbuch S, Ochs J, Morris C et al. (2002) ZD1839, a selective oral epidermal growth factor receptor-tyrosine kinase inhibitor, is well tolerated and active in patients with solid, malignant tumors: results of a phase I trial. *J Clin Oncol* 20: 2240–2250

59 Marty M, Fumoleau P, Adenis A, Rousseau Y, Merrouche Y, Robinet G, Senac I, Puozzo C (2001) Oral vinorelbine pharmacokinetics and absolute bioavailability study in patients with solid tumors. *Ann Oncol* 12: 1643–1649

60 Variol P, Nguyen L, Tranchand B, Puozzo C (2002) A simultaneous oral/intravenous population pharmacokinetic model for vinorelbine. *Eur J Clin Pharmacol* 58: 467–476

61 Freyer G, Delozier T, Lichinister M, Gedouin D, Bougnoux P, His P, Imadalou K, Trillet-Lenoir V (2003) Phase II study of oral vinorelbine in first-line advanced breast cancer chemotherapy. *J Clin Oncol* 21: 35–40

62 Schellens JHM, Creemers GJ, Beijnen JH, Rosing H, de Boer-Dennert M, McDonald M, Davies B, Verweij J (1996) Bioavailability and pharmacokinetics of oral topotecan: a new topoisomerase I inhibitor. *B J Cancer* 73: 1268–1271

63 von Pawel J, Gatzemeier U, Pujol JL, Moreau L, Bildat S, Ranson M, Richardson G, Steppert C, Riviere A, Camlett I et al. (2001) Phase II comparator study of oral *versus* intravenous topotecan in patients with chemosensitive small-cell lung cancer. *J Clin Oncol* 19: 1743–1749

64 Gore M, Oza A, Rustin G, Malfetano J, Calvert H, Clarke-Pearson D, Carmichael J, Ross G, Beckman RA, Fields SZ (2002) A randomised trial of oral *versus* intravenous topotecan in patients with relapsed epithelial ovarian cancer. *Eur J Cancer* 38: 57–63

Drugs Affecting Growth of Tumours
Edited by Herbert M. Pinedo and Carolien H. Smorenburg
© 2006 Birkhäuser Verlag/Switzerland

Anti-angiogenesis agents

Bart C. Kuenen

VU Medical Center, Department Medical Oncology, De Boelelaan 1117, 1081 HV Amsterdam, The Netherlands

Introduction

Angiogenesis, the formation of new blood vessels from the existing vasculature, is a physiological process in wound healing and the menstrual cycle, but has a pathological role in several diseases, such as retinopathy, rheumatoid arthritis and cancer. One of the hallmarks of cancer is the presence of sustained angiogenesis after tumors have made the angiogenic switch [1, 2]. Angiogenesis is a multi-step process in which endothelial cells have to become activated, subsequently proliferate and migrate in the direction of the angiogenic stimulus, which will only be possible after breakdown and remodeling of the extracellular matrix (ECM) so that tube formation can take place. Finally, the newly formed blood vessel has to be stabilized by pericytes and fibroblasts. This complicated and highly orchestrated process is regulated by multiple factors, such as growth factors, proteases (matrix metalloproteases) and integrins. Important driving forces in tumor-induced angiogenesis are hypoxia and oncogenes [3–6]. Hypoxia induces via hypoxia inducible factor 1α (HIF-1α) the expression of several angiogenic growth factors, whereas mutated and/or activated oncogenes are also capable of inducing the upregulation of angiogenesis stimulating factors and/or the downregulation of angiogenesis inhibiting genes and proteins.

Several lines of evidence indicate that vascular endothelial growth factor (VEGF) is one of the most important and potent stimulators of the angiogenic process. VEGF, first discovered as vascular permeability factor, is a proliferation and migration factor for endothelial cells [7–9]. At present the family of VEGFs consist of six members, designated VEGF-A, -B, -C, -D, -E and placenta growth factor (PlGF). VEGF-A has several isoforms (121, 145, 165, 183, 189, and 206 amino acids residues), which are all encoded by the same gene by alternative splicing. $VEGF_{165}$ appears to be the most biologically active isoform in both physiological and pathological angiogenesis. The VEGFs exert their effects via receptors (VEGFRs) which are almost exclusively expressed by endothelial cells. The family of VEGFRs consists of three members, named VEGFR-1 (Flt-1), VEGFR-2 (KDR/Flk-1), and VEGFR-3 (Flt-4). All VEGFRs, except soluble (s)VEGFR-1, are characterized by seven extracellular immunoglobulin (Ig)-like domains, of which the second and third are critical for ligand binding. The receptor consists furthermore of a transmembrane hydrophobic domain and an intracellular protein tyrosine kinase (TK) catalytic domain. The intracellular TK domain is characteristic for the family of receptor tyrosine kinases (RTKs), which consist of 20 members including furthermore the receptors for epidermal growth factor (EGF), fibroblast growth factor (FGF), and platelet derived growth factor (PDGF). RTKs are present as inactive diffusible monomers in the plane of the plasma membrane. Transmembrane signal transduction occurs when ligand binding induces receptor dimerization, which subsequently induces a conformational change in the catalytic TK domain resulting in autophosphorylation of the receptor and the subsequent activation of intracellular pathways. sVEGFR-1 lacks the seventh Ig-like domain, transmembrane sequence and cytoplasmatic TK domain, but binds the ligand VEGF with the same affinity as full length VEGFR-1, thereby probably acting as a physiological negative regulator. VEGF-A/VEGFR-2 signaling appears to be responsible for the most changes in endothelial cells, such as differentiation, proliferation, migration and sprouting, whereas involvement of VEGFR-1 in contrast might diminish some of these cellular responses [10].

Overwhelming evidence is present confirming the essential role of VEGF/VEGFR-signaling pathway in tumor development and growth, and the significance for prognosis and survival of cancer patients. Whereas angiogenesis occurs infrequently during adult life and most endothelial cells are quiescent, except during wound healing, inflammation, ovulation, pregnancy and ischemia, the endothelium in tumors is active, immature and proliferating. Therefore, disruption of the VEGF/VEGFR-pathway represents an attractive target for anti-cancer therapy. The approach of modifying the tumor environment by affecting endothelial and supporting cells includes furthermore the promise that, because these cells are in contrast to tumor cells genetically stable, they do not become resistant to the therapy. Proof of principle came from xenograft mouse models, in which treatment with a retrovirus encoding a dominant-interfering form of the VEGFR-2 was successful in inhibiting tumor growth and moreover induced a reduction in tumor size [11, 12].

However, VEGF is not the only factor in inducing the angiogenic switch and sustaining angiogenesis. Also FGF which *in vitro* has shown to act synergistically with VEGF in angiogenesis assays and PDGF which plays a role in the recruitment of pericytes and stabilization of vessels have a positive contribution in the angiogenic process [13–17]. Remodeling of the ECM, which also results in the release of VEGF, by MMPs can also activate the angiogenic switch, which has been shown for MMP-9 [18]. Alternatively, a decrease in concentration or lack of naturally occurring angiogenesis inhibitors, such as thrombospondin-1, platelet factor-4, angiostatin, and endostatin can change the balance between pro- and anti-angiogenic factors and activate the switch.

Consequently, several possibilities of anti-angiogenic treatment are conceivable and at present in pre- and clinical development. The most promising anti-angiogenic strategy seems to be the disruption of the VEGF/VEGF-receptor pathway with tyrosine kinase (TK) inhibitors or recombinant humanized monoclonal antibodies (rhuMAb). In addition, the combined inhibition of VEGF and other growth factors, such as FGF and PDGF, with TK inhibitors is practically possible and might result in a powerful anti-angiogenic treatment [19, 20]. Another potential anti-angiogenic treatment is the administration of endogenous anti-angiogenic proteins, such as endostatin and angiostatin. Compounds that interfere with remodeling of the ECM, such as MMP-inhibitors, have been developed and tested in clinical practice. Potentially interesting are compounds that block integrins, which are involved in adhesion and migration of both endothelial and tumor cells [21].

An important advantage that anti-angiogenesis agents offer is the possibility to combine them with classical chemotherapy. Whereas chemotherapy affects genetically instable tumor cells which emerges resistance, the angiogenesis inhibitor affects genetically stable endothelial cells. Besides their different way of action and non-overlapping toxicity patterns other theoretical advantages of combined treatments might be foreseen, which might result in synergistic or even additive effects. In tumors for instance a high interstitial fluid pressure might result in insufficient drug penetration, which can be improved by blocking of VEGF [22]. In preclinical models this phenomenon has been shown to occur in cases of inhibition of the PDGFR [23]. Anti-angiogenic compounds that affect the VEGF/VEGFR pathway and endogenous anti-angiogenesis proteins, which are in preclinical and clinical development, and the results of combination treatments with classical chemotherapy will be discussed in this chapter.

Disruption of the VEGF/VEGF-receptor pathway

Several strategies of interfering with the VEGF/VEGF-receptor pathway have been developed. Blocking of the intracellular tyrosine kinase domain of the VEGFR with specially designed small molecules (TK inhibitors) thereby preventing subsequent downstream signaling after activation of the receptor, represents an attractive strategy. Antibodies directed to VEGF have been developed of which bevacizumab is at present the most successful anti-angiogenic strategy in the clinic. Another way of targeting and inactivating the VEGFR is the administration of antibodies directed to the VEGFR. A strategy in development is called VEGF-Trap, which is the highest-affinity VEGF blocker described to date.

Tyrosine kinase inhibitors

The mechanism of action of all the TK inhibitors is more less the same and consists of binding in the vicinity of the ATP-binding site of their target tyro-

Table 1. Examples of TK inhibitors and their targets

Agent	Target(s)	Developmental status
SU5416	VEGFR-1 and -2, KIT	Discontinued
SU6668	VEGFR-2, PDGFRβ, FGFR1, KIT	Discontinued
SU11248	VEGFR-1, -2, PDGFRα/β, KIT, Flt-3	Phase II/III
SU14813	VEGFR-1, -2, PDGFRα/β, KIT, Flt-3	Phase I
PTK787/ZK 222584	VEGFR-2, PDGFRβ, KIT	Phase II
CP-547,632	VEGFR-2, FGFR1	Phase I/II
AG013736	VEGFR-2, PDGFRβ	Phase I
AZD2171	VEGFR-1–3	Phase I
AMG 706	VEGFR-1–3	Phase I
CEP-7055	VEGFR-1–3, Flt-3, Mlk1-3	Phase I
CEP-701	Flt-3, Trk kinases	Phase II
PKC-412	PKC, VEGFR-2, PDGFR, c-kit, Flt-3	Phase II
MLN-518	PDGFRβ, c-kit, Flt-3	Phase I
GW-572016	EGFR, Her2	Phase I
EKB-569	EGFR, Her2 (irreversible inhibitor)	Phase I
PKI-166	EGFR, Her2	Phase II
CI-1033	EGFR, Her2 (irreversible inhibitor)	Phase II
OSI-774/Erlotinib	EGFR	Phase II/III
ZD1839/Gefitinib	EGFR	Registered
STI571/Imatinib	Bcr/Abl, PDGFRβ, c-kit	Registered

sine kinase thereby preventing phosphorylation of tyrosine residues of the receptor and subsequent intracellular signaling. Because of the large homology of the TK-domains it is possible that one compound targets several receptors at the same time. Many TK inhibitors have been developed which all target one or more specific receptors (Tab. 1). One important group of TK inhibitors that will not be discussed in this chapter are those TK inhibitors which block the family of the epidermal growth factor receptors, thereby targeting tumor cells which express these receptors. Another famous TK inhibitor that falls beyond the scope of this chapter is Imatinib, which blocks the bcr-abl fusion protein, PDGFR and KIT (stem cell factor receptor). Imatinib is highly successful in the treatment of chronic myeloid leukemia (CML) and gastrointestinal stroma cell tumors (GIST). Its success is explained by the fact that it blocks the dominant driving force in CML and GIST which in CML is the tyrosine kinase activity of the bcr-abl fusion protein and in GIST gain of function mutations in KIT [24–26].

In this chapter we will only focus on specific anti-angiogenic TK inhibitors, which predominantly block the VEGFRs. Some compounds however have a broader range of activity and also block the PDGFRs, FGFRs, and/or KIT. These compounds furthermore differ in their affinity for the TK domain of the receptors and as a consequence might have totally different clinical effects.

These compounds predominantly target the micro-environment of tumors, but because some tumors also express the PDGFR, not only the tumor micro-environment but also the tumor cells might be directly affected by those compounds that also targets the PDGFR. Examples of TK inhibitors, which among other growth factor receptors target the VEGFR(s), in different stages of clinical development, are mentioned in Table 1. SU5416, an inhibitor of VEGFR-1, VEGFR-2, and KIT, is one of the first compounds with which clinical experience has been obtained. SU5416 has been investigated in Phase II and III clinical trials as a single agent and in combination with chemotherapy. Compared to other TK inhibitors, which are all oral compounds, the disadvantage of SU5416 is that it has to be administered intravenously twice weekly over at least 90 min to prevent severe headache. Furthermore prior administration of dexamethasone to prevent allergic reactions on the solvent Cremophor is necessary. Plasma levels of SU5416 at the maximum tolerated dose of 145 mg/m^2 were in the range which in preclinical models resulted in long lasting inhibition of the VEGFR-2 [27, 28]. Dose limiting toxicities consisted of projectile vomiting, nausea, headache, pheblitis, diarrhea and fatigue. In Phase II trials in patients with advanced renal cell carcinoma (RCC), soft tissue sarcoma (STS), and melanoma (M) treatment with SU5416 was well tolerated. However the efficacy was low with response rates, which consisted mainly of short-lived stable disease, of 21% for RCC and 19% for STS, while no responses at all were observed in the M group [29]. The low efficacy is remarkable in view of the substantial evidence that VEGF is definitively involved in the development and progress of these three tumor types and the efficacy of SU5416 in human melanoma and sarcoma xenograft models. Which mechanisms could explain this lack of efficacy? One possible explanation is the fact that large tumors have established mature vessels instead of immature VEGF-dependent vessels. The observation, however, that almost 50% of the patients developed new lesions during treatment indicates that these tumors were able to generate neovascularization despite blockage of the VEGF/VEGFR pathway with SU5416. Also insufficient bioavailability of SU5416 in human tumors and the short plasma half-life could be significant contributing factors. Other important explanations for lack of efficacy could be the presence of many different pro-angiogenic factors, such as FGF, PDGF, hepatocyte growth factor (HGF) and the redundancy of these pro-angiogenic factors [30]. In conclusion, inhibition of the VEGF/VEGFR pathway with SU5416 single agent seems not to be enough to inhibit angiogenesis and tumor growth entirely. Compounds which target multiple growth factors at the same time and combination treatment of those new compounds with classical chemotherapy or immunotherapy might hopefully result in larger clinical benefit.

A Phase II trial investigating the combination of SU5416 twice weekly plus α-interferon 1 million units subcutaneously twice daily in patients with advanced RCC however revealed that the efficacy was low with an 1 year event free survival of 6%, whereas toxicity was substantial in particular fatigue [31]. The feasibility and pharmacokinetics of the combination of cisplatin 80 mg/m^2

on day 1 and gemcitabine 1,250 mg/m^2 on days 1 and 8 every 3 weeks, in combination with SU5416 (85 and 145 mg/m^2) was investigated in a Phase I trial [32]. No significant pharmacological interaction between the three drugs was observed and most toxicities observed were those previously reported for SU5416 alone and for this chemotherapy regimen. Anti-tumor activity was similar to that expected in the patient population selected for this study. In the small heterogeneous patient population, no clear conclusion could be drawn on possible additive or synergistic effects between chemotherapy and SU5416. This Phase I study was however terminated because almost 50% of the patients experienced a thromboembolic event. Eight out of 19 patients developed nine events (three transient ischemic attacks, two cerebrovascular events, four deep venous thrombosis which in two was complicated by pulmonary emboli). As this exceeds the incidence observed with this type of chemotherapy alone and SU5416 alone, this is likely the result of the combination. Additional analysis of endothelial cell and coagulation parameters revealed that SU5416 alone had no influence on the coagulation cascade, but induced endothelial cell perturbation [33]. The combination treatment induced endothelial cell activation as well as activation of the coagulation cascade in cyclic pattern which was opposite to the change in platelet number. It was hypothesized that endothelial cells deprived of VEGF after exposure to SU5416 became activated and more susceptible to damage during treatment with cisplatin/gemcitabine, which was aggravated by a transient decrease in platelets, which are amongst others carriers of VEGF [34]. The unexpected high rate of serious adverse events presses for caution in trials investigating new compounds and combination treatments. Also a large Phase III trial comparing standard chemotherapy with standard chemotherapy plus SU5416 in patients with advanced colorectal cancer has been conducted [35]. The results of this trial have not been published yet, but no difference in efficacy was observed between both arms.

At present, preliminary experience has been obtained with compounds that target multiple growth factor receptors. Several Phase I trials exploring the optimal dosing regimen have been conducted with SU6668, an oral compound which targets the VEGFR-2, PDGFR-β, FGFR-1 and KIT [36–38]. Once daily doses up to 2,000 mg/m^2 in fasted conditions were well tolerated but did not reach steady state levels. Preclinical studies in dogs demonstrated that twice daily dosing resulted in increased steady state trough levels, and a nearly five-fold increase in oral bioavailability in fed compared to fasted animals. The maximum tolerated dose of SU6668 orally thrice daily dosing under fed conditions was 100 mg/m^2, which resulted in plasma concentrations of SU6668 in the range of 1 μg/ml, the level which in xenograft models was associated with inhibition of VEGFR phosphorylation [39]. The plasma level at which receptor phosphorylation in humans is inhibited by SU6668 is however unknown. The plasma levels of SU6668 decreased during treatment at all dose levels probably due to induction of metabolic liver enzymes. The dose limiting toxicities were unexpected and consisted of serositis-like pains, fatigue, and anorexia. Less severe grades of serositis-like pains as well as flu-like complaints were also frequently (respectively, 53% and 47%) observed. These adverse events in combination with an acute phase response mediated by IL-6 as assessed in serial blood samples is indicative for the induction of an inflammatory reaction. Possibly, SU6668 modifies inhibitory components of the inflammatory process. However, the exact mechanism is unclear and complex since SU6668 inhibits at least four different receptors. No anti-tumor activity of SU6668 was observed. In conclusion, SU6668 has an unfavorable pharmacological profile and probably a small therapeutic index.

Another compound in the same category is SU11248 which is also an oral compound but has a slightly different targeting profile. SU11248 blocks the VEGFR-1/2, PDGFRα/β, KIT and Flt-3 and has shown activity in several human xenograft models [40–42]. SU11248 is a potentially promising compound in the treatment of acute myeloid leukemia (AML), because activating mutations of Flt-3 occur in up to 30% of patients with AML. Inhibitory effects of SU11248 on the intracellular signaling cascade of leukemic cells after a single dose have been shown [43]. In a Phase I trial entering 28 patients with advanced solid tumors, the recommended Phase II dose has been defined as 50 mg/day for 28 days followed by 14 days rest [44]. Grade 3 fatigue and hypertension, reversible upon treatment discontinuation, were the dose limiting toxicities (DLTs). Pharmacokinetics showed good oral bioavailability with modest SU11248 intra/inter-patient variability. At higher doses, tumor responses were associated with reduced intratumoral vascularization and central tumor necrosis resulting in tumor perforation in one patient and fistula formation in another patient. In 6 out of 23 evaluable patients (renal cell carcinoma, neuro-endocrine tumors) tumor responses were observed. Remarkably, hair depigmentation has been noted with strikingly bands of depigmentation and pigmentation that correspond, respectively, to periods of treatment and dosing rest periods. This phenomenon was also observed in mice receiving SU11248, which demonstrate that hair pigmentation can serve as a biological readout for treatment with SU11248 [45]. SU11248 showed furthermore clinical activity in patients with objectively progressing Imatinib-resistant GIST [46]. In conclusion, evidence of activity and the manageable toxicity profile of SU11248 support further studies. A successor of SU11248 is SU14813, which is an inhibitor of the same growth factor receptors, but has possibly a better pharmacokinetic profile and therapeutic index. SU14813 is currently being investigated in a Phase I trial.

Monoclonal antibodies

At present the most famous anti-angiogenic treatment is the humanized monoclonal antibody bevacizumab (Avastin) which is directed to all VEGF isoforms. It is the first anti-angiogenic compound of which direct and rapid anti-vascular effects in human tumors has been demonstrated [47]. A single infusion of this compound in patients with rectal carcinoma induced a decrease of tumor perfusion, vascular volume, microvascular density, interstitial fluid pressure and the number of viable, circulating endothelial and progenitor cells, and an increase of the fraction of vessels with pericyte coverage. It is also the first anti-angiogenic compound of which clinical efficacy, both as single agent and in combination with chemotherapy, has been demonstrated.

In a randomized, double-blind Phase II trial comparing placebo *versus* a low dose (3 mg/kg) and a high dose (10 mg/kg) bevacizumab every 2 weeks in patients with advanced RCC time to disease progression and response rate were primary endpoints [48]. Although crossover from placebo to bevacizumab was allowed, survival was a secondary endpoint. A significant prolongation of the time to disease progression in the high-dose antibody group as compared with the placebo group (hazard ratio, 2.55; P < 0.001) was observed, which met the criteria for early stopping after the interim analysis of 116 patients and resulted in termination of the trial. A small difference in the time to disease progression of borderline significance, was observed in the low-dose antibody group as compared with the placebo group (hazard ratio, 1.26; P = 0.053). The probability of being progression-free for patients given high-dose antibody, low-dose-antibody, and placebo was 64%, 39%, and 20%, respectively, at 4 months and 30%, 14%, and 5% at 8 months. Response rates were low (10%) and occurred only in those patients receiving high-dose bevacizumab. There were two patients, one with a long lasting partial and one with a long lasting minor response, who relapsed after treatment discontinuation, but responded again upon retreatment. No significant differences in overall survival between the three groups were observed. The treatment was well tolerated with minimal toxic effects, predominantly hypertension and asymptomatic proteinuria. No increased incidences of thromboembolic events and/or bleeding complications were observed during treatment with bevacizumab as compared to placebo. In conclusion, bevacizumab can, although response rates were low, significantly prolong the time to progression of disease in patients with metastatic renal-cell cancer.

There are now two studies, one small randomized Phase II and a large randomized Phase III, which have shown a longer disease-free survival in patients with advanced colorectal cancer of standard chemotherapy plus bevacizumab *versus* standard chemotherapy alone [49, 50]. In the Phase II trial 104 patients were randomly assigned to the standard arm 5-fluorouracil/leucovorin (FU/LV) (500 mg/m^2 weekly for 6 weeks of each 8-week cycle), or FU/LV plus high dose bevacizumab (10 mg/kg every 2 weeks), or FU/LV plus low dose bevacizumab (5 mg/kg every 2 weeks). Although the three treatment arms were not equally balanced, which may have affected treatment outcome, both arms with bevacizumab resulted in higher response rates, longer median time to disease progression, and longer median survival as compared with FU/LV alone (Tab. 2). When the data for bevacizumab-treated patients were pooled, there was a 55% reduction in the hazard of progressing compared with the control arm (P = 0.003). Statistically significant more grade 3 and 4 adverse events were observed in the bevacizumab arms (P = 0.042), which might be confounded by patients longer on study in these arms. Bevacizumab therapy

Table 2. Response rate, progression free survival and overall survival of chemotherapy plus placebo or bevacizumab in metastatic colorectal cancer

Treatment	RR (ci)	PFS (ci)	OS (ci)
5-FU/LV	17% (7–34%)	5.2 months (3.5–5.6)	13.8 months (9.1–23.0)
5-FU/LV plus high dose bevacizumab (10 mg/kg)	24% (12–43%)	7.2 months (3.8–9.2)	16.1 months (11.0–20.7)
5-FU/LV plus low dose bevacizumab (5 mg/kg)	40% (24–58%)	9.0 months (5.8–10.9)	21.5 months (17.3–nd)
Irinotecan/5-FU/LV plus placebo	35% (na)	6.2 months (na)	15.6 months (na)
Irinotecan/5-FU/LV plus bevacizumab (5 mg/kg)	45% (na)	10.6 months (na)	20.3 months (na)

RR = response rate; PFS = progression free survival; OS = overall survival; ci = confidence interval

was associated, besides generally mild-to-moderate, headache, fever, rash, and chills, with an increased incidence of bleeding, thrombosis, hypertension, and proteinuria. Bleeding consisted mostly of transient epistaxis, but three patients in the high-dose arm had a grade 3 or 4 gastrointestinal hemorrhage. Thrombosis was the most significant adverse event with nine events (arterial as well as venous) in the low dose arm and four in the high dose arm, of which one was fatal. This incidence of 19% (13 of 67 patients) is clearly elevated compared to 9% (3 of 35 patients) in the control arm. Hypertension was reported in 19 patients, of which nine had a preexisting history of hypertension, and 16 required oral antihypertensive therapy. Adverse events known to be associated with FU/LV (diarrhea, leucopenia, and stomatitis) were not increased in incidence and severity when bevacizumab was added to the regimen.

In the Phase III study, over 800 patients were randomized to receive irinotecan/5FU/leucovorin (IFL, respectively 125 mg/m^2, 500 mg/m^2, 20 mg/m^2 given 4 of 6 weeks) plus placebo or IFL plus bevacizumab 5 mg/kg every 2 weeks. The primary efficacy endpoint was survival; secondary efficacy endpoints included progression free survival (PFS), objective response rate (ORR), duration of response and quality of life. The bevacizumab arm resulted in significantly longer median survival, PFS, and ORR (Tab. 2). The duration of response was also significantly longer in the bevacizumab arm with a gain of 3.3 months (P = 0.0014). More or less the same toxicities were observed in this trial as compared to the Phase II trial. Almost 20% of the patients in the bevacizumab arm experienced thromboembolic events, which was however not statistically different as compared to the control arm since a remarkably high incidence of 16.2% occurred in this arm. Grade III hypertension, which was easily manageable with oral medications, was clearly increased in the bevacizumab arm with an incidence of 10.9%. Grade III proteinuria and grade III/IV bleeding occurred in both arms equally, respectively about 1 and 3%.

Bevacizumab added to standard chemotherapy regimens has also been investigated in other tumor types. Patients with advanced non-small cell lung cancer (NSCLC) were in a Phase II trial randomly assigned to bevacizumab 7.5 mg/kg or 15 mg/kg plus carboplatin (area under the curve 6) and paclitaxel (200 mg/m^2) every 3 weeks or carboplatin and paclitaxel alone [51]. Combination treatment with high dose bevacizumab resulted in a higher response rate (31.5% *versus* 18.8%), longer median time to progression (7.4 *versus* 4.2 months) and a modest increase in survival (17.7 *versus* 14.9 months) compared to the control arm. There was no difference between the control arm and the low dose bevacizuamb arm. A major concern in this trial was sudden and life-threatening hemoptysis in 6 (4 fatal) out of 67 patients. This severe hemoptysis was associated with squamous cell histology, tumor necrosis and cavitation, and disease location close to major blood vessels.

Although it is thought that angiogenesis is needed for the growth of all tumor types and that inhibition of angiogenesis should work irrespectively of the tumor type, there seems to be a difference in efficacy per tumor type. Bevacizumab was evaluated in a Phase III trial comparing capecitabine (CAP) alone to CAP plus BV in 462 patients with metastatic breast cancer (MBC) who had previously been treated with both an anthracycline and a taxane. Although there was a statistically significant increase in objective response rate (9.1% *versus* 19.8%), no improvement in progression-free survival was observed [52]. A possible explanation for the difference in efficacy could be the fact that these breast cancer patients were heavily pretreated as compared to the colorectal cancer patients.

Other monoclonal antibodies have been developed and one of them is IMC-1C11, which is directed to VEGFR-2 and has been investigated in a Phase I trial [53]. The advantage of targeting VEGFR-2 might be the selectivity because of the fact that this receptor is primarily found on activated endothelial cells. The treatment with IMC-1C11 was well tolerated over a dose range of 0.2–4 mg/kg weekly with an obvious dose dependent pharmacokinetics. The dose levels of 2 and 4 mg/kg resulted in plasma levels above 5 μg/ml, the concentration which prevented VEGFR-2 phosphorylation *in vitro*. None of the patients had objective tumor regression. A possible disadvantage of IMC-1C11 was its immunogenicity since 50% of the patients developed a human antibody to chimeric antibody response (HACA). Whether HACAs affect the efficacy of this compound remains to be established. Other monoclonal antibodies directed to VEGFR-2 are IMC-2C6 and IMC-1121. They seem to have a higher affinity for VEGFR-2 and are in preclinical development [54].

VEGF-Trap

The highest-affinity VEGF blocker described to date is VEGF-Trap, a composite decoy receptor based on VEGFR-1 and VEGFR-2 fused to an Fc segment of IgG1. This VEGF-Trap abolishes mature, preexisting vasculature in established xenografts, resulting in stunted and almost completely avascular tumors subsequently followed by marked tumor regression and suppressed tumor growth [55, 56]. Compared to two other anti-VEGF agents (an anti-human VEGF(165) RNA-based fluoropyrimidine aptamer and a monoclonal anti-human VEGF antibody) high dose VEGF-Trap caused the greatest inhibition of tumor growth in a neuroblastoma xenograft model [57]. In this model persistence of co-option of host vasculature might represent a novel mechanism by which neuroblastoma can partly evade anti-angiogenic therapy. More effective VEGF blockade, as achieved by VEGF-Trap, can lead to regression of co-opted vascular structures. However, results of VEGF-Trap in the clinic have to be awaited, but may be superior to that achieved by other agents, such as monoclonal antibodies targeted against VEGF or the VEGF receptor.

Endogenous angiogenesis inhibitors

Endostatin

Revolutionary results have been obtained with endostatin in preclinical models [58]. Repeated treatment with endostatin in mice bearing Lewis lung carcinoma, T241 fibrosarcoma or B16F10 melanoma, which were allowed to regrow after discontinuation of endostatin treatment, resulted in repeated disappearance of the tumor lesions and even in prolonged tumor dormancy without further therapy after 6, 4 or 2 treatment cycles, respectively. The mechanism of action of endostatin, a carboxy-terminal fragment of collagen type XVIII, is still not completely elucidated. Endostatin probably affects endothelial cell migration by interfering with integrins, especially $\alpha V\beta 1$ [59, 60]. Via a heparan sulfate proteoglycan-dependent mechanism endostatin recruits $\alpha V\beta 1$ integrin into lipid rafts and subsequently induces Src-dependent activation of p190RhoGAP with concomitant decrease in RhoA activity resulting in disassembly of actin stress fibers and focal adhesions [61]. Another proposed mechanism of action is the interaction of endostatin with tropomyosin resulting in disruption of microfilament integrity leading to inhibition of cell motility, induction of apoptosis, and ultimately inhibition of tumor growth [62].

The revolutionary results obtained in the preclinical models however have not yet been followed in the clinic. The main conclusion of three different Phase I trials is that endostatin is generally well tolerated and can be administered safely in doses ranging from 15 to 600 mg/m^2 [63–65]. Although in two of the three trials, recombinant endostatin plasma levels achieved area under the concentration-time curves associated with activity in preclinical models, no objective tumor responses were observed. In one of these trials the results of additional quantitative analysis of biomarkers showed significant increases in endothelial cell death and decreases in tumor microvessel density with maximal effects of endostatin at a dose of $\approx$250 mg/m^2 [66]. Tumor cell death however was uniformly low and did not correlate with endostatin dose. These data suggest that endostatin might have a bell-shaped biological activity dose response curve and that endostatin single agent fails to induce tumor regression due to its lack of induction of tumor cell death.

Angiostatin

A fragment of plasminogen, called angiostatin and containing 3–4 N-terminal kringle domains, is another endogenous potent angiogenesis inhibitor. Human recombinant angiostatin induced an almost complete inhibition of tumor growth without detectable toxicity or resistance in preclinical models [67]. Although the exact anti-angiogenic mechanism of action of angiostatin is unclear, currently three different possible mechanisms have been elucidated. Angiostatin, which is a ligand for $\alpha V\beta 3$-integrin, prevents the binding of plasmin to this integrin, thereby inhibiting plasmin-induced migration of endothelial cells [68]. Furthermore, it has been shown that angiostatin binds to annexin II, a protein that acts as a regulator of cell surface plasmin generation, and that impaired endothelial cell fibrinolytic activity constitutes a barrier to effective neoangiogenesis [69, 70]. Angiostatin seems also to antagonize the effects of VEGF-A by inducing apoptosis via the modulation of two distinct signaling pathways, one involving p53 and the other the Fas-mediated apoptotic pathway [71].

A clinical Phase I study investigating recombinant human angiostatin administered twice daily by s.c. injection in 24 patients has been performed [72]. Three groups of 8 patients received 7.5, 15, or 30 mg/m^2/day divided in two s.c. injections for 28 consecutive days followed by a 7-day washout period. Treatment was continued in absence of toxicity or a 100% increase in tumor size. Pharmacokinetics showed a linear relation between dose and area under the curve and Cmax, which were at all three dose levels within the range of drug exposure that has biological activity in preclinical models. Treatment was well tolerated with erythema at injection sites being the most frequent side effect. Serious adverse events with an uncertain relationship to the study drug were hemorrhage in brain metastases in two patients and deep venous thrombosis in two other patients. No objective responses were observed. Although long-term (>6 months) stable disease (<25% growth of measurable uni- or bidimensional tumor size) was observed in 6 of 24 patients, the design of the study does not allow to draw conclusions on antitumor effects of angostatin.

Tumstatin

Several lines of evidence suggests that tumstatin, a cleavage fragment of the $\alpha 3$ chain of type IV collagen, which is present in the circulation plays an important role in pathological angiogenesis [59, 73]. Deletion of the $\alpha 3$ chain of collagen type IV in mice induces accelerated tumor growth associated with enhanced pathological angiogenesis, while angiogenesis associated with development and tissue repair remain unaffected. Administration of recombinant tumstatin to a normal physiological concentration to these collagen type IV-α3-deficient mice abolishes the increased rate of tumor growth. Matrix metalloproteinase-9 (MMP-9) seems to be essential for the release of tumstatin, since mice deficient in MMP-9 have decreased levels of circulating tumstatin and exhibit accelerated tumor growth. The effects of tumstatin on pathological angiogenesis are probably explained by binding of tumstatin to αVβ3 integrin expressed on pathological angiogenic blood vessels. At present only preclinical data regarding tumstatin are available and clinical efficacy have to be awaited.

Summary/conclusion

With the introduction of anti-angiogenic agents in the clinic a new era in the treatment of cancer has begun. Although this new class of compounds was introduced very recently in the clinic one compound (bevacizumab) has already become a standard part of first line treatment of colorectal cancer. The results obtained with bevacizumab demonstrate that inhibition of angiogenesis via blocking the effects of VEGF is not only theory but works in the clinic. This observation is exciting and promises a lot for the future regarding all potent compounds in pre- and early clinical development. However a lot of questions remain to be answered. First of all is there a difference between monoclonal antibodies and TK inhibitors in interfering with VEGF signaling. Thus far the monoclonal antibody is more successful in the clinic and seems to have a slightly different toxicity pattern compared to TK inhibitors. Another question is how to explain the disappointing clinical effects of the administered recombinant endogenous angiogenesis inhibitors endostatin and angiostatin, whereas both were very successful in preclinical experiments. It is very important to learn from failures of potentially interesting compounds in the clinic. As mentioned above, these compounds offer the opportunity to combine them with classical chemotherapy and moreover with other biological compounds, such as for instance EGFR inhibitors. The number of conceivable combinations however is without number. It is therefore necessary to develop preclinical and clinical tools to predict which combination could have efficacy in the clinic. This issue encloses also the need to identify which patient would have benefit from what kind of treatment. Although experience with specific characteristics and side effects of anti-angiogenic agents is rapidly growing, unexpected side effects of these compounds alone and in combination may occur frequently. Especially the TK inhibitors with there broad range of activity on several known, but maybe also unknown receptors, may have unexpected effects in patients. Furthermore it is not exactly known how and which biological networks are being changed and modulated by those multiple receptor targeting compounds. However, one conclusion to be drawn for certain is that the treatment of cancer will change dramatically and rapidly in the near future.

References

1 Hanahan D, Folkman J (1996) Patterns and emerging mechanisms of the angiogenic switch during tumorigenesis. *Cell* 86: 353–364
2 Hanahan D, Weinberg RA (2000) The hallmarks of cancer. *Cell* 100: 57–70
3 Carmeliet P, Dor Y, Herbert JM, Fukumura D, Brusselmans K, Dewerchin M, Neeman M, Bono F, Abramovitch R, Maxwell P et al. (1998) Role of HIF-1alpha in hypoxia-mediated apoptosis, cell proliferation and tumour angiogenesis. *Nature* 394: 485–490
4 Rak J, Yu JL, Klement G, Kerbel RS (2000) Oncogenes and angiogenesis: signaling three-dimensional tumor growth. *J Investig Dermatol Symp Proc* 5: 24–33
5 Rak J, Filmus J, Finkenzeller G, Grugel S, Marme D, Kerbel RS (1995) Oncogenes as inducers of tumor angiogenesis. *Cancer Metastasis Rev* 14: 263–277
6 Blancher C, Moore JW, Robertson N, Harris AL (2001) Effects of ras and von Hippel-Lindau (VHL) gene mutations on hypoxia-inducible factor (HIF)-1alpha, HIF-2alpha, and vascular endothelial growth factor expression and their regulation by the phosphatidylinositol 3'-kinase/Akt signaling pathway. *Cancer Res* 61: 7349–7355
7 Neufeld G, Cohen T, Gengrinovitch S, Poltorak Z (1999) Vascular endothelial growth factor (VEGF) and its receptors. *FASEB J* 13: 9–22
8 Robinson CJ, Stringer SE (2001) The splice variants of vascular endothelial growth factor (VEGF) and their receptors. *J Cell Sci* 114: 853–865
9 Veikkola T, Karkkainen M, Claesson-Welsh L, Alitalo K (2000) Regulation of angiogenesis via vascular endothelial growth factor receptors. *Cancer Res* 60: 203–212
10 Waltenberger J, Claesson-Welsh L, Siegbahn A, Shibuya M, Heldin CH (1994) Different signal transduction properties of KDR and Flt1, two receptors for vascular endothelial growth factor. *J Biol Chem* 269: 26988–26995
11 Kim KJ, Li B, Winer J, Armanini M, Gillett N, Phillips HS, Ferrara N (1993) Inhibition of vascular endothelial growth factor-induced angiogenesis suppresses tumour growth *in vivo*. *Nature* 362: 841–844
12 Millauer B, Shawver LK, Plate KH, Risau W, Ullrich A (1994) Glioblastoma growth inhibited *in vivo* by a dominant-negative Flk-1 mutant. *Nature* 367: 576–579
13 Gerwins P, Skoldenberg E, Claesson-Welsh L (2000) Function of fibroblast growth factors and vascular endothelial growth factors and their receptors in angiogenesis. *Crit Rev Oncol Hematol* 34: 185–194
14 Pepper MS, Mandriota SJ, Jeltsch M, Kumar V, Alitalo K (1998) Vascular endothelial growth factor (VEGF)-C synergizes with basic fibroblast growth factor and VEGF in the induction of angiogenesis *in vitro* and alters endothelial cell extracellular proteolytic activity. *J Cell Physiol* 177: 439–452
15 Goto F, Goto K, Weindel K, Folkman J (1993) Synergistic effects of vascular endothelial growth factor and basic fibroblast growth factor on the proliferation and cord formation of bovine capillary endothelial cells within collagen gels. *Lab Invest* 69: 508–517
16 Asahara T, Bauters C, Zheng LP, Takeshita S, Bunting S, Ferrara N, Symes JF, Isner JM (1995) Synergistic effect of vascular endothelial growth factor and basic fibroblast growth factor on angiogenesis *in vivo*. *Circulation* 92: II365–II371
17 Heldin CH, Westermark B (1999) Mechanism of action and *in vivo* role of platelet-derived growth factor. *Physiol Rev* 79: 1283–1316
18 Bergers G, Brekken R, McMahon G, Vu TH, Itoh T, Tamaki K, Tanzawa K, Thorpe P, Itohara S, Werb Z et al. (2000) Matrix metalloproteinase-9 triggers the angiogenic switch during carcinogenesis. *Nat Cell Biol* 2: 737–744
19 Bergers G, Song S, Meyer-Morse N, Bergsland E, Hanahan D (2003) Benefits of targeting both pericytes and endothelial cells in the tumor vasculature with kinase inhibitors. *J Clin Invest* 111: 1287–1295
20 Laird AD, Cherrington JM (2003) Small molecule tyrosine kinase inhibitors: clinical development of anticancer agents. *Expert Opin Investig Drugs* 12: 51–64
21 Westlin WF (2001) Integrins as targets of angiogenesis inhibition. *Cancer J* 7 Suppl 3: S139–S143
22 Jain RK (1994) Barriers to drug delivery in solid tumors. *Sci Am* 271: 58–65
23 Pietras K, Rubin K, Sjoblom T, Buchdunger E, Sjoquist M, Heldin CH, Ostman A (2002) Inhibition of PDGF receptor signaling in tumor stroma enhances antitumor effect of chemotherapy. *Cancer Res* 62: 5476–5484

24 Demetri GD, von Mehren M, Blanke CD, Van den Abbeele AD, Eisenberg B, Roberts PJ, Heinrich MC, Tuveson DA, Singer S, Janicek M et al. (2002) Efficacy and safety of imatinib mesylate in advanced gastrointestinal stromal tumors. *N Engl J Med* 347: 472–480

25 Druker BJ, Talpaz M, Resta DJ, Peng B, Buchdunger E, Ford JM, Lydon NB, Kantarjian H, Capdeville R, Ohno-Jones S et al. (2001) Efficacy and safety of a specific inhibitor of the BCR-ABL tyrosine kinase in chronic myeloid leukemia. *N Engl J Med* 344: 1031–1037

26 O'Brien SG, Guilhot F, Larson RA, Gathmann I, Baccarani M, Cervantes F, Cornelissen JJ, Fischer T, Hochhaus A, Hughes T et al. (2003) Imatinib compared with interferon and low-dose cytarabine for newly diagnosed chronic-phase chronic myeloid leukemia. *N Engl J Med* 348: 994–1004

27 Mendel DB, Schreck RE, West DC, Li G, Strawn LM, Tanciongco SS, Vasile S, Shawver LK, Cherrington JM (2000) The angiogenesis inhibitor SU5416 has long-lasting effects on vascular endothelial growth factor receptor phosphorylation and function. *Clin Cancer Res* 6: 4848–4858

28 Stopeck A, Sheldon M, Vahedian M, Cropp G, Gosalia R, Hannah A (2002) Results of a Phase I dose-escalating study of the antiangiogenic agent, SU5416, in patients with advanced malignancies. *Clin Cancer Res* 8: 2798–2805

29 Kuenen BC, Tabernero J, Baselga J, Cavalli F, Pfanner E, Conte PF, Seeber S, Madhusudan S, Deplanque G, Huisman H et al. (2003) Efficacy and toxicity of the angiogenesis inhibitor SU5416 as a single agent in patients with advanced renal cell carcinoma, melanoma, and soft tissue sarcoma. *Clin Cancer Res* 9: 1648–1655

30 Sengupta S, Gherardi E, Sellers LA, Wood JM, Sasisekharan R, Fan TP (2003) Hepatocyte growth factor/scatter factor can induce angiogenesis independently of vascular endothelial growth factor. *Arterioscler Thromb Vasc Biol* 23: 69–75

31 Lara PN Jr, Quinn DI, Margolin K, Meyers FJ, Longmate J, Frankel P, Mack PC, Turrell C, Valk P, Rao J et al. (2003) SU5416 plus interferon alpha in advanced renal cell carcinoma: a Phase II California Cancer Consortium study with biological and imaging correlates of angiogenesis inhibition. *Clin Cancer Res* 9: 4772–4781

32 Kuenen BC, Rosen L, Smit EF, Parson MR, Levi M, Ruijter R, Huisman H, Kedde MA, Noordhuis P, van der Vijgh WJ et al. (2002) Dose-finding and pharmacokinetic study of cisplatin, gemcitabine, and SU5416 in patients with solid tumors. *J Clin Oncol* 20: 1657–1667

33 Kuenen BC, Levi M, Meijers JC, Kakkar AK, van Hinsbergh VW, Kostense PJ, Pinedo HM, Hoekman K (2002) Analysis of coagulation cascade and endothelial cell activation during inhibition of vascular endothelial growth factor/vascular endothelial growth factor receptor pathway in cancer patients. *Arterioscler Thromb Vasc Biol* 22: 1500–1505

34 Kuenen BC, Levi M, Meijers JC, van Hinsbergh VW, Berkhof J, Kakkar AK, Hoekman K, Pinedo HM (2003) Potential role of platelets in endothelial damage observed during treatment with cisplatin, gemcitabine, and the angiogenesis inhibitor SU5416. *J Clin Oncol* 21: 2192–2198

35 DePrimo SE, Wong LM, Khatry DB, Nicholas SL, Manning WC, Smolich BD, O'Farrell AM, Cherrington JM (2003) Expression profiling of blood samples from an SU5416 Phase III metastatic colorectal cancer clinical trial: a novel strategy for biomarker identification. *BMC Cancer* 3: 3

36 Laird AD, Vajkoczy P, Shawver LK, Thurnher A, Liang C, Mohammadi M, Schlessinger J, Ullrich A, Hubbard SR, Blake RA et al. (2000) SU6668 is a potent antiangiogenic and antitumor agent that induces regression of established tumors. *Cancer Res* 60: 4152–4160

37 Laird AD, Christensen JG, Li G, Carver J, Smith K, Xin X, Moss KG, Louie SG, Mendel DB, Cherrington JM (2002) SU6668 inhibits Flk-1/KDR and PDGFRbeta *in vivo*, resulting in rapid apoptosis of tumor vasculature and tumor regression in mice. *FASEB J* 16: 681–690

38 Hoekman K (2001) SU6668, a multitargeted angiogenesis inhibitor. *Cancer J* 7 Suppl 3: S134–S138

39 Kuenen BC, Giaccone G, Ruijter R, Kok A, Schalkwijk C, Hoekman K, Pinedo HM (2005) Dose finding study of the multi-targeted tyrosine kinase inhibitor SU6668 in patients with advanced malignancies. *Clin Cancer Res* 11: 6240–6246

40 Abrams TJ, Murray LJ, Pesenti E, Holway VW, Colombo T, Lee LB, Cherrington JM, Pryer NK (2003) Preclinical evaluation of the tyrosine kinase inhibitor SU11248 as a single agent and in combination with 'standard of care' therapeutic agents for the treatment of breast cancer. *Mol Cancer Ther* 2: 1011–1021

41 Abrams TJ, Lee LB, Murray LJ, Pryer NK, Cherrington JM (2003) SU11248 inhibits KIT and platelet-derived growth factor receptor beta in preclinical models of human small cell lung cancer. *Mol Cancer Ther* 2: 471–478

42 Mendel DB, Laird AD, Xin X, Louie SG, Christensen JG, Li G, Schreck RE, Abrams TJ, Ngai TJ, Lee LB et al. (2003) *In vivo* antitumor activity of SU11248, a novel tyrosine kinase inhibitor targeting vascular endothelial growth factor and platelet-derived growth factor receptors: determination of a pharmacokinetic/pharmacodynamic relationship. *Clin Cancer Res* 9: 327–337

43 O'Farrell AM, Foran JM, Fiedler W, Serve H, Paquette RL, Cooper MA, Yuen HA, Louie SG, Kim H, Nicholas S et al. (2003) An innovative Phase I clinical study demonstrates inhibition of FLT3 phosphorylation by SU11248 in acute myeloid leukemia patients. *Clin Cancer Res* 9: 5465–5476

44 Raymond E, Faivre S, Vera K, Delbaldo C, Robert C, Spatz A, Bello C, Brega N, Scigalla P, Armand JP (2003) Final results of a Phase I and pharmacokinetic study of SU11248, a novel multi-target tyrosine kinase inhibitor, in patients with advanced cancers. *Proc Am Soc Clin Oncol* 22: 192, abstract 796

45 Moss KG, Toner GC, Cherrington JM, Mendel DB, Laird AD (2003) Hair depigmentation is a biological readout for pharmacological inhibition of KIT in mice and humans. *J Pharmacol Exp Ther* 307: 476–480

46 Demetri GD, George S, Heinrich MC, Fletcher JA, Fletcher CDM, Desai J, Cohen DP, Scigalla P, Cherrington JM, Van den Abbeele AD (2003) Clinical activity and tolerability of the multi-targeted tyrosine kinase inhibitor SU11248 in patients (pts) with metastatic gastrointestinal stromal tumor (GIST) refractory to imatinib mesylate. *Proc Am Soc Clin Oncol* 22: 814, abstract 3273

47 Willett CG, Boucher Y, di Tomaso E, Duda DG, Munn LL, Tong RT, Chung DC, Sahani DV, Kalva SP, Kozin SV et al. (2004) Direct evidence that the VEGF-specific antibody bevacizumab has antivascular effects in human rectal cancer. *Nat Med* 10: 145–147

48 Yang JC, Haworth L, Sherry RM, Hwu P, Schwartzentruber DJ, Topalian SL, Steinberg SM, Chen HX, Rosenberg SA (2003) A randomized trial of bevacizumab, an anti-vascular endothelial growth factor antibody, for metastatic renal cancer. *N Engl J Med* 349: 427–434

49 Hurwitz H, Fehrenbacher L, Novotny W, Cartwright T, Hainsworth J, Heim W, Berlin J, Baron A, Griffing S, Holmgren E et al. (2004) Bevacizumab plus irinotecan, fluorouracil, and leucovorin for metastatic colorectal cancer. *N Engl J Med* 350: 2335–2342

50 Kabbinavar F, Hurwitz HI, Fehrenbacher L, Meropol NJ, Novotny WF, Lieberman G, Griffing S, Bergsland E (2003) Phase II, randomized trial comparing bevacizumab plus fluorouracil (FU)/leucovorin (LV) with FU/LV alone in patients with metastatic colorectal cancer. *J Clin Oncol* 21: 60–65

51 Johnson DH, Fehrenbacher L, Novotny WF, Herbst RS, Nemunaitis JJ, Jablons DM, Langer CJ, DeVore RF, III, Gaudreault J, Damico LA et al. (2004) Randomized phase ii trial comparing bevacizumab plus carboplatin and paclitaxel with carboplatin and paclitaxel alone in previously untreated locally advanced or metastatic non-small-cell lung cancer. *J Clin Oncol* 22: 2184–2191

52 Hillan KJ, Koeppen HKW, Tobin P, Pham T, Landon TH, Miller KD, Holmes FA, Cobleigh MA, Reimann JD, Langmuir VK (2003) The role of VEGF expression in response to bevacizumab plus capecitabine in metastatic breast cancer (MBC). *Proc Am Soc Clin Oncol* 22: 191, abstract 766

53 Posey JA, Ng TC, Yang B, Khazaeli MB, Carpenter MD, Fox F, Needle M, Waksal H, LoBuglio AF (2003) A Phase I study of anti-kinase insert domain-containing receptor antibody, IMC-1C11, in patients with liver metastases from colorectal carcinoma. *Clin Cancer Res* 9: 1323–1332

54 Zhu Z, Hattori K, Zhang H, Jimenez X, Ludwig DL, Dias S, Kussie P, Koo H, Kim HJ, Lu D et al. (2003) Inhibition of human leukemia in an animal model with human antibodies directed against vascular endothelial growth factor receptor 2. Correlation between antibody affinity and biological activity. *Leukemia* 17: 604–611

55 Holash J, Davis S, Papadopoulos N, Croll SD, Ho L, Russell M, Boland P, Leidich R, Hylton D, Burova E (2002) VEGF-Trap: a VEGF blocker with potent antitumor effects. *Proc Natl Acad Sci USA* 99: 11393–11398

56 Huang J, Frischer JS, Serur A, Kadenhe A, Yokoi A, McCrudden KW, New T, O'Toole K, Zabski S, Rudge JS et al. (2003) Regression of established tumors and metastases by potent vascular endothelial growth factor blockade. *Proc Natl Acad Sci USA* 100: 7785–7790

57 Kim ES, Serur A, Huang J, Manley CA, McCrudden KW, Frischer JS, Soffer SZ, Ring L, New T, Zabski S et al. (2002) Potent VEGF blockade causes regression of coopted vessels in a model of neuroblastoma. *Proc Natl Acad Sci USA* 99: 11399–11404

58 Boehm T, Folkman J, Browder T, O'Reilly MS (1997) Antiangiogenic therapy of experimental cancer does not induce acquired drug resistance. *Nature* 390: 404–407

59 Sudhakar A, Sugimoto H, Yang C, Lively J, Zeisberg M, Kalluri R (2003) Human tumstatin and human endostatin exhibit distinct antiangiogenic activities mediated by alpha v beta 3 and alpha

5 beta 1 integrins. *Proc Natl Acad Sci USA* 100: 4766–4771

60 Rehn M, Veikkola T, Kukk-Valdre E, Nakamura H, Ilmonen M, Lombardo C, Pihlajaniemi T, Alitalo K, Vuori K (2001) Interaction of endostatin with integrins implicated in angiogenesis. *Proc Natl Acad Sci USA* 98: 1024–1029

61 Wickstrom SA, Alitalo K, Keski-Oja J (2003) Endostatin associates with lipid rafts and induces reorganization of the actin cytoskeleton via down-regulation of RhoA activity. *J Biol Chem* 278: 37895–37901

62 MacDonald NJ, Shivers WY, Narum DL, Plum SM, Wingard JN, Fuhrmann SR, Liang H, Holland-Linn J, Chen DH, Sim BK (2001) Endostatin binds tropomyosin. A potential modulator of the antitumor activity of endostatin. *J Biol Chem* 276: 25190–25196

63 Eder JP Jr, Supko JG, Clark JW, Puchalski TA, Garcia-Carbonero R, Ryan DP, Shulman LN, Proper J, Kirvan M et al. (2002) Phase I clinical trial of recombinant human endostatin administered as a short intravenous infusion repeated daily. *J Clin Oncol* 20: 3772–3784

64 Herbst RS, Hess KR, Tran HT, Tseng JE, Mullani NA, Charnsangavej C, Madden T, Davis DW, McConkey DJ, O'Reilly MS et al. (2002) Phase I study of recombinant human endostatin in patients with advanced solid tumors. *J Clin Oncol* 20: 3792–3803

65 Thomas JP, Arzoomanian RZ, Alberti D, Marnocha R, Lee F, Friedl A, Tutsch K, Dresen A, Geiger P, Pluda J et al. (2003) Phase I pharmacokinetic and pharmacodynamic study of recombinant human endostatin in patients with advanced solid tumors. *J Clin Oncol* 21: 223–231

66 Davis DW, Shen Y, Mullani NA, Wen S, Herbst RS, O'Reilly M, Abbruzzese JL, McConkey DJ (2004) Quantitative analysis of biomarkers defines an optimal biological dose for recombinant human endostatin in primary human tumors. *Clin Cancer Res* 10: 33–42

67 O'Reilly MS, Holmgren L, Chen C, Folkman J (1996) Angiostatin induces and sustains dormancy of human primary tumors in mice. *Nat Med* 2: 689–692

68 Tarui T, Majumdar M, Miles LA, Ruf W, Takada Y (2002) Plasmin-induced migration of endothelial cells. A potential target for the anti-angiogenic action of angiostatin. *J Biol Chem* 277: 33564–33570

69 Tuszynski GP, Sharma MR, Rothman VL, Sharma MC (2002) Angiostatin binds to tyrosine kinase substrate annexin II through the lysine-binding domain in endothelial cells. *Microvasc Res* 64: 448–462

70 Ling Q, Jacovina AT, Deora A, Febbraio M, Simantov R, Silverstein RL, Hempstead B, Mark WH, Hajjar KA (2004) Annexin II regulates fibrin homeostasis and neoangiogenesis *in vivo. J Clin Invest* 113: 38–48

71 Chen YH, Wu HL, Chen CK, Huang YH, Yang BC, Wu LW (2003) Angiostatin antagonizes the action of VEGF-A in human endothelial cells via two distinct pathways. *Biochem Biophys Res Commun* 310: 804–810

72 Beerepoot LV, Witteveen EO, Groenewegen G, Fogler WE, Sim BKL, Sidor C, Zonnenberg BA, Schramel F, Gebbink MFBG, Voest EE (2003) Recombinant human angiostatin by twice-daily subcutaneous injection in advanced cancer: A pharmacokinetic and long-term safety study. *Clin Cancer Res* 9: 4025–4033

73 Hamano Y, Zeisberg M, Sugimoto H, Lively JC, Maeshima Y, Yang C, Hynes RO, Werb Z, Sudhakar A, Kalluri R (2003) Physiological levels of tumstatin, a fragment of collagen IV alpha3 chain, are generated by MMP-9 proteolysis and suppress angiogenesis via alphaV beta3 integrin. *Cancer Cell* 3: 589–601

Drugs Affecting Growth of Tumours
Edited by Herbert M. Pinedo and Carolien H. Smorenburg
© 2006 Birkhäuser Verlag/Switzerland

Signal transduction inhibitors

Ferry A.L.M Eskens

Erasmus University Medical Center Rotterdam, Department of Medical Oncology, PO Box 2040, 3000 CA Rotterdam, The Netherlands

Introduction

As a result of significant advances in fundamental research over the past 20 years, the process of signal transduction pathways involving receptor tyrosine kinases is now recognised to play a key role in the regulation of several physiological processes such as cell cycle, metabolism, growth, differentiation and proliferation. In addition, and more recently acknowledged, abnormal activation of these signal transduction pathways may lead to increased and uncontrolled cellular proliferation and a decrease in apoptosis, thus playing an important role in the development and growth of many human epithelial tumours.

While the epidermal growth factor receptor (EGFR) family is among the most important group of receptor tyrosine kinases and is functionally active in physiological and pathophysiological conditions, other receptor tyrosine kinases such as the BCR-Abl fusion protein and the c-Kit receptor play important roles in the development and growth of acute myeloid leukaemia and gastrointestinal stromal tumours (GIST), respectively.

In this chapter, we describe the physiology and pathophysiology of receptor tyrosine kinase activity. As abnormal tyrosine kinase activity has become an important target for the development of a completely new group of targeted anticancer agents, we will describe the various approaches that have been developed, and will summarise some important results that have been achieved in clinical studies.

During preclinical and early clinical development of specific tyrosine kinase inhibitors cell growth inhibition was most frequently observed in *in vitro* models. Dose-dependent tumour growth inhibition was also seen in tumour xenograft models for most of these agents, with only sporadic cases of tumour regressions. In these models, prolonged or even continuous administration, often feasible without dose-limiting toxicity, was necessary to obtain optimal growth inhibition. Therefore, in designing early clinical studies with these agents, it must be considered that endpoints used in studies with so-called classical cytotoxic agents will probably not be suitable, and new endpoints will have to be defined in order to be able to accurately assess antitumour activity. As this has major consequences for the design of these studies, these issues will be discussed briefly.

Physiology and pathophysiology of tyrosine kinase activity

The ErbB family, including EGFR, was the first family of receptor tyrosine kinases to be identified. The ErbB family consists of four closely related members; erbB-1/EGFR/HER1, ErbB-2/HER2, ErbB-3/HER3 and ErbB-4/HER4. These receptors share a common structure that consists of an extracellular ligand-binding domain, a single transmembrane helix that anchors the receptor to the cell, and a cytoplasmic domain that consists of a protein kinase domain and a tightly attached regulatory carboxyl terminal segment or domain that can be phosphorylated. The number of ligands that can activate a specific erbB receptor is variable (Fig. 1). Whereas the erbB-1 receptor has various different high-affinity ligands such as EGF, amphiregulin and transforming growth factor-α (TGF-α), no specific ligand for the structurally identical erbB-2 receptor has as yet been identified. It has been demonstrated, however, that ErbB-2 acts as a coreceptor to other erbB receptors and as such can initiate a wide variety of intracellular signal transduction pathways. As the erbB-3 receptor has no tyrosine kinase domain, this receptor also needs to heterodimerise with other members of the erbB family to induce intracellular signalling [1–4].

Ligand-receptor interaction induces receptor dimerisation, which can either be homodimerisation (binding between two identical receptors) or heterodimerisation (binding between two different receptors such as erbB-1 and

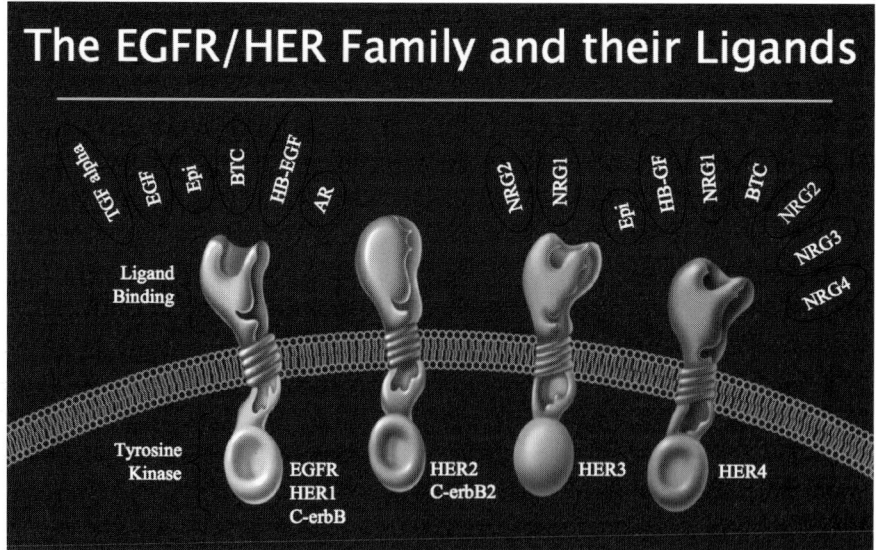

Figure 1. The EGFR family of receptors and their ligands.

erbB-2). Various ligands can induce specific dimerisation reactions that can be tissue or even tumour specific [5]. Receptor dimerisation induces the activation of protein tyrosine kinase activity, which induces tyrosine autophosphorylation of the cytoplasmic regulatory segment of the kinase domain. Finally, tyrosine autophosphorylation activates multiple downstream effector pathways that ultimately lead to various cellular responses.

The two most important downstream signal transduction pathways activated by the erbB family are the Ras-Raf-MAPK and the PI3K/AKT pathways. The MAPK pathway leads to activation of ERK1 and ERK2 that regulate cell transcription and are linked to cell survival and proliferation, and the phosphatidylinositol-3 (PI3K) and serine/threonine Akt kinase pathway plays an important role in cell survival. In addition, and illustrating the enormous diversity of effects and interactions, the activation of other, less defined pathways through increased tyrosine kinase activity is responsible for an increase in the production and secretion of various proangiogenic factors such as interleukin 8 (IL-8) and Vascular Endothelial Growth Factor (VEGF) (Fig. 2).

Under physiological conditions, negative regulation of tyrosine kinase activity is exerted through various different mechanisms. Inhibitory protein tyrosine phosphatases can dephosphorylate the regulatory segment related to the tyrosine kinase domain, whereas ligand-receptor interaction induces a rapid endocytosis and subsequent degradation of both receptor and ligand leading to turning off the erbB signalling pathways.

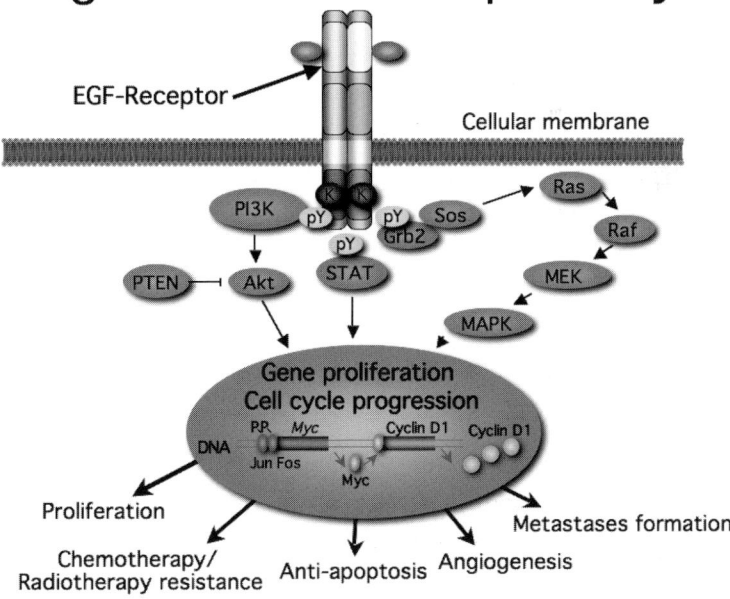

Figure 2. EGFR activation and downstream receptor pathways.

   The essential role of the erbB receptor tyrosine kinase family in physiological development and growth processes has been established through studies in
genetically modified mouse models, in which mutations of erbB receptors
resulted in early fetal or perinatal death and multiple abnormalities in various
organ systems.

   In contrast, the pathological role of abnormal tyrosine kinase activity in the
development and growth of human cancer was established in the 1980s following the observation that overexpression of EGFR in fibroblasts and other
cell lines resulted in the development of a malignant phenotype. More recently, and confirming these *in vitro* observations, the association between EGFR
overexpression and increased tumour aggressiveness and poor clinical outcome in various human epithelial cancers, such as colorectal cancer, breast
cancer, ovarian cancer, squamous head and neck cancer, gastric carcinoma and
gliomas has been established. In addition, in many other frequently occurring
epithelial cancers such as renal cell cancer, esophageal cancer, pancreatic can-
cer, bladder cancer, cervical cancer and prostate cancer, increased EGFR
expression has been demonstrated. Furthermore, EGFR overexpression has
been found to be correlated with tumour size and loss of tumour differentiation. In order to illustrate the magnitude of 'overexpression' of EGFR, it has
been found that some tumour cells express as many as two million EGFRs at
the cell surface, while under physiological conditions cells usually express not
more than 100,000 EGFRs. These findings may explain the fact that EGFR
overexpression is correlated with increased resistance to chemotherapy, hormonal therapy and radiation therapy.

   Several theories concerning the pathophysiological mechanisms that under-
lie the relationship between EGFR overexpression and increased autonomous
tumour growth have been postulated (Fig. 3).

• EGFR overexpression as such leads to a constitutively increased intracellular signalling activity even in the presence of normal concentrations of ligands.
• Overexpression of some classes of erbB receptors can lead to specific pref-
  erential heterodimerisations that result in increased activity of downstream
  effector pathways leading to constitutive or autonomous cell proliferation.
  Of note is that some of these receptor heterodimerisations can occur in the
  absence of specific ligands.
• Tumour cells are often able to produce both receptor and ligand, predominantly EGF and TGF-α. Therefore, an autocrine loop can be formed that
  leads to constitutive receptor activation.
• Finally, mutations within the receptors have been described that lead to constitutive receptor activation, decreased endocytosis and reduced lysosomal
  degradation.

   Apart from the ErbB family of receptors, two other receptor tyrosine kinases have been recognised to play a crucial role in the development and growth
of human cancer. Chronic myelogenous leukemia (CML) and gastrointestinal

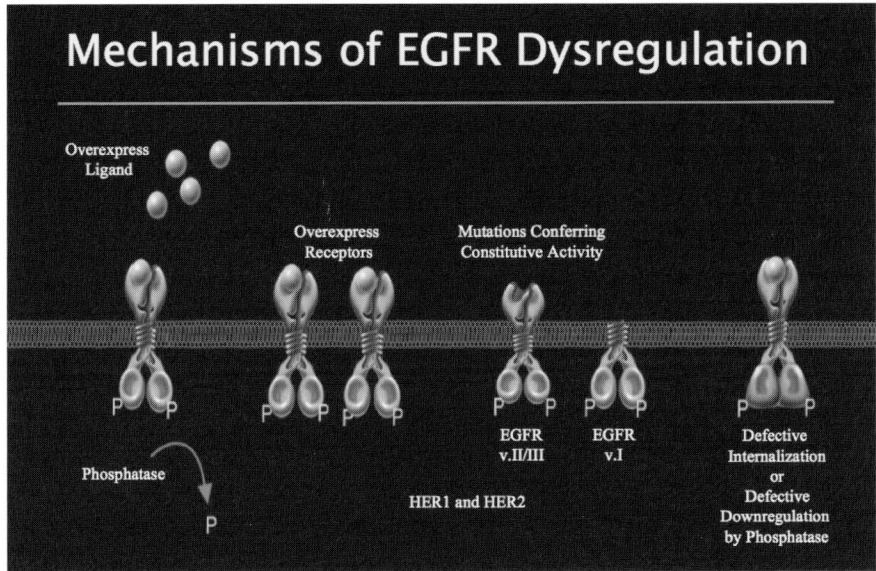

Figure 3. Abnormalities in EGFR activation.

stromal tumours (GIST) are two separate disease entities whose development and growth are almost completely dependent on one single genetic defect, the Philadelphia chromosome as a result of a t(9,22) reciprocal translocation and the *KIT* proto-oncogene, respectively. These genetic defects each alter the function of a specific transmembrane receptor, the BCR-ABL fusion protein and the KIT receptor, respectively. Both receptors harbour a constitutively activated tyrosine kinase.

Assessing the status of ErbB receptors in human tumours

As it is now considered to have potential therapeutic consequences, determining EGFR overexpression in human epithelial tumours is increasingly becoming part of the diagnostic work-up, although until now only within the framework of well-designed clinical studies. In theory, the EGFR status can be assessed via a number of different assays using separate tumour cells, human tumour tissue samples or even plasma. Gene amplification or gene mutations can be detected by Southern blot techniques, fluorescence *in situ* hybridisation (FISH), or polymerase chain reaction (PCR), while mRNA expression can be detected by, among others, Northern blot techniques, RNAse protection assays, RNA *in situ* hybridisation and RT-PCR. However, the direct detection of EGFR expression in tumour samples by means of direct immunohistochemical analysis is most frequently used. This technique is technically easy and widely available, making routine assessments, even in the absence of gene mutations, easy to perform. Unfortunately, these assays thus far lack a standardised method, hampering comparisons between studies.

Immunohistochemical staining of HER2 overexpression in breast cancer tissue is also becoming part of the standard diagnostic work-up. This staining is scored semiquantitatively, with a 3+ positive staining being considered to be positive. In many centres, a 2+ positive staining in combination with a positive FISH test, however, is also considered to be positive.

Receptor tyrosine kinases as target for anticancer drug development

Considering the crucial role of constitutive activated receptor tyrosine kinases in the development and growth of a large number of frequently occurring human epithelial tumour types, inhibiting the activity of these receptors (including EGFR, BCR-ABL and KIT) is an attractive and rational approach in developing specific anticancer agents.

In theory, a number of different strategies can be considered to target and inhibit EGFR tyrosine kinase activity:
• Antisense oligonucleotides or rybozymes blocking mRNA transcription and thus prohibiting protein translation of EGFR or one of its ligands
• Vaccines stimulating the formation of inhibitory antibodies targeting EGFR
• Immunotoxins containing EGFR ligands or antibodies targeting EGFR coupled to radioactive isotopes or toxins
• Anti-EGFR antibodies that competitively inhibit EGFR by binding to the extracellular domain (Fig. 4)
• Small molecule tyrosine kinase inhibitors that bind to the intracellular domain of EFGR (Fig. 5)

As currently the last two approaches have been studied most extensively, both preclinically and clinically, and numerous specific target inhibitory compounds are in advanced stages of clinical testing, with some of them now being approved and licensed, we will focus on these developments.

ErbB-2/HER2 antibodies

ErbB-2/HER2 is overexpressed in 25–30% of breast cancers. HER2 overexpression is correlated with hormone and chemotherapy insensitivity, and poor prognosis. Trastuzumab is a monoclonal IgG antibody that specifically targets the extracellular domain of HER2. Trastuzumab is administered intravenously, with the first administration usually given as a loading dose, followed by weekly maintenance treatment.

In single-agent Phase II studies trastuzumab has shown antitumour activity in patients with HER2 overexpressing metastatic breast cancer when it was given as first-line treatment, and in patients progressing after chemotherapy [6,

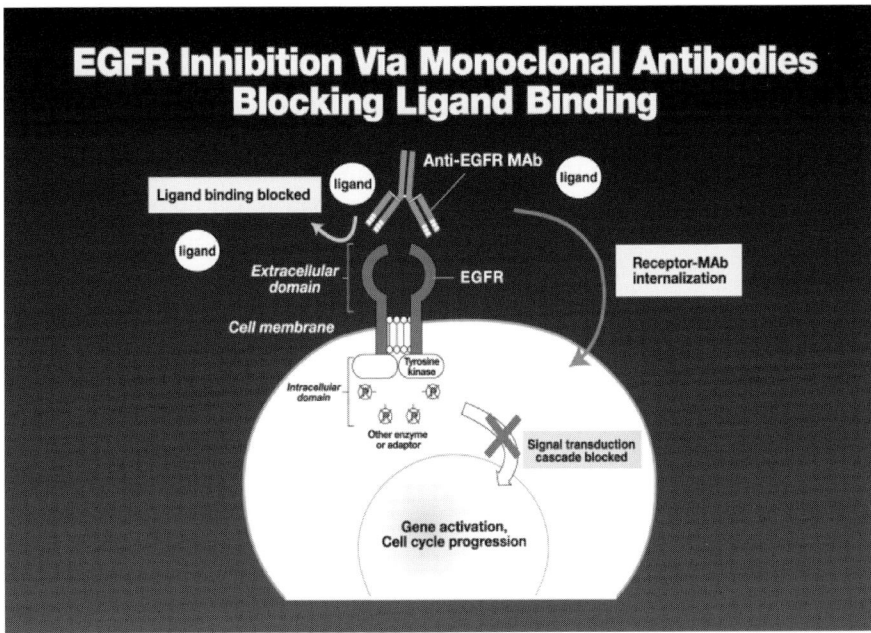

Figure 4. Mechanism of action of anti-EGFR antibodies.

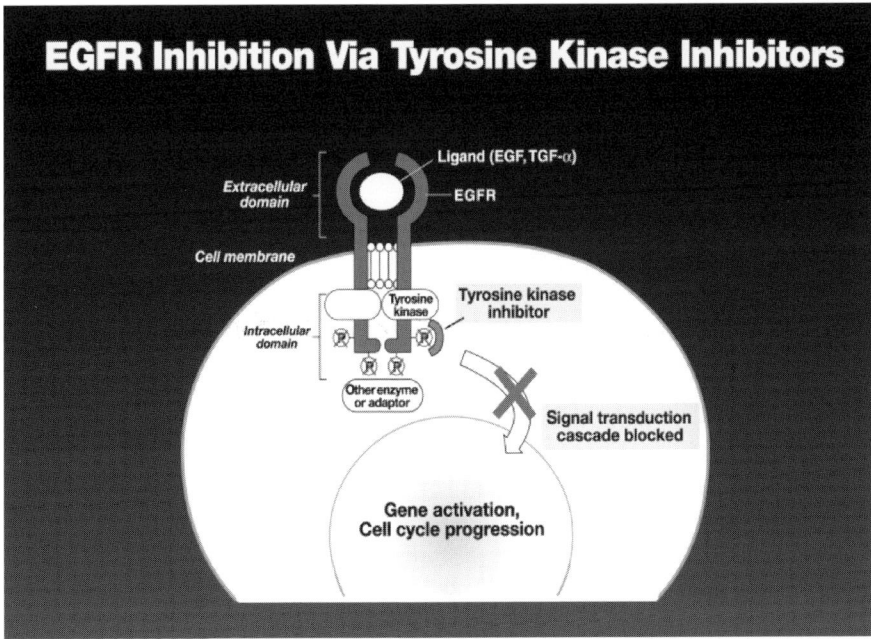

Figure 5. Mechanism of action of EGFR tyrosine kinase inhibitors.

7]. In randomised studies comparing chemotherapy to a combination of chemotherapy and trastuzumab in patients with metastatic breast cancer over-expressing HER2, increased response rates and increased time to disease progression were observed with the combination [8]. Studies exploring the role of trastuzumab in the adjuvant disease setting have recently demonstrated impressive clinical benefit in HER2-neu overexpressing tumors [9–11].

The most prominent side effect of trastuzumab is cardiotoxicity, which potentially hampers the co-administration with antracyclines. Studies are currently analysing whether 3-weekly administration schedules of trastuzumab are as efficacious as weekly schedules and induce less cardiotoxicity. Other studies are investigating the efficacy and safety of the combination of trastuzumab and liposomal formulations of anthracyclines.

Trastuzumab is currently approved for treatment in patients with metastatic breast cancer, either as a single agent or in combination with non-antracycline cytotoxic chemotherapy, both as first-line and second-line treatment.

ErbB-1/HER1 antibodies

EGFR antibodies are directed against the extracellular domain of EGFR, where they compete with natural ligands for binding. Following antibody binding, the receptor dimerises and is subsequently downregulated (Fig. 4).

Murine monoclonal antibody 225 (Mab 225) was among the first antibodies to be tested, and was found to have profound growth inhibitory effects in tumour xenografts overexpressing EGFR. IMC-C225 is a chimerical human:murine antibody that was subsequently developed to obviate human immune response following repeated exposure to Mab 225. IMC-C225 binds to EGFR with higher affinity than the natural ligands EGF and TGF-α, and blocks receptor dimerisation by these natural ligands. In preclinical studies in EGFR overexpressing head and neck, colorectal, renal and prostate cancer cell lines, IMC-C225 profoundly inhibited tumour growth, and synergistic antitumour activity was seen when C225 was combined with various frequently used cytotoxic anticancer agents. Early clinical studies with C225 yielded some promising tumour responses, while pharmacodynamic analyses revealed inhibition of EGFR tyrosine kinase activation, profound inhibition of MAPK activation and decreased cell proliferation. These effects can be observed in sequentially taken tumour biopsies, although in practice this procedure will be quite cumbersome for patients. Fortunately, the skin, where intense EGFR expression occurs, has been found to be a very useful and more easy to reach surrogate tissue for these measurements.

Side effects of IMC-C225 in these studies included diarrhoea and reversible skin rash, which is considered to be potentially predictive for a greater response rate in patients with colorectal cancer [12].

A large number of Phase II studies with IMC-C225, currently known as cetuximab, have meanwhile been performed, either as single-agent or in combination therapy. Based on the results of a large randomised Phase II study in patients with metastatic irinotecan-refractory colorectal carcinoma, where cetuximab was given either as single agent or in combination with irinotecan, cetuximab has recently been approved as second-line treatment option for this group of patients [13].

The efficacy of cetuximab in the treatment of squamous head and neck cancer has recently been demonstrated in a large randomised Phase III study where this antibody given in combination with radiotherapy improved progression free survival and, although to a lesser extent, overall survival [14].

Results of early clinical studies with fully humanised EGFR antibodies such as EMD 72000 and ABX-EGF have recently been published, demonstrating safety, biological and preliminary antitumour activity of these agents. As in the case of cetuximab, a relation between the severity of drug-induced rash and clinical benefit was suggested [15, 16].

Due to their large molecular size, erbB-1 antibodies have to be administered intravenously. In most clinical studies, these agents are given on a weekly basis, often preceded by a loading dose.

EGFR tyrosine kinase inhibitors

Small-molecule tyrosine kinase inhibitors of EGFR and other members of the ErbB family are also in advanced stage of clinical testing. These agents block the ATP binding site of the intracellular domain of the EGFR and have the advantage that they can be given orally, potentially improving patient convenience and compliance with long-term therapy (Fig. 5).

Gefitinib was the first compound to be tested clinically after it had shown growth inhibitory activity towards a range of human tumour cell lines expressing EGFR. In addition to this single-agent activity, additive and even synergistic antitumour activity with various frequently used cytotoxic antitumour agents and radiation therapy was demonstrated in these models. Gefitinib demonstrated promising clinical activity and patient benefit in Phase I and II trials in patients with non-small-cell lung cancer (NSCLC). Based on these results, gefitinib has meanwhile been approved and licensed for the third-line treatment of this disease. To determine the potential role of gefitinib in combination with chemotherapy in first-line treatment in patients with metastatic NSCLC, two large Phase III trials with gefitinib in combination with two of the most frequently used regimens in these patients have been performed. Somewhat surprisingly, both studies failed to show any clinical benefit of gefitinib [17, 18]. The fact that patients in these two studies were not selected based on having EGFR overexpressing tumours was postulated as a possible explanation for these disappointing results. However, in a retrospective molecular analysis of tumour biopsies from patients responding to gefitinib treatment, it was found that almost all of these patients harboured specific activating somatic mutations in the EGFR gene. This finding will undoubtedly have enormous consequences for future trials, as detection of these gene defects will likely enable a better selection of patients that may benefit from these agents [19, 20].

Erlotinib is a second oral EGFR tyrosine kinase inhibitor that has undergone extensive clinical testing. As the safety and efficacy results in Phase I and II trials of erlotinib were highly comparable to those of gefitinib, two large randomised Phase III trials in patients with advanced NSCLC were performed. In these studies, erlotinib was combined with two frequently used cytotoxic chemotherapy schedules. Patients were not selected for EGFR expression and, not surprisingly, these studies yielded comparable disappointing results with regard to patient benefit [21, 22]. With regard to single-agent EGFR tyrosine kinase activity, a large placebo controlled randomised Phase III trial in patients with advanced NSCLC failing previous chemotherapy has recently demonstrated significant clinical patient benefit with a relief of tumour-related symptoms as well as a, although modest, survival benefit in patients treated with erlotinib [23].

A large number of new, often dual specific EGFR tyrosine kinase inhibitors (i.e., targeting both erbB-1 and erbB-2) are currently undergoing clinical testing (Tab. 1).

Diarrhoea and skin rash have been described as the most prominent side effects of these compounds, being often dose limiting in early clinical studies. Whether this skin rash correlates with EGFR inhibition and/or might predict antitumour activity is an ongoing matter of debate. A typical example of this rash is shown in Fig. 6.

Despite the proven synergistic antitumour activity of these two EGFR inhibitors with chemotherapy seen in models, it can be argued that the combination of an EGFR inhibitor and chemotherapy might render proliferating human tumour cells less sensitive to the cytotoxic effects of chemotherapy. It therefore seems likely that in order to optimally benefit from EGFR tyrosine

Typical examples of drug-induced rash

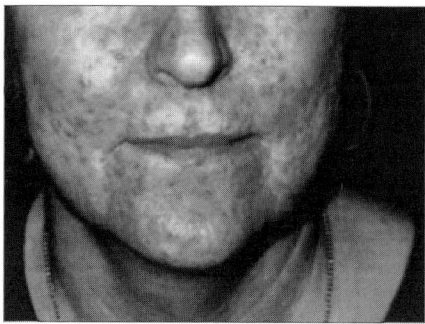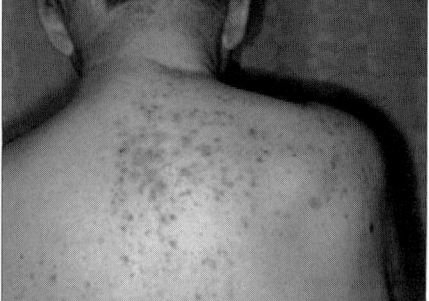

Figure 6. Typical rash following treatment with EGFR tyrosine kinase inhibitors.

Table 1. Signal transduction inhibitors in clinical trials

Drug	Target	Clinical development
	Receptor antibodies	
Cetuximab	ErbB1/EGFR	Approved
ABX-EGF	ErbB1/EGFR	Phase I, II
EMD 72000	ErbB1/EGFR	Phase I, II
h-R3	ErbB1/EGFR	Phase I, II
MDX-447	ErbB1/EGFR	Phase I, II
Trastuzumab	ErbB2/HER-2	Approved
Pertuzumab	ErbB2/HER-2	Phase I, II
	Tyrosine kinase inhibitors (reversible)	
Gefitinib	ErbB1/EGFR	Approved
Erlotinib	ErbB1/EGFR	Phase III
PKI-166	ErbB1/EGFR ErbB2	Phase I
GW2016	ErbB1/EGFR ErbB2	Phase I
BIBW 2992	ErbB1/EGFR ErbB2	Phase I
Imatinib	KIT, PDGFR, BCR-ABL	Approved
	Tyrosine kinase inhibitors (irreversible)	
EKB569	ErbB1/EGFR ErbB2	Phase I
CI-1033	ErbB1/ErbB2 ErbB3/ErbB-4	Phase III

kinase inhibitors, these agents should preferably be given sequentially rather than concomitantly.

BCR-Abl and c-KIT tyrosine kinase inhibitors

Imatinib mesylate is a specific inhibitor of the tyrosine kinase activity of ABL kinase, the BCR-ABL fusion protein, KIT and the PDGF receptor (PDGFR) and therefore has been considered a rational treatment option for CML and GIST. In a Phase I study in GIST, imatinib yielded surprisingly high response rates that were confirmed in a multicentre Phase II study [24, 25]. With the absence of any other standard therapy, imatinib is now considered standard of care for unresectable GIST.

In patients with CML, imatinib was superior to the combination of interferon-α and low-dose cytarabine with regard to several cytogenetic and clinical endpoints, and therefore has already been referred to as the new gold standard for treatment patients [26, 27].

Future clinical studies with EGFR inhibiting agents; challenges for design

Phase I studies with so-called classical cytotoxic anticancer agents are designed to describe the acute toxicity profile of these agents and to define the maximum tolerated dose. As cytotoxic agents most often cause a steep dose dependent cell kill that induces dose dependent tumour regressions, it makes sense to dose these agents at their highest possible dose. Usually, this dose is recommended for subsequent Phase II and III trials (Tab. 2).

In contrast to this 'cytotoxic paradigm', however, many preclinical studies with targeted anticancer agents such as EGFR antibodies and small molecule EGFR tyrosine kinase inhibitors have shown that these agents often induce tumour growth inhibition rather than tumour regression, and that optimal growth inhibition often can be obtained at doses that do not induce dose-limiting toxicity. Thus the endpoint of Phase I studies with these targeted agents should be defining an optimal biologic effect dose, rather than the maximum tolerated dose. When trying to define such an optimal biologic effect dose, it is important to realise that biological and antitumour activity often correlate with certain threshold concentrations in preclinical studies. Therefore, it is conceivable that pharmacokinetic parameters such as the area under the plasma concentration-time curve or the time above a certain threshold concentration may become important new endpoints in these Phase I studies. In addition, pharmacodynamic analyses showing target inhibition within the tumour

Table 2. Endpoints in the design of cytotoxic and growth inhibitory anticancer agents

	Cytotoxic agents	Growth inhibitory agents
Phase I studies	1: Acute toxicity 2: Maximum tolerated dose Defined by: Toxicity	1: Acute and chronic toxicity 3: Optimal biologic effect dose Defined by: Target AUC Inhibition of cellular target Inhibition of surrogate marker
Phase II studies	1: Antitumour activity Defined by: Tumour regression rate Surrogate marker inhibition 2: Delayed toxicity	1: Antitumour activity Defined by: Time to progression Symptom relief
Phase III studies	1: Antitumour efficacy Defined by: Cure rate Time to progression Disease free survival Overall survival Quality of life	1: Antitumour efficacy Defined by: Cure rate Time to progression Disease free survival Overall survival Quality of life

or within a surrogate tissue can also be of help to determine a biological effective dose.

When interpretating these findings, however, one must take into account that preclinical models of EGFR inhibiting agents often have shown to be poor predictors for the clinical situation and that several EGFR tyrosine kinase inhibitors have caused dose-limiting toxicities in Phase I clinical trials.

The focus of Phase II studies with classical cytotoxic anticancer agents is to define the percentage of tumour regressions in a group of patients treated with the recommended dose. This 'cytotoxic paradigm' assumes that tumour regression will correlate with patient benefit in the long-term. If this paradigm is applied to single-agent Phase II studies with EGFR antibodies or EGFR receptor tyrosine kinase inhibitors, which may induce growth inhibition rather than tumour shrinkage, many agents would be considered to be of no clinical benefit to patients, leading to premature 'pharmacoptosis'. Finally, randomised Phase III trials remain pivotal to definitely proof the efficacy of EGFR antibodies or EGFR receptor tyrosine kinase inhibitors with regard to such endpoints as improvement of time to progression, overall survival and quality of life.

Conclusions

The development and clinical introduction of new classes of signal transduction inhibitors, such as tyrosine kinase inhibitors and receptor antibodies, has significantly changed our way of thinking of cancer. Nowadays, cancer can increasingly be regarded as a disease for which specific and rationally designed growth inhibiting agents exist.

In addition, the recent recognition that some target receptors as a result of genetic mutations can be expected to play a more essential role in maintaining cancer growth, may facilitate the rational application of these growth-inhibiting agents.

Although the results from some large randomised studies have been somewhat disappointing, the lessons learned from these studies will undoubtedly further improve the rational application of this new group of anticancer agents.

After decades of non-selective trial-and-error treatment, these new insights and the availability of specifically targeted anticancer agents such as signal transduction inhibitors finally makes rational anticancer treatment a realistic option.

References

1 Yarden Y, Sliwkowski M (2001) Untangling the ErbB signalling network. *Nat Rev Mol Cell Biol* 2: 127–137
2 Hackel P, Zwick E, Prenzel N, Ullrich A (1999) Epidermal growth factor receptors: critical mediators of multiple receptor pathways. *Current Opin Cell Biol* 11: 184–189

3 Klapper L, Glathe S, Vaisman N, Hynes N, Andrews G, Sela M, Yarden Y (1999) The ErbB-2/HER2 oncoprotein of human carcinomas may function solely as a shared coreceptor for multiple stroma-derived growth factors. *Proc Natl Acad Sci USA* 96: 4995–5000

4 Graus-Porta D, Beerli R, Daly J, Hynes N (1997) ErbB-2, the preferred heterodimerization partner for all erbB receptors, is a mediator of lateral signalling. *EMBO J* 16: 1647–1655

5 Olayoiye M, Neve R, Lane H, Hynes N (2000) The ErbB signalling network: receptor heterodimerization in development and cancer. *EMBO J* 19: 3159–3167

6 Cobleigh M, Vogel C, Tripathy D, Robert N, Scholl S, Fehrenbacher L, Wolter J, Paton V, Shak S, Lieberman G et al. (1999) Multinational study of the efficacy and safety of humanized anti-HER2 monoclonal antibody in women who have HER2-overexpressing metastatic breast cancer that has progressed after chemotherapy for metastatic disease. *J Clin Oncol* 17: 2639–2648

7 Vogel C, Cobleigh M, Tripathy D, Gutheil J, Harris L, Fehrenbacher L, Slamon D, Murphy M, Novotny W, Burchmore M et al. (2002) Efficacy and safety of trastuzumab as a single agent in first-line treatment of HER2-overexpressing metastatic breast cancer. *J Clin Oncol* 20: 719–726

8 Slamon D, Leyland-Jones B, Shak S, Fuchs H, Paton V, Bajamonde A, Fleming T, Eiermann W, Wolter J, Pegram M et al. (2001) Use of chemotherapy plus a monoclonal antibody against HER2 for metastatic breast cancer that overexpresses HER2. *N Engl J Med* 344: 783–792

9 Romond EH, Perez EA, Bryant J, Suman VJ, Geyer CE Jr, Davidson NE, Tan-Chiu E, Martino S, Paik S, Kaufman PA et al. (2005) Trastuzumab plus adjuvant chemotherapy for operable HER2-positive breast cancer. *N Engl J Med* 353: 1673–1684

10 Piccart-Gebhart MJ, Procter M, Leyland-Jones B, Goldhirsch A, Untch M, Smith I, Gianni L, Baselga J, Bell R, Jackisch C et al.; Herceptin Adjuvant (HERA) Trial Study Team (2005) Trastuzumab after adjuvant chemotherapy in HER2-positive breast cancer. *N Engl J Med* 353: 1659–1672

11 Joensuu H, Kellokumpu-Lehtinen PL, Bono P, Alanko T, Kataja V, Asola R, Utriainen T, Kokko R, Hemminki A, Tarkkanen M et al.; FinHer Study Investigators (2006) Adjuvant docetaxel or vinorelbine with or without trastuzumab for breast cancer. *N Engl J Med* 354: 789–790

12 Saltz L, Kies M, Abbruzzese J, Azarnia N, Needle M (2003) The presence and intensity of the cetuximab-induced acne-like rash predicts increased survival in studies across multiple malignancies. *Proc Amer Soc Clin Oncol* 23: abstr 204

13 Cunningham D, Humblet Y, Siena S, Khayat D, Bleiberg H, Santoro A, Bets D, Mueser M, Harstrick A, Verslype C et al. (2004) Cetuximab monotherapy and cetuximab plus irinotecan in irinotecan-refractory metastatic colorectal cancer. *N Engl J Med* 22: 337–345

14 Bonner J, Harari P, Giralt J, Azarnia N, Cohen R, Raben D, Jones C, Kies M, Baselga J, Ang K (2004) Cetuximab prolongs survival in patients with locoregionally advanced squamous cell carcinoma of head and neck: A phase III study of high dose radiation therapy with or without cetuximab. *Proc Amer Soc Clin Oncol* 23: abstr 7022

15 Vanhoefer U, Tewes M, Rojo F, Dirsch O, Schleucher N, Rosen O, Tillner J, Kovar A, Braun AH, Trarbach T et al. (2004) Phase I study of the humanized antiepidermal growth factor receptor monoclonal antibody EMD72000 in patients with advanced solid tumours that express the epidermal growth factor receptor. *J Clin Oncol* 22: 175–184

16 Rowinsky E, Schwartz G, Gollob J, Thompson J, Vogelzang N, Figlin R, Bukowski R, Haas N, Lockbaum P, Li Y et al. (2004) Safety, Pharmacokinetics, and Activity of ABX-EGF, a Fully Human Anti-Epidermal Growth Factor Receptor Monoclonal Antibody in Patients With Metastatic Renal Cell Cancer. *J Clin Oncol* 22: 3003–3015

17 Giaccone G, Herbst R, Manegold C, Scagliotti G, Rosell R, Miller V, Natale R, Schiller J, Pawel J, Pluzanska A et al. (2004) Gefitinib in combination with gemcitabine and cisplatin in advanced non-small-cell lung cancer: A Phase III trial – INTACT 1. *J Clin Oncol* 22: 777–784

18 Herbst R, Giaccone G, Schiller J, Natale R, Miller V, Manegold C, Scagliotti G, Rosell R, Oliff I, Reeves J et al. (2004) Gefitinib in combination with paclitaxel and carboplatin in advanced non-small-cell lung cancer: A Phase III trial – INTACT 2. *J Clin Oncol* 22: 785–794

19 Lynch T, Bell D, Sordella R, Gurubhagavatula S, Okimoto R, Brannigan B, Harris P, Haserlat S, Supko J, Haluska F et al. *(2004)* Activating mutations in the epidermal growth factor receptor underlying responsiveness of non-small-cell lung cancer to gefitinib. N Engl J Med 20: 2129–2139

20 Paez J, Janne P, Lee J, Tracy S, Greulich H, Gabriel S, Herman P, Kaye F, Lindeman N, Boggon T et al. (2004) EGFR mutations in lung cancer: correlation with clinical response to gefitinib therapy. *Science* 304: 1497–1500

21 Gatzemeier U, Pluzanska A, Szczesna A, Kaukel E, Roubec J, Brennscheidt U, De Rosa U, Mueller B, Von Pawel J (2004) Results of a phase III trial of erlotinib (OSI-774) combined with cisplatin and gemcitabine (GC) chemotherapy in advanced non-small cell lung cancer (NSCLC). *Proc Amer Soc Clin Oncol* 24: abstr 7010

22 Herbst R, Prager D, Hermann R, Miller V, Fehrenbacher L, Hoffman P, Johnson B, Sandler A, Mass R, Johnson D (2004) TRIBUTE – A phase III trial of erlotinib HCl (OSI-774) combined with carboplatin and paclitaxel (CP) chemotherapy in advanced non-small cell lung cancer (NSCLC). Proc Amer Soc Clin Oncol 24: abstr 7011

23 Shepherd F, Pereira J, Ciuleanu T, Tan E, Hirsh V, Thongprasert S, Bezjak A, Tu D, Santabárbara P, Seymour L (2004) A randomized placebo-controlled trial of erlotinib in patients with advanced non-small cell lung cancer (NSCLC) following failure of 1st line or 2nd line chemotherapy. A National Cancer Institute of Canada Clinical Trials Group (NCIC CTG) trial. *Proc Amer Soc Clin Oncol* 24: abstr 7022

24 Van Oosterom A, Judson I, Verweij J, Stroobants S, Donato di Paola E, Dimitrijevic S, Martens M, Webb A, Sciot R, Van Glabbeke M et al. (2001) Safety and efficacy of imatinib (STI571) in metastatic gastrointestinal stromal tumours: a phase I study. *Lancet* 358: 1421–1423

25 Demetri G, von Mehren M, Blanke C, Van den Abbeele C, Eisenberg S, Roberts P, Heinrich M, Tuveson D, Singer S, Janicek M et al. (2002) Efficacy and safety of imatinib mesylate in advanced gastrointestinal stromal tumours. *N Engl J Med* 347: 472–480

26 O'Brien S, Guilhot F, Larson R, Gathmann I, Baccarani M, Cervantes F, Cornelissen J, Fischer T, Hochhaus A, Hughes T et al. (2003) Imatinib compared to interfron and low-dose cytarabine for newly diagnosed chronic-phase chronic myeloid leukaemia. *N Engl J Med* 348: 994–1004

27 Peggs K, Mackinnon S (2003) Imatinib mesylate – the new gold standard for treatment of chronic myeloid leukaemia. *N Engl J Med* 348: 1048–1050

Endocrine therapy of breast cancer

Rosalba Torrisi[1], Alessandra Balduzzi[2] and Aron Goldhirsch[3]

[1, 2] *Research Unit of Medical Senology,* [1, 2, 3] *Department of Medicine, European Institute of Oncology, via Ripamonti 435, 20141 Milano, Italy,* [3] *Oncology Institute of Southern Switzerland, Bellinzona & Lugano, Switzerland*

Introduction

Endocrine therapies have been used for more than a century to treat breast cancer. A huge body of evidence from preclinical and clinical studies substantiates the outstanding role of estrogens in initiation and in promotion of almost two-thirds of breast cancer [1, 2]. Since the initial observations of Beatson who, in 1896 reported of dramatic tumor responses after bilateral oophorectomy in premenopausal women with advanced breast cancer, the suppression of the estrogen activity has represented the rationale for the manipulations aimed to affect the growth of breast cancer [3, 4]. Initially, surgical procedures as oophorectomy and adrenalectomy were shown to induce some tumor regres-

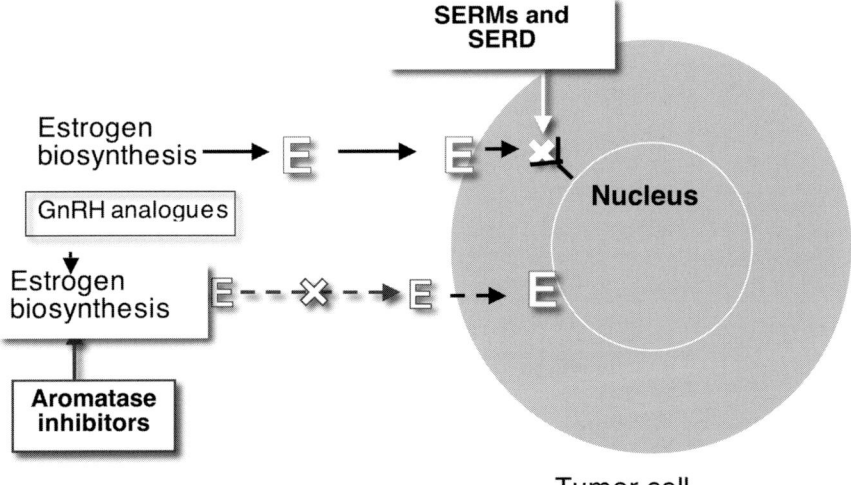

Figure 1. Sites of interaction of the different endocrine agents with the estrogen synthesis and the estrogen-receptor. SERMs: Selective Estrogen Receptor Modulators; SERD: Selective Estrogen Receptor Downregulator; GnRH: Gonadotropin Releasing Hormone; E: Estrogens

sion; since then, the development of drugs which negatively affect the production and/or the activity of estrogens led to a widespread use of hormonal manipulations [4]. Developed in the late 1960s, tamoxifen has represented the gold standard of endocrine therapy in pre- and postmenopausal women, either in the advanced or the adjuvant setting [5]. The determination of the estrogen receptor, a most sensitive predictive factor of response to treatments, allows for the identification of patients that are most likely to benefit from endocrine therapy [6]. Recently, the availability of drugs which differently affect the estrogen signaling and do not show cross-resistance with tamoxifen has extended the options of endocrine therapies in advanced and in early breast cancer [7] (Fig. 1).

Since estrogen receptor represents the ultimate target of all endocrine manipulations, we will briefly discuss the biology of the receptor, focusing on the ligand specific activity and the potential mechanisms of *de novo* and induced resistance. We will then review the pharmacological and clinical results of the most commonly used endocrine agents.

The estrogen receptor

Physiology

The estrogen receptors (ER), α and β, belong to a superfamily of nuclear hormone receptors including those for other steroid hormones, thyroid hormones, vitamin D and retinoic acid [8]. These receptor proteins function as transcription factors in the nucleus when they are bound to their respective ligands [9]. The receptor has a ligand binding domain, several transcription activation domains and a DNA-binding domain which interacts with specific regions in the promoter of target genes, known as estrogen-responsive elements (ERE) [5, 10]. Upon binding an agonist, these receptors form heterodimers in cells expressing both subtypes or homodimers in cells expressing a single subtype [11]. ERα and ERβ are coded on different chromosomes and share similar but not identical structure, but they appear to play different roles in estrogen action, with ERα being a more robust activator of transcription while ERβ moderates the agonist activity of estradiol and has been involved in the mechanisms of resistance to tamoxifen [11]. The ERs possess two major transcriptional activation domains residing in their NH2 and the COOH terminal-domain, which harbor, respectively, the constitutively active, hormone-independent AF-1 and the hormone-dependent AF-2 functions [12]. Estrogen regulates the expression of genes which are crucial for cell proliferation, inhibition of apoptosis, stimulation of invasion and metastasis and promotion of angiogenesis [10]. It is becoming more clear that the molecular pharmacology of the estrogen receptor is extremely complex, affected by the expression of the two receptor subtypes, the ligand-specific effect on the receptor structure, the availability of receptor interacting proteins as cofactors, corepressor and the ERE [9].

When estrogen binds to the receptor it induces phosphorylation of the receptor, triggers receptor dimerization and activates DNA binding to ERE in the promoter regions of target genes [12]. Promoter-bound ER dimer forms a complex with coregulatory proteins (coactivators) with acetyltransferase activity as AIB1 which helps to unwind the chromatin and facilitates transcription of estrogen-responsive genes [13]. This transcriptional activity of ER is called genomic activity (Fig. 2). On the other hand, when a selective estrogen receptor modulator (SERM) binds the receptor it induces conformational changes which prevents binding of co-activators and blocks AF-2-induced transcription [14]. It has been hypothesized that individual SERMs may induce specific and unique changes in receptor conformation which account for their pharmacologic properties in target tissues [15]. The complex ER–tamoxifen activates different coregulatory proteins (corepressors) with histone deacetylase activity, resulting in a condensation of chromatin and repression of transcriptional activity [16]. The balance between coactivators and corepressors in the target tissues is the major determinant of the agonist/antagonist activities of tamoxifen [10]. In fact, in the presence of high concentrations of AIB1 and other coactivators, the complex SERM-ER may result in enhanced estrogenic activity of tamoxifen [17]. As for the mechanism of action of pure antagonists, when fulvestrant binds to the ER the subsequent conformational changes, which prevents dimerization and

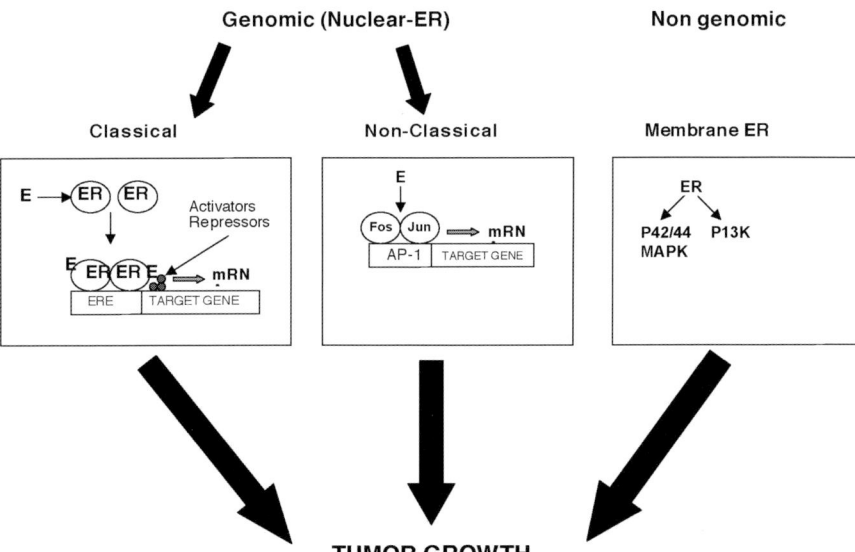

Figure 2. Mechanisms of activation of estrogen receptor (ER): on binding of its estrogen ligands (E), nuclear ER transcription is activated (genomic action) either by direct DNA binding (classical mode) or by other components (non-classical mode). Membrane ER activity (non-genomic action) through direct interaction with different signaling intermediates.

ERE binding, leads to the destabilization and degradation of the receptor. In this case the relative presence of coregulatory proteins is not relevant to the agent activity and this explains the pure antagonistic properties of fulvestrant [10].

An alternative to this 'classical' nuclear transcriptional activity on ERE, ER have been shown to modulate gene expression at alternative regulatory DNA sequences such AP-1, SP-1, thus regulating transcription of IGF-1R, cyclin D1, myc and the antiapoptotic factor Bcl-2 [18] (Fig. 2).

More recently, a so-called non genomic activity has been established for the ER. A small pool of ERs, with a partially different structure from the nuclear form, is located outside of the nucleus, bound to plasma membrane (Fig. 2). This membrane-bound ER can mediate signals originating in the membrane or in the cytoplasm, which occur within a few minutes, by directly interacting or activating growth factor signalling pathways as IGF-1R, EGFR, HER2, the p85 subunit of P13k [10]. The membrane effect of ER may be cell, receptor subtype and ligand specific and it may also be influenced by the presence of growth factor signaling, being more relevant when EGFR and HER2 are over-expressed [10]. Since tamoxifen behaves as an agonist on the membrane mediated activity, while fulvestrant is not able to activate membrane ER, the non genomic activity of ER may contribute to explain the *de novo* resistance to SERMs [10].

Mechanism of resistance to endocrine agents and clinical implications

Evidence from clinical trials indicates that almost 50% of tumors do not respond to first line endocrine therapy and that 40% of patients receiving tamoxifen experience tumor relapse, suggesting the occurrence of either *de novo* and acquired mechanisms of resistance [19].

Multiple potential mechanisms of resistance to endocrine agents have been postulated. Findings derived from preclinical and clinical studies suggest that acquired resistance to SERMs and to aromatase inhibitors arises through mechanisms that are partially distinct.

Since ER is the ultimate target of endocrine agents, it seems plausible that the loss of expression of ERα might be responsible for acquired resistance to tamoxifen. However, this represents only one of the potential mechanisms since loss of ERα has been demonstrated in less than 30% of breast cancers and that aromatase inhibitors induce approximately 30% objective response rate after tamoxifen failure [20, 21]. The occurrence of a mutated ER is even more uncommon and cannot be claimed as a major cause of resistance to antie-strogens [19].

Similarly, contradictory data have emerged on the role of ERβ in the resist-ance to endocrine agents with some studies showing increased levels of ERβ in tumors from tamoxifen resistant patients, while another study failed to cor-relate ERβ mRNA with response to toremifene [22, 23].

The intricate modulation of the receptor:coregulator ratio in different cells and tissues is also implicated in determining either response or resistance to endocrine therapy [10].

Cumulative evidence supports the existence of a cross-talk between activated EGFR/HER-2 and ER signalling pathways with reciprocal upregulation. In estrogen-dependent breast cancer cell lines overexpression of EGFR/HER-2 leads to an increase of intracellular kinase as mitogen-activated protein kinase (MAPK) and PI3/Akt, which in turn increase ER phosphorylation and may promote its binding with coactivators rather than corepressors, inducing ER-dependent gene transcription [24, 25]. Conversely, activated ER can increase EGFR dependent transcription, responsible for a positive feedback loop which enhances the cross-talk between growth factor and hormone receptors [25]. A high expression of AIB1 is frequently associated with HER-2 overexpression and both are involved in tamoxifen resistance [10]. It has been shown that in HER-2 overexpressing tumors, the tamoxifen-ER complex is able to recruit coactivators as AIB1 rather than corepressors, switching tamoxifen into an agonist [17].

Hyperactivity of MAPK has been reported also as a consequence of chronic estrogen deprivation [26]. Estrogen receptor positive breast cancer cells grown in estrogen-depleted conditions exhibit increased MAPK activity, which, in turn, makes cells more sensitive to low concentrations of estrogens [26]. Moreover, long-term estrogen deprivation may enhance the non-genomic ER activity, increasing the levels of membrane ERα and, consequently, the cross-talk with growth factor signaling pathways [27]. It may thus be hypothesized that increased MAPK activity may be involved also in resistance to aromatase inhibitors. In addition to a supersensitive phenotype, which may be abrogated by treatment with fulvestrant, prolonged activation of growth factor signaling pathways may also lead to transcriptional repression of ERα as an ultimate step of resistance to endocrine agents [19].

The proof of principle of the involvement of EGFR/HER-2 pathway in the *de novo* and acquired resistance to antiestrogens has been sustained by clinical evidence suggesting:

a) HER-2 overexpression predicts poor clinical outcome and a lower response rate in patients with hormone receptor positive breast cancer treated with tamoxifen [28–30]

b) the blockade of the EGFR/HER-2 signaling pathway is able to restore sensitivity to antiestrogens [17, 31]

Finally, recent evidence suggests that after long-term treatment with SERMs tumor cells undergo spontaneous growth and estrogens, rather than stimulating growth, may induce apoptosis [32]. Preclinical studies have demonstrated that tumors recurring after estrogen-induced apoptosis are sensitive to treatment with SERMs or aromatase inhibitors [15]. These findings support the potential of an estrogen purge as a means to restore sensitivity to hormonal agents and should be explored in clinical trials [15].

Selective estrogen receptor modulators

Selective estrogen receptor modulators (SERMs) are synthetic agents that bind the ER and act as either agonists or antagonists, depending on the balance between coregulatory (coactivators and corepressors) molecules in the tissue [5, 10]. Chemically, they lack the steroid structure of estrogens but possess a tertiary structure which allows them to bind to the ER (Fig. 3).

Tamoxifen is a non steroidal triphenylethylene derivative which was first developed in the 1970s and since then has represented the gold standard of treatment of endocrine responsive breast cancer at all stages [5].

The 1998 Oxford meta-analysis reported the results of 55 randomized trials of tamoxifen in early breast cancer. Tamoxifen significantly reduced risk of recurrence by 18%, 25% and 42%, for 1, 2 and 5 years of treatment, respectively. For mortality, the proportional reductions in the death rates in the trials of 1 year, 2 years and about 5 years of tamoxifen were 10%, 15% and 22% [33].

The benefit of tamoxifen was independent of nodal status, although in terms of 10-year outcome, the same proportional benefit for node-positive as for node-negative disease would generally imply a greater absolute benefit for women with node-positive disease. For the trials of 1 or 2 years of tamoxifen the absolute improvements in this 10-year recurrence risk appear larger for women with node-positive disease than for those with node-negative disease. In the trials of about 5 years of tamoxifen, the absolute improvement in this

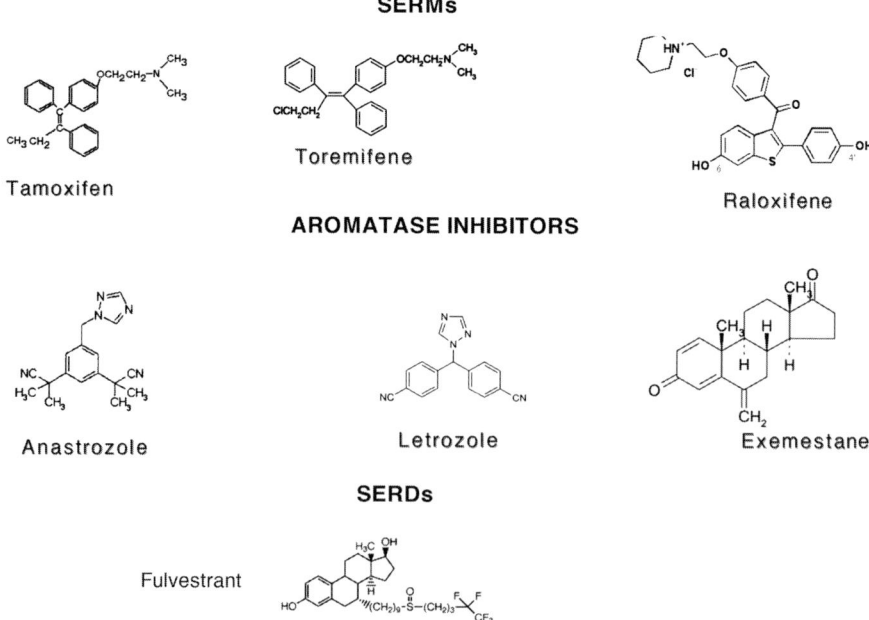

Figure 3. Chemical structure of the most common endocrine agents. SERMs: Selective Estrogen Receptor Modulators; SERDs: Selective Estrogen Receptor Downregulators

10-year recurrence risk appears to be about as great for women with node-negative disease (absolute improvement 14.9%) as for those with node-positive
disease (absolute improvement 15.2%). For patients with node-negative disease in the trials of 1, 2 and 5 years of tamoxifen the improvement in 10-year
survival are 3.4%, 2.3% 5.6%, respectively; for those with node-positive disease it is 4.5%, 7.2% and 10.9% [33].

Tamoxifen was active irrespective of age and menopausal status. The recurrence of reductions produced by about 5 years of tamoxifen are substantial and
highly significant both in the women aged under 40 (54% reduction) and in
those aged between 40–49 (41% reduction) [33].

The principal determinant of tamoxifen activity was ER expression. In fact,
tamoxifen was shown to have a small if any effect (6%) on ER poor (<10%)
with no evidence of greater benefit with longer treatment. This finding is
somewhat consistent with a non-significant trend towards a detrimental effect
(HR = 1.21, 95% CI 0.89–1.67 p = 0.21) on the occurrence of ER-negative
tumors observed in an overview of the tamoxifen prevention trials, while ER-
positive cancers were dramatically decreased by 48% [34]. No dose-dependent effect was demonstrated, and the benefit appeared to be about as big in the
trials of 20 mg/day as in the trials of 30–40 mg/day, in terms both of recurrence and of mortality [33].

The 2000 Overview, published in 2005, confirmed, with comparable figures, these results, showing that 5-years tamoxifen almost halved annual recurrence rate (recurrence rate ratio = 0.59) and reduced breast cancer mortality by
one-third (death rate ratio = 0.66). The advantage of 5 years *versus* 1–2 years
was confirmed although a non-significant increase in mortality rate from other
causes was observed with longer treatment, mostly attributable to an excess of
deaths for thromboembolisms [35].

The role of progesterone receptor (PgR) in determining endocrine responsiveness is not fully elucidated. While the overviews failed to show any additional information by PgR in assessing the benefit of adjuvant endocrine therapy, with ER+/PgR+ patients performing comparably to ER+/PgR– patients
[33], some recent reports suggest that patients with PgR negative tumors have
a different behavior, showing a worse prognosis and less benefit from tamoxifen therapy as compared with PgR positive. The Swedish trial on adjuvant
tamoxifen showed that patients with ER+/PgR+ tumors performed better than
patients with ER+/PgR– primaries [36].

The additional prognostic value for PgR was supported from a recent retrospective analysis of two large independent databases, the PP and the SPORE
databases, containing information on more than 50,000 patients with early
breast cancer, of whom more than 15,000 were untreated or had received adjuvant endocrine therapy [37]. All specimens were assessed for ER and PgR status by ligand-binding assay by central pathology laboratories with standardized assays and quality-controlled procedures. The databases yielded similar
results showing that ER+/PgR+ patients did better than ER+/PgR patients as
for 5-years disease free survival (DFS) and overall survival (OS) either in

untreated and in endocrine-treated patients. The additional predictive value of PgR was confirmed either in untreated patients or in patients receiving endocrine therapies (mostly tamoxifen).

It has been hypothesized that PgR loss may be related to a cross-talk between hormone receptors and EGFR/Her-2 pathways and is the consequence of the upregulation of the growth factor receptor signalling [38]. In a recent retrospective analysis among tamoxifen-treated patients included in a large database of more 40,000 patients, ER+/PR– tumors expressed higher levels of HER-1 and HER-2, displayed more aggressive features than ER+/PR+ tumors and were associated with a higher likelihood of recurrence [39]. On the other hand, PgR loss does not appear to be predicitive to resistance to all endocrine agents. Both preclinical and clinical data show that aromatase inhibitors may be active also in breast cancer cell lines and in PgR negative tumors (see paragraph on *Aromatase inhibitors*).

It is evident that the risk of recurrence of breast cancer maintains fairly high (2–4% per year) even up to 15 years after the diagnosis, particularly for patients with ER positive tumors [40]. The long-term results of some trials (NSABP B-14, Scottish trial) have shown that 5-year tamoxifen lowers the risk of recurrence and death up to 15 years after diagnosis, and the recent overview has confirmed a prolonged benefit of survival after treatment discontinuation [35, 41, 42]. The issue of prolonging the duration of the endocrine manipulation beyond 5 years has been addressed in several trials. Firstly, the optimal duration of tamoxifen has been addressed in randomized trials comparing 5 years with longer durations [41, 43]. The question has not been answered definitively yet, awaiting for the results of two huge ongoing trials (aTTom and ATLAS), although the results from the Scottish and the NSABP B-14 trials clearly show the lack of benefit from extending tamoxifen therapy, due to the higher rate of endometrial cancer, and cerebrovascular events in the prolonged treatment arm [41, 43]. The availability of a new class of agents, with different mechanism of action and substantially different toxic profile, has prompted the attempt of extending endocrine manipulation using a different agent, which is shown to be active after tamoxifen failure. The results of these trials will be discussed in the paragraph on *Aromatase inhibitors*.

The toxicity profile of tamoxifen is attributable mostly to its agonistic properties and has been extensively looked into by large individual trials and by the overviews [33, 43, 44]. A recent meta-analysis estimated the effects of tamoxifen on vascular and neoplastic outcomes including data from more than 30 large randomized controlled trials [44]. Tamoxifen is associated with an increased risk of endometrial cancer (RR = 2.7 95% CI 1.94–3.75), gastrointestinal cancer (RR = 1.31 95% CI 1.01–1.69), stroke (RR = 1.49 95% CI 1.16–1.9), pulmonary embolism (RR = 1.88 95% CI 1.77–3.01) and deep venous thrombosis (RR = 1.87 95% CI 1.33–2.64). In contrast, tamoxifen significantly decreased myocardial infarction deaths (RR = 0.62 95% CI 0.41–0.93) [44]. These results were superimposable to those of the 1998 EBCTCG overview except for the protective effect on myocardial infarction deaths, which was not specifically reported in the overview although a non-statistically significant effect on all causes of cardiac mortality was observed [33]. It has been argued that a large part of this effect may be due to a single trial (the Scottish trial) since significance was not maintained if this trial was excluded from the meta-analysis [41, 33]. Tamoxifen has a known positive effect on blood lipids and on C reactive protein, all considered intermediate biomarkers of cardiovascular risk and a protective effect on death for myocardial infarction in patients with active coronary disease was observed in the prevention trial with tamoxifen [45].

Toremifene is the only SERM other than tamoxifen which has shown some activity in the treatment of breast cancer (Fig. 3). Toremifene is a chlorinated derivative with similar site specific activity of tamoxifen and in preclinical studies it has shown equivalent ER binding and anti-tumor efficacy [46]. Clinical studies confirm the cross-resistance beween tamoxifen and toremifene; in fact, no response was observed with either drug as second-line treatment of metastatic disease [47]. Randomized Phase III trials either in the advanced or in the adjuvant setting have confirmed the similar activity of toremifene and tamoxifen [48–50]. A potential advantage of toremifene could be related to a less uterotrophic effect than tamoxifen [51]. However, results from adjuvant studies do not support this hypothesis, with a rate of endometrial cancer which was similar in two randomized large studies [49–50].

Two other SERMs structurally related to tamoxifen are **droloxifene** and **idoxifene.** Both drugs did show less agonist effects than tamoxifen in preclinical studies. However, Phase III studies in patients with advanced breast cancer showed an advantage for tamoxifen as compared to droloxifene and a similar activity and toxicity when compared to idoxifene. The development of both drugs was stopped [46].

The second-generation SERMs share a different structure than the triphenylethylene derivative, the so-called 'fixed-ring' (Fig. 3). The most developed drugs of this group are raloxifene and arzoxifene (SERM3). **Raloxifene** was tested unsuccessfully in advanced breast cancer, but it is gaining a role in the prevention setting [52]. After the results of the MORE trial, showing a dramatic decrease of breast cancer incidence in postmenopausal women treated with raloxifene for osteoporosis [53], the drug is currently compared to tamoxifen in a large randomized trial (STAR) including about 20,000 patients [54]. Results are awaited in 2006.

Arzoxifene, a benzothiophene derivative of raloxifene completely devoid of estrogenic activity, has not shown significant clinical activity against advanced disease and it is currently being investigated as a chemopreventive agent [55].

Ovarian ablation and GnRH analogs

Since the early data of Beatson more than a century ago, oophorectomy represented the first means of endocrine manipulation for the treatment of breast cancer [3]. Randomized trials of ovarian ablation, obtained by surgery or radiation therapy, as adjuvant therapy were started in the 1940s [56].

Surgical ovarian ablation has the advantage of obtaining rapid and irreversible decrease of estradiol and to decrease the risk of an ovarian cancer. Moreover, the advent of laparoscopic surgery has significantly reduced the morbidity related to the procedure [57]. A former alternative to surgery was ovarian irradiation, which can be achieved by a single dose of 450 cGy, or, more commonly, with a total dose of 10–20 Gy administered in 5–6 fractions [56]. However, this procedure has a longer time to achievement of full effect, and may be unsuccessful in up to 35% of patients, depending on the age of the patients [58]. Since the 1990s the availability of synthetic analogs of gonadotropin releasing hormone (GnRH), which mimicks the structure of the hypothalamic peptide, led to an extended use of medical castration, preferred for the reversibility of its effect [59].

GnRH is a decapeptide synthesized in the diencephalon of the brain [60]. It is packaged into granules and released in synchronized pulses into the capillaries of the hypophyseal-portal circulation [60]. GnRH then binds selectively to highly specific receptors consisting of seven transmembrane domains located in the anterior pituitary gonadotrophic cells, thus stimulating in a pulsatile fashion, the synthesis and the release of LH and FSH, which, in turn, induces sex hormone secretion from ovaries and testis [60]. After the binding of GnRH to the receptor, an intracellular signal is triggered which ultimately leads to the activation of MAPkinase and to gonadotropin release [61]. **GnRH analogs** or **agonists** share structural homology with the natural decapeptide but they differ for the amino acid in positions 6 and 10 [59]. This difference protects the molecule from enzymatic degradation and brings a 100–200-fold higher affinity for the GnRH receptor. Synthetic GnRH analogs lead initially to an intense release of stored LH and FSH (flare-up effect), while prolonged administration desensitize the gonadotrophic cells by downregulation of GnRH receptors and dysregulation of the intracellular signalling, resulting in a decrease of gonadotropins and consequently of ovarian and testicular hormones [61]. This effect commonly occurs after 2–3 weeks. GnRH receptors have been found in breast and ovary tissues and also in some tumor cells (prostate, breast, endometrium) and a direct antitumor effect of GnRH analogs has been hypothesized but it has not been fully elucidated [62].

Since the half-life of the majority of the molecules when delivered in circulation is of a few hours, the most common delivery system of the GnRH analogs is represented by subcutaneous or intramuscular implants, conjugated in polymeric, which allow a 28-day administration [61].

Synthetic **GnRH antagonists**, which competitively bind to the receptor, have been less extensively investigated. Although devoid of the flare up effect, the clinical use of the first compounds was limited by the occurrence of a severe histaminic skin reaction. More recent drugs, such as cetrorelix, have shown no significant histamine-releasing effect and are under investigation in prostate and in breast cancer [62].

Few data have been reported on the comparison between different methods of ovarian ablation. A couple of randomized studies have compared surgical with medical ablation in premenopausal women with metastatic breast cancer; the results were inconsistent but both trials were closed prematurely because of poor accrual, thus leaving the question unresolved [63]. The reversibility of the effect upon discontinuation of treatment represents the major advantage but also a potential limit of medical castration, questioning the optimal duration of the treatment. Studies using GnRH analogs as a means of obtaining ovarian ablation have investigated treatment for 2–5 years in different populations, by age and risk of recurrence and concomitant treatments, but no direct comparison between different duration has yet been performed [63].

In premenopausal women with advanced ER positive breast cancer, ovarian ablation yielded a response rate up to 60%. A meta-analysis of the small randomized trials which compared ovarian ablation and tamoxifen did not find significant differences in response rate, time to progression and death between the two treatments. However, the meta-analysis of four randomized studies comparing the combination of tamoxifen plus GnRH analogs with single agent therapy showed a significantly increased survival, higher objective response rate and longer duration of response for the combination [64].

In 1995 the Oxford Overview of 12 randomized trials of adjuvant therapy, including more than 2,000 women obtaining ovarian ablation through oophorectomy or ovarian irradiation, reported a 25% reduction either in mortality and recurrence risk in women aged less than 50 years, a figure similar with that obtained with chemotherapy and tamoxifen [33, 56, 65]. Interestingly, hormone receptor studies was available in a minority of patients and that when used in addition with chemotherapy, ovarian ablation yielded a non-significant advantage, presumably because of the endocrine effects of chemotherapy [56]. The 2000 EBCTCG overview included for the first time more than 3,400 women under 50 years old receiving GnRH analogs, confirmed either the benefit of ovarian ablation *versus* no treatment and a lack of benefit when used after chemotherapy [35]. The fundamental role of ovarian suppression in the treatment of premenopausal women with endocrine responsive breast cancer was highlighted also by the retrospective analysis performed by the International Breast Cancer Study Group of IBCSG studies I, II, V and VI, showing that patients under 35 years old with hormone receptor (HR) positive tumors treated with chemotherapy had a worse disease free survival when compared either to the same age group with ER negative tumors or to older premenopausal patients (aged > 35 years) with HR positive tumors receiving the same treatment. In contrast, older patients with HR positive tumors fared better than those with HR negative tumors. The lower incidence of chemotherapy induced amenorrhea in younger patients may account for the worse prognosis observed in this subset of patients [66].

Since the 1990s a number of trials have compared the activity of ovarian suppression, obtained mainly through GnRH analogs with chemotherapy. The design of these trials substantially differed by experimental arm, which includ-

ed GnRH agonists alone or in combination with tamoxifen, patient characteristics, either for nodal status and hormone status with studies allowing patients
with hormone receptor negative tumors to be enrolled, and duration of ovarian
ablation, ranging from 2–5 years [67]. The results of the principal studies have
been reported in the past few years and are summarized in Table 1 [68–77].
Ovarian ablation was similar to chemotherapy in patients with endocrine
responsive breast cancer, while in patients with receptor negative tumors
chemotherapy was superior as for DFS and OS [68, 69]. The combination of
ovarian suppression plus tamoxifen was at least equivalent to chemotherapy,
with an advantage for relapse free survival in one study [71, 72]. It has been
argued that chemotherapy did not include taxane-based regimens, but was represented principally by CMF either classical or intravenous, although two
French studies which used anthracyclins containing regimens obtained comparable results [74, 75].

The addition of ovarian suppression to chemotherapy did not seem of benefit, as previously shown by the overview [69, 77]. However, in subgroup
analyses of both the IBCSG study VIII and the Intergroup 0101, the addition
of GnRH analogs resulted in an improved DFS in patients younger than 40
years or in patients not achieving permanent amenorrhea after chemotherapy
[69, 77].

Although these studies have enrolled thousands of patients, several questions remain unresolved:
a) the utility of ovarian suppression in women receiving chemotherapy especially in those maintaining or resuming ovarian activity
b) the optimal duration of ovarian suppression
c) the role of ovarian suppression in addition to tamoxifen
d) the role of chemotherapy in addition to an optimal endocrine therapy
e) the role of aromatase inhibitors in premenopausal women

A set of randomized studies, specifically addressing some of these issues is
currently being conducted by the Breast International Group and the North
American Intergroup. The BIG 02–02 SOFT (Ovarian Function Suppression
Trial) is designed to evaluate the role of ovarian function suppression in
women who remain premenopausal after surgery or after chemotherapy and in
addition to tamoxifen. The three arms include tamoxifen alone and the combination of ovarian suppression with either tamoxifen or exemestane. The BIG
04–02 PERCHE trial (CHemotherapy in Premenopausal Endocrine
Responsive trial) will evaluate the addition of chemotherapy to optimal
endocrine therapy (ovarian ablation + tamoxifen or exemestane). Finally, the
BIG 03–02 TEXT (Tamoxifen and Exemestane Trial) trial will evaluate the
role of aromatase inhibitors in comparison to tamoxifen in the adjuvant treatment of premenopausal women.

Table 1. Overview of the studies of ovarian suppression ± tamoxifen *versus* chemotherapy

Trial	Patients	Treatments	Results
ZEBRA [68]	1640 N+; ER+/ER–	Goserelin × 24 months *versus* CMF 1, 8 × 6 cycles	ER +ve: CMF *versus* Gos HR = 1.05 (0.88–1.24) p = NS ER–ve: CMF *versus* Gos HR = 1.83 (1.33–2.52) p = .0001
IBCSG VIII [69]	1063 N–; ER+/ER–	Goserelin × 24 months *versus* CMF *per os* × 6 cycles *versus* CMF × 6 cycles → Goserelin × 18 months	ER+ve: CMF *versus* Gos HR = 0.97 (0.66–1.92) p = NS CMF→ Gos *versus* Gos HR = 0.84 (0.56–1.26) p = NS Age ≤ 39 CMF→ Gos *versus* Gos HR = 0.34 (0.16–0.89) p = .02 ER–ve CMF *versus* Gos HR = 1.52 (0.89–2.58) p = NS
TABLE [70]	600 N+; ER+ve	Leuprorelin × 24 months *versus* CMF 1, 8 × 6 cycles	No difference
ABCSG 5 [71]	1099 N–ve/N+; ER or PgR+	CMF *per os* × 6 cycles *versus* Goserelin × 3 years + TAM 30 mg × 5 years	CMF *versus* Gos + TAM HR = 1.4 (1.06–1.87) p = .017
GROCTA 2 [72]	244 N–/+; ER+	CMF × os × 6 cycles *versus* Goserelin × 2 years + TAM 20 mg × 5 years	CMF *versus* Gos + TAM HR = 0.98 (0.66–1.47) p = NS
ZIPP [73]	2631 N–/N+; ER–ve/ER+ve	Goserelin × 2 years *versus* TAM × 2 years *versus* Goserelin × 2 years + TAM × 2 years No hormonal therapy [1]	Goserelin > no Goserelin HR = 0.77 (0.66–0.89) p < . 001
FRENCH [74]	162 N+/ER+ve	FAC × 6 cycles *versus* OA + TAM 30 mg × 2 years	DFS 55% *versus* 83% p = NS
FASG 06 [75]	333 N+/ER+ve	FEC × 6 cycles *versus* Triptorelin × 3 years + TAM 30 mg × 3 years	DFS 81% *versus* 92% p = NS

(Continued on next page)

Table 1. (Continued)

Trial	Patients	Treatments	Results
IBCSG XI [76]	174 N+, ER+ve	AC × 4 cycles + OA + TAM 20 mg × 5 years *versus* OA + TAM 20 × 5 years	DFS 88% *versus* 87% p = NS
INT- 0101 [77]	1504 N+/ER+ve	CAF × 6 cycles *versus* CAF × 6 cycles + Goserelin × 5 years *versus* CAF + Gos + TAM 20 mg × 5 years	CAF + Gos HR = 0.93 (0.76–1.14) p = NS CAF + Gos + TAM = 0.73 (0.59–0.90) p < .01

N+ node positive; N– node negative; ER–ve estrogen receptor negative; ER+ve estrogen receptor positive; HR = hazard ratio (95% Confidence Intervals); Gos = goserelin; TAM = tamoxifen; OA = ovarian ablation (oophorectomy/radiotherapy or by GnRH analog); [1] 43% of patients received chemotherapy

Aromatase inhibitors

An alternative mean to interfere with the estrogen receptor is represented by the inhibition of estrogen biosynthesis (Fig. 1). In postmenopausal women the primary source of estrogens is represented by peripheral tissues, mainly adipose tissue. Breast tissue, as well, has been found to have several-fold higher levels of estrogen than those in plasma [78].

The classical pathway of estrogen biosynthesis starts with cholesterol and comprises a series of steps till the transformation in estrogens. The last step in this sequence is catalyzed by an enzyme called aromatase, which converses androgen substrate (Δ-4-androstenedione, testosterone) in estrogens (estrone, estradiol). Aromatase is an enzyme of the cytochrome P-450 family and is coded by the CYP19 gene [7]. The inhibition of aromatase is most specific and does not affect the biosynthesis of other steroid classes. Since androstenedione is the preferred substrate for aromatization, full estrogenic activity requires the conversion of estrone in estradiol through the enzyme 17-β-dehydrogenase [79]. Increased aromatase activity was found in approximately 60% of breast tumors [79, 80].

Aromatase inhibitors (AIs) are classified according either to different stages of development or to mechanism of action [81, 82]. According to the first classification, three generations of aromatase inhibitors have been developed in clinical practice. The first compound, which was proven active in breast cancer in the 1960s, was aminoglutethimide. This drug leads to a medical adrenalectomy, since it inhibits, along with aromatase other steps of steroidogenesis as the 11-β-hydroxylase which mediates the synthesis of cortisol and other P-450 enzymes. Because of this lack of selectivity the drug, albeit it's antitumor activity causes a number of toxicities (lethargy, drowsiness, skin rash) which were considered unacceptable for large use [81]. The second generation of AI included formestane and fadrozole. These two drugs, although more selective than aminoglutethimide (achieving an inhibition of aromatase activity by 90%), had a limited use. Formestane, as a result of first-pass metabolism, cannot be administered orally and has to be given twice monthly as intramuscular injection, leading to reports of local reactions in 17% of patients [81, 83]. Conversely, fadrozole partially inhibited 11- and 18-β-hydroxylase with decrease in serum cortisol and aldosterone [7]. In addition no advantage was shown in comparison with megestrole acetate, and currently this drug is available only in Japan [84]. The third generation of AIs includes the triazoles **anastrozole**, **letrozole**, vorozole and the steroidal **exemestane**; all compared favorably with earlier AIs for aromatase inhibition and oestrogen suppression, showing higher selectivity (Fig. 3).

According to the mechanism of action, AIs are classified in 2 groups. Type 1 inhibitors are steroidal analogs of androstenedione which bind irreversibily to the catalytic site on the aromatase molecule, causing loss of the enzyme activity. They are known also as enzyme inactivators or suicide inhibitors. Formestane and exemestane share this mechanism of action. Because of their steroidal structure, exemestane and its metabolite 17-hydroexemestane have the potential for androgenic effects. The affinity of the metabolite for the androgen receptor is about 100 times that of the parent compound [82]. Type 2 inhibitors include the triazoles and reversibly interact with the cytochrome P-450 moiety of the enzyme and their activity is dependent on the continued presence of the drug [82].

Some differences in pharmacokinetic properties of the three drugs have been reported with anastrozole and exemestane attaining steady-state after 7 days, while letrozole takes 60 days to achieve steady-state plasma levels. In addition, half-life is longer for the non-steroidal AIs [84].

All third-generation AIs compared favorably with earlier AIs for aromatase inhibition and oestrogen suppression [85, 86]. Significant differences in the extent of aromatase inhibition, estrone and estrone sulphate have been reported favoring letrozole over anastrozole, while estrogen suppression was only marginally greater with letrozole [87].

However, no dose response effect of two different doses of letrozole on aromatase inhibition and estradiol suppression were observed, although an improved clinical activity for the higher dose was reported only in one Phase III trial [21, 88]. Conversely, anastrozole may be more selective for the enzyme. In fact, no impact on cortisol and aldosterone has been reported in patients receiving anastrozole while contradictory data have been reported for letrozole, showing a significant decrease in plasma cortisol although not below normal levels [84].

It is currently unknown whether the different mechanisms of action and different potencies have any clinical implication. However, this represents the rationale for the sequential use of the different type of AIs. Some activity has been shown with either sequence (irreversible $\rightarrow$ reversible and the opposite sequence) in the advanced setting. Preliminary results of a crossover trial showed that women receiving a non-steroidal AI after failure with exemestane achieved a 10% of objective responses (OR) and a 47% of clinical benefit (CB) defined as response rate + stable disease lasting $\geq$ 24 weeks, while exemestane obtained a 4% of OR and a 25% of clinical benefit after anastrozole or letrozole. Studies are ongoing to prospectively address the issue of the optimal sequencing of AIs [89].

In premenopausal women the use of AIs leads to an increase in gonadotropin secretion because of the reduced feedback of estrogens on hypothalamus and pituitary and a subsequent stimulation of ovarian activity. Short-term letrozole has been successful for the induction of ovulation in women with infertility [90]. Thus, up to now the clinical development of AIs has been limited to postmenopausal women [81].

Aromatase inhibitors have shown significant clinical activity on breast cancer in all settings. As second-line treatment in tamoxifen-resistant breast cancer, the third-generation AIs showed improved activity in at least one of the clinical outcome measures (response rates [RR], time to progression [TTP] and time to treatment failure) as compared to megestrol acetate in a series of Phase III studies [21, 91, 92].

The results of the randomized studies of AIs as front-line therapy in comparison with tamoxifen in metastatic disease seem to substantiate the superiority of AIs. Letrozole was significantly superior in terms of TTP (9.4 months *versus* 6.0 months, P = .0001), RR (32% *versus* 21%, P = .0002), and CB (50% *versus* 38%, P = .0004), independent of disease site, receptor status, or prior adjuvant anti-oestrogen therapy. No significant difference in OS was observed (34 months for letrozole and 30 months for tamoxifen) [93]. The results of the two studies, which compared anastrozole with tamoxifen, are somewhat conflicting. A recent combined analysis of the two trials confirmed overall the lack of difference between the two treatments, while anastrozole was superior only in a retrospective subgroup analysis of patients with hormone receptor positive tumors (time to progression 10.7 months *versus* 6.4 months p = 0.0.22) [94]. Exemestane also has been compared with tamoxifen in a smaller Phase III double-blind randomized study. Results showed a significantly higher response rate (46% *versus* 31%) and a longer TTP (9.9 *versus* 5.8 months) for exemestane *versus* tamoxifen, although, similarly with other AIs, no advantage in OS was observed [95].

Given the results in the advanced disease, the activity of AIs as adjuvant treatment of postmenopausal patients with HR-positive breast cancer was compared to tamoxifen in a number of large randomized double-blind trials.

The first trial showing an advantage for an AI as compared to tamoxifen is the Arimidex Tamoxifen Alone or in Combination (ATAC) trial, first reported in 2002 and updated in 2005 [96]. The trial compared 5-year treatment with tamoxifen, anastrozole or the combination of both in postmenopausal women mostly with HR positive tumors. However, the combination arm was stopped after the first analysis at 33 months due to the lack of benefit as compared to tamoxifen.

The late results of this study at a median follow-up of 68 months are summarized in Table 2. Importantly, in HR positive tumors the absolute differences in recurrence rate appears to increase with time (1.7% at 2 years and 3.7% at 6 years) suggesting the occurence of a carry over effect beyond treatment discontinuation which appears to be greater than that observed with tamoxifen, although the observation is more limited in time [96]. Interestingly, a retrospective analysis showed that the subpopulation of patients with ER+/PgR– tumors who received anastrozole experienced a 57% reduction (HR = 0.43 95% CI 0.31–0.61 p < .0001) in breast cancer events as compared to tamoxifen treated patients, while in the ER+/PgR+ subpopulation a 16% reduction was observed (HR = 0.84 95% CI 0.69–1.02 p = .07) [97]. Although highly provoking, these data should be considered with caution firstly since they derived from a retrospective unplanned analysis, and secondly, they were not confirmed by the results of the other trials comparing AIs with tamoxifen; last, but not least, the determination of HR was performed by a variety of assays. However,

R. Torrisi et al.

Table 2. Principal results of the adjuvant trials with aromatase inhibitors

TRIAL	pts	DFS (95% CI)	DDFS (95% CI)	Absolute RR (time)	TTR (95% CI)	OS (95% CI)	Subgroup analyses
ATAC [96, 97]	9366 61% N-ve	HR = 0.87 (0.78–0.97) p = .01	HR = 0.86[1] (0.74–0.99) p = .04	3.3% (6 years)	HR = 0.79 (0.70–0.90) p = .0005	HR = 0.97 (0.85–1.12) p = 0.7	ER+/PgR-ve TTR HR = 0.43 (0.31–0.61)
BIG 01-98 [98]	8010 57% N-ve	HR = 0.81 (0.70–0.93) p = .003	HR = 0.73 (0.60–0.88) p = .001	2.6% (5 years)	HR = 0.72 (0.61–0.86) p < .001	P = 0.16	N+ve DFS HR = 0.71 (0.59–0.85) p < .0001
IES [99]	4742 51% N-ve	HR = 0.68 (0.56–0.82) p < .001	HR = 0.66 (0.52–0.83) p = .0004	4.7% (3 years)	NR	HR = 0.88 (0.67–1.16) p = 0.37	No difference
ABCSG 08/ ARNO [101]	224 74% N-ve	HR = 0.60 (0.44–0.81) p = .0009	HR = 0.61 (0.42–0.87) p = .0067	3.1% (3 years)	NR	P = 0.16	ER+/PgR -ve DFS HR = 0.42 (0.19–0.92) p < .03
ITA [102]	448 0% N-ve	HR = 0.35 (0.18–0.68) p = .001	HR = 0.49 (0.22–1.05) p = .06	5.8% (3 years)	NR	P = 0.1	Not reported
MA.17 [103, 104]	5187 50% N-ve	HR = 0.58 (0.45–0.76) p < .001	HR = 0.60 (0.43–0.84) p = .002	4.6% (4 years)	NR	HR = 0.82 (0.57–1.19) p = 0.3	N+ve OS HR = 0.61 (0.38–0.98) p = .04

DFS = disease free survival; DDFS = distant disease free survival; RR = risk reduction; time = time from randomization; TTR = time to recurrence; OS = overall survival; HR = hazard ratio; N-ve = node negative; ER+ = estrogen receptor positive; PgR-ve = progesterone receptor negative; [1] results were reported as time to distant recurrence (see text for treatment description) this finding is consistent with clinical and preclinical data which have associated the PgR loss to higher HER expression and resistance to tamoxifen [39].

A second trial, the Breast International Group (BIG) 1–98, reported recently, has compared upfront tamoxifen and letrozole for 5 years, the sequence of 2-year tamoxifen followed by letrozole for 3 years and the reciprocal sequence of the two agents [98]. The primary core analysis included patients from the four study arms, excluding events occurring after 30 days after crossover in the sequential treatment arms. Median follow up was 25.8 months and the principal results are reported in Table 2. Noticeably, a greater effect of letrozole was observed in reducing recurrence at distant sites, in node positive patients and in patients who received chemotherapy. An increase in grade 3–5 cardiac events was observed in patients receiving letrozole, possibly related at least in part to the cardioprotective effect of tamoxifen [98].

The theoretical rationale of investigating the switching to an AI after 2–3 years of tamoxifen is based on the knowledge that resistance to tamoxifen, due to its agonistic activity, usually arises after 18 months; moreover the safety concerns on endometrial cancer and thromboembolic events increase for longer treatment duration as shown in the NSABP B14 study [43]. Conversely, the agonistic effects of tamoxifen on blood lipids and bone resorption may reduce the concern on the effects for prolonged treatment with aromatase inhibitors on cardiac event risk and fractures. The sequential trials include the Intergroup Exemestane Study (IES), the Austrian Breast Cancer Study Group (ABCSG) 8-ARNO 95 and the smaller Italian Tamoxifen Anastrozole (ITA) trial [99–102].

The IES is the largest of these trials and compared 5-year tamoxifen with the sequence tamoxifen for 2–3 years followed by exemestane. Results are described in Table 2. Interestingly, all patient subgroups equally benefited by switching to exemestane. Patients receiving exemestane had a significantly reduced risk of developing a new non-breast primary cancer but a 2.5-fold increase in myocardial infarction was observed in this group in the updated analysis at 42 months [99, 100].

In the combined analysis of the ABCSG study 8 and ARNO 95, patients receiving anastrozole after 2 years of tamoxifen had a 40% decrease of a breast event (local or distant recurrence and contralateral breast cancer) (see Tab. 2). Differently from other trials, two-thirds of the patients included in this study were node negative and the vast majority (94%) had low-intermediate grade tumors. Similarly to the ATAC trial, a subgroup analysis showed a trend for an increased efficacy of anastrozole on ER+/PgR– tumors [101].

The ITA trial, which included 448 patients with node positive ER positive breast cancer who were switched to anastrozole after 2–3 years of tamoxifen, showed similarly after a follow up of 36 months a clear benefit for the switching arm (Tab. 2) [102].

The rationale of extending treatment with AIs beyond 5 year tamoxifen has been reported above. The MA.17 trial was designed to assess the activity of letrozole after the completion of adjuvant tamoxifen and randomized patients to receive 5 years of letrozole or placebo. The trial was closed prematurely after a median follow up of 2.4 years based on the significant improved 4-years DFS observed in the letrozole arm (93% *versus* 87% HR = .0.57 95% CI 0.43–0.75 p = .00008) [103]. The final results at a median follow up of 30 months also showed an improvement in OS, which was limited to node-positive patients. According to the results of this trial, letrozole was recently approved by the US Food and Drug Administration (FDA) for extended adjuvant treatment in patients completing 5 years of tamoxifen [104]. A second trial, with a similar design but considering exemestane after 5 years of tamoxifen, conducted by the NSABP (NSABP B-33) suspended its accrual after the early disclosure of the MA.17 results. Adverse events and major toxicities reported in the larger five trials are summarized in Table 3.

Which aromatase inhibitor has the greater activity and better tolerability? It may be speculated that the known structural and pharmacokinetic differences between AIs may translate in different potencies and safety profiles. Only one study up to now has compared letrozole and anastrozole in second-line therapy of advanced breast cancer, showing a higher response rate for letrozole without any advantage on other outcome measures (TTP, duration of response) [105]. However, no other direct comparison between AIs is available and data are extrapolated from comparisons with tamoxifen. Moreover, differences in study populations, treatment duration, previous 'priming' with tamoxifen and also endpoint definition should be considered when comparing activity and tolerability of the AIs. A large Phase III randomized trial (MA.27) is currently active in North America and in Europe comparing anastrozole with exemestane and is expected to enroll about 8,000 patients.

A great debate is ongoing in the scientific community as to whether the upfront use of an AI is superior to the switch to an AI after 2–3 years of tamoxifen. Although the size of benefit appears greater in the sequential studies, these studies include an exquisite endocrine responsive population, who did

Table 3. Principal adverse events reported in the adjuvant trials with aromatase inhibitors

TRIAL	Thombo-embolism (%)	Cardiac events (grade 3–5) (%)	Fractures (%)	Musculoskeletal symptoms (%)	Vasomotor symptoms (%)
ATAC [96]	2.8	4.1	11	35.6	35.7
BIG 01-98 [98]	1.5	3.7	5.7	26.7	22.7[*]
IES [99, 100]	1.3	42.6[1]	3.1	38.6	24[*]
ABCSG 8/ ARNO95 [101]	0.3	0.2[2]	2	19	48
MA.17 [104]	NR	5.8	5.3	45	35*

Musculoskeletal symptoms included bone pain, arthralgia and myalgia; vasomotor symptoms included hot flashes and sweating. [*] only grade ≥ 2 were considered; [1] any grade cardoiovascular disease except myocardial infarction were included; [2] only myocardial infarction rate was reported not relapse early during endocrine treatment and direct comparisons of the results with upfront treatment are inappropriate. The definitive results of the four-arm comparison of the BIG 1-98 trial will help to untangle this dilemma.

Finally the ASCO Technology Assessment stated in 2004 that all post-menopausal women with hormone receptor positive breast cancer should receive an AI as part of the adjuvant treatment [106]. However, some concern should be raised on the non-significant increase in non-breast cancer related deaths observed in some trials in patients treated with AIs and the balance with the expected gain in breast cancer related events should be taken into account individually in treatment decisions. In addition, it should be remembered that no survival advantage has been observed up to now for either AI.

The above-reported results have prompted the investigation of the activity and the safety of AIs in premenopausal patients. A series of studies (BIG 02–02 SOFT, BIG 03–02 TEXT, BIG 04–02 PERCHE and ABCSG 12) comparing exemestane and anastrozole, both in association with a GnRH analog, with tamoxifen as adjuvant treatment of premenopausal women with HR breast cancer are currently active.

Selective estrogen receptor downregulator

Fulvestrant (ICI 182780) is a selective estrogen receptor downregulator (SERD), which behaves as a pure estrogen receptor antagonist. It binds competitively to the ER with high affinity, which is 89% that of estradiol, but much greater than that of tamoxifen, which in turn is only 2.5% that of estradiol [107]. The binding to the ER downregulates the receptor by preventing dimerization, the binding to the ERE, and the uptake into the nucleus of the ER. In any case, the fulvestrant-ER complex is transcriptionally inactive because both AF-1 and AF-2 are disabled [107]. The fulvestrant-ER complex is unstable, resulting in accelerated degradation of the receptor [108]. The downregulation of the cellular levels of ER protein leads to a complete abrogation of the transcription of the estrogen-regulated genes as PgR and pS2 [10]. Moreover, fulvestrant showed an antiestrogenic effect on the endometrium. The disruption of both AF-1 and AF-2 sites implies that, differently from tamoxifen, fulvestrant is completely devoid of agonist activity and thus it is considered a pure antiestrogen [107].

Due to the prolonged presence of active plasma concentrations the drug may be administered at a 28-day interval as an intramuscular injection. Pharmacokinetics of the drug is not affected by liver or renal impairment [109].

In experimental models, both *in vitro* and *in vivo*, fulvestrant proved to be more active than tamoxifen in inhibiting breast cancer growth and showed antitumor effect after tamoxifen failure [107, 110]. A dose ranging trial of three different doses of fulvestrant compared also with tamoxifen and placebo was conducted in 201 postmenopausal women with untreated ER-positive or unknown breast cancer, evaluating intermediate endpoints as the effect on ER and PgR expression and the proliferative acitivity measured by the Ki67 antigen. Fulvestrant induced a significant dose-dependent reduction of ER and PgR expression and of ki67 as compared with placebo, while only the highest dose (250 mg) was more effective than tamoxifen in reducing ER histochemical scores [111]. The dose of 250 mg every 4 weeks as an intramuscular injection was identified as the standard dose to be used in Phase II trials.

Activity of fulvestrant in tamoxifen-resistant advanced breast cancer was investigated in a Phase II study [112]. A clinical benefit was obtained in 69% of patients with a 37% partial clinical response. Most interestingly, five out of seven responding patients still maintained remission after 30–33 months, with a median duration of response of 26 months which was significantly higher than that reported with megestrol acetate in the same patient population [112].

The drug is well tolerated and shows a side effect profile which is consistent with estrogen deprivation as vasomotor symptoms, although hot flushes were less frequently reported than tamoxifen and other symptoms as nausea, asthenia, and headache with a frequency comparable to anastrozole [113]. Injection site reaction occurred in about 7% of patients. The effect of fulvestrant on bone density *in vivo* needs to be clarified while clinical data suggest that fulvestrant does not significantly affect blood lipids [112].

Clinical efficacy of fulvestrant in comparison with anastrozole was investigated in two randomized trials, conducted in several countries primarily in Europe, North America and Australia, involving 851 postmenopausal women with ER-positive advanced breast cancer progressing on adjuvant or first-line antiestrogen therapy [113]. The planned combined analysis of the two trials reported a TTP of 5.5 *versus* 4.1 months, and a RR of 19.2% *versus* 16.5% for fulvestrant and anastrozole, respectively, and a median duration of response which still favored fulvestrant (16.7 *versus* 13.7 months) [113]. A subgroup analysis performed according to the site of metastatic disease, showed no difference in the rate of objective response in both visceral and non-visceral sites. Preliminary analysis of survival, at an extended follow-up of 27 months showed no difference between the two treatments (74.5% *versus* 76.1% of patients dead in the two arms) [114].

In a Phase III trial comparing fulvestrant and tamoxifen as first-line therapies in postmenopausal patients with advanced breast cancer, the TTP favored, although not significantly, tamoxifen (8.3 *versus* 6.8 months) [115]. However, a relevant clinical benefit was observed after treatment with AIs, demonstrating the lack of cross-resistance with other endocrine agents [112]. Moreover, *in vitro* data suggest that fulvestrant may be more effective when estrogen levels are maintained at low levels, providing a rationale for the combination with AIs, also in patients progressing on these agents [112]. These data support an important role for fulvestrant in the panel of endocrine tools available for postmenopausal women with endocrine responsive advanced breast cancer. The proper sequence is unknown and is currently investigated in a number of ongoing Phase III trials.

Conclusions

Approximately two-thirds of all breast cancers are hormone receptor positive. However, until the early 1990s tamoxifen, and to a lesser extent oophorectomy, represented the only endocrine options that were offered to these patients and there were very few chances of response with further endocrine manipulations after progression on these agents.

In the past 10 years the armamentarium of endocrine agents has expanded with the identification of more selective drugs devoid of agonistic properties as third-generation aromatase inhibitors and SERDs. The sequential use of these non-cross-resistant agents has improved and prolonged the chance of manipulating growth of hormone responsive advanced tumors. On the other hand, the differences in toxicity profile and biological targets exhibited by each drug will allow, in the future, to tailor adjuvant therapies according to tumor biology and individual risks of coexisting morbidities in early breast cancer. Moreover SERMs, such as tamoxifen and raloxifene, have proven to be effective in preventing the occurrence of hormone-dependent tumors and third-generation AIs represent a most promising option in this setting.

The development of gene expression profiling techniques attempts to better characterize the molecular patterns of immunohistochemical hormone-receptor breast cancers with the aim to identify molecular predictors of endocrine responsiveness [116]. A first step has been the definition of a 21-gene assay which proved to be effective in predicting which ER-positive breast cancer patients are adequately treated with tamoxifen within the NSABP studies B-14 and B-20 [117, 118].

The increased knowledge of the ER biology and complexity allows a better understanding of the mechanisms of *de novo* and acquired resistance to endocrine agents and to depict the molecular targeted approach to overcome this resistance. For example, new drugs such as anti-receptor antibodies and small tyrosine kinase inhibitors, interfering with the growth factor pathway, appear as promising strategies to restore or delay the resistance to endocrine agents [19, 31].

The availability of multiple drugs raises the issue of determining the optimal sequence and timing of each agent. Preclinical data and *ad hoc* designed studies will help to resolve this question.

Finally, the last two St Gallen Consensus Conferences have recognized endocrine responsiveness as a crucial criterion within all risk categories for deciding adjuvant therapy for early breast cancer [119, 120]. Endocrine therapy, once considered adequate for patients not suitable for receiving more aggressive and 'active' treatments, has gained a leading role among the therapeutic tools available for the management of hormone receptor positive breast cancer at all stages.

References

1 Yue W, Santen RJ, Wang JP, Li Y, Verderame MF, Bocchinfuso WP, Korach KS, Devanesan P, Todorovic R, Rogan EG et al. (2003) Genotoxic metabolites of estradiol in breast: potential mechanism of estradiol induced carcinogenesis. *J Steroid Biochem Mol Biol* 86: 477–486

2 Key TJ, Appleby, Barnes I, Reeves G, Endogenous hormones and breast cancer collaborative group (2002) Endogenous sex hormones and breast cancer in postmenopausal women: reanalysis of nine prospective studies. *J Natl Cancer Inst* 94: 606–616

3 Beatson GT (1896) On the treatment of inoperable cases of carcinoma of the mamma: suggestions for a new method of treatment with illustrative cases. *Lancet* 2: 104–107

4 Goldhirsch A, Colleoni M, Gelber RD (2002) Endocrine theray of breast cancer. *Ann Oncol* 13 S4: 61–68

5 Osborne CK, Zhao H, Fuqua SA (2000) Selective estrogen receptor modulators: structure, function, and clinical use. *J Clin Oncol* 18: 3172–3186

6 Harvey JM, Clark GM, Osborne CK, Allred DC (1999) Estrogen receptor status by immunochemistry is superior to the ligand-binding assay for predicting response to adjuvant endocrine therapy in breast cancer. *J Clin Oncol* 17: 1474–1481

7 Miller WR (2003) Aromatase Inhibitors: mechanism of action and role in the treatment of breast cancer. *Sem Oncol* 30 (S14): 3–11

8 Beato M, Herrlich P, Schutz G (1995) Steroid hormone receptors: many actors in search of a plot. *Cell* 83: 851–857

9 Osborne CK, Schiff R (2005) Estrogen-receptor biology: continuing process and therapeutic implications. *J Clin Oncol* 23: 1616–1622

10 Osborne CK, Shou J, Massarweh S, Schiff R (2005) Crosstalk between estrogen receptor and growth factor receptor pathways as a cause for endocrine therapy resistance in breast cancer. *Clin Cancer Res* 11: 865S–870S

11 Hall JM, McDonnell DP (1999) The estrogen receptor β-isoform (ERβ) of the human estrogen receptor modulates ERα transcriptional activity and is a key regulator of the cellular response to estrogens and antiestrogens. *Endocrinology* 140: 5566–5578

12 Smith CL, O'Malley BW (1999) Evolving concepts of selective estrogen receptor action: from basic science to clinical applications. *Trends Endocrinol Metab* 10: 299–300

13 McKenna NJ, Lanz RB, O'Malley BW (1999) Nuclear receptor coregulators: cellular and molecular biology. *Endocrine Rev* 20: 321–344

14 Shiau AK, Barstad D, Loria PM, Cheng L, Kushner PJ, Agard DA, Greene GL (1998) The structural basis of estrogen receptor/coactivator recognition and the antagonism of this interaction with tamoxifen. *Cell* 95: 927–937

15 Lewis JS, Jordan VC (2005) Selective estrogen receptor modulators (SERMs): mechanisms of anticarcinogenesis and drug resistance. *Mut Res* 591: 247–263

16 Smith CL, Nawaz Z, O'Malley BW (1997) Coactivator and corepressor regulation of the agonist/antagonist activity of the mixed antiestrogen 4-hydroxytamoxifen. *Mol Endocrinol* 11: 657–666

17 Shou J, Massarweh S, Osborne CK, Wakelin AE, Ali S, Wiss H, Schiff R (2004) Mechanisms of tamoxifen resistance: increased estrogen receptor-HER2/neu cross-talk in ER/HER2-positive breast cancer. *J Natl Cancer Inst* 96: 926–935

18 Kushner PJ, Agard DA, Greene GL, Scanlan TS, Shiau AK, Uht RM, Webb P (2000) Estrogen receptor pathway to AP-1. *J Steroid Biochem Mol Biol* 74: 311–317

19 Normanno N, Di Maio M, De Maio E, De Luca A, de Matteis A, Giordano A, Perrone F (2005) Mechanisms of endocrine resistance and novel therapeutic strategies in breast cancer. *Endocrine Related Cancer* 12: 721–747

20 Gutierrez MC, Detre S, Johnston S, Mohsin SK, Shou J, Allred DC, Schiff R, Osborne CK, Dowsett M (2005) Molecular changes in tamoxifen-resistant breast cancer. Relationship between estrogen receptor, HER-2 and p38 mitogen-activated protein kinase. *J Clin Oncl* 23: 2469–2476

21 Dombernowsky P, Smith I, Falkson G, Leonard R, Panasci R, Bellmunt J, Bezwoda W, Gardin G, Gudgeon A, Morgan M et al. (1998) Letrozole, a new oral aromatase inhibitor for advanced breast cancer: double-blind randomized trial showing a dose effect and improved efficacy and tolerability compared with megestrol acetate. *J Clin Oncol* 16: 453–456

22 Speirs V, Malone C, Walton DS, Kerin MJ, Atkin SL (1999) Increased expression of estrogen receptor-beta mRNA in tamoxifen-resistant breast cancer patients. *Cancer Res* 59: 5421–5424

23 Cappelletti V, Celio L, Bajetta E, Allevi A, Longarini R, Miodini P, Villa R, Fabbri A, Mariani L, Giovannazzi R et al. (2004) Prospective evaluation of estrogen receptor-beta in predicting response to neoadjuvant antiestrogen therapy in elderly breast cancer patients. *Endocrine Related Cancer* 11: 761–770

24 Lavinsky RM, Jepsen K, Heinzel T, Torchia J, Mullen TM, Schiff R, DelRio AL, Ricote M, Ngo S, Gemsch J et al. (1998) Diverse signaling pathways modulate nuclear receptor recruitment of N-CoR and SMRT complexes. *Proc Natl Acad Sci USA* 95: 2920–2925

25 Yarden Y, Sliwkowski MX (2001) Untangling the ErbB signalling network. *Nat Rev Mol Cell Biol* 2: 127–137

26 Shim WS, Conaway M, Masamura S, Yue W, Wang JP, Kmar R, Santen RJ (2000) Estradiol hypersensitivity and mitogen-activated protein kinase expression in long-term estrogen deprived human breast cancer cells *in vivo*. *Endocrinology* 141: 396–405

27 Santen RJ, Song RX, Zhang Z, Kumar R, Jeng MH, Masamura S, Yue W, Berstein L (2003) Adaptive hypersensitivity to estrogen: mechanism for superiority of aromatase inhibitors over selective estrogen receptor modulators for breast cancer treatment and prevention. *Endocrine Related Cancer* 10: 111–130

28 Osborne CK, Bardou V, Hopp TA, Chamness GC, Hilsenbeck SG, Fuqua SA, Wong J, Allred DC, Clark GM, Schiff R (2003) Role of the estrogen receptor coactivator AIB1 (SRC-3) and HER-2/neu in tamoxifen resistance in breast cancer. *J Natl Cancer Inst* 95: 353–361

29 De Placido S, De Laurentiis M, Carlomagno C, Gallo C, Perrone F, Pepe S, Ruggiero A, Marinelli A, Pagliarulo C, Panico L et al. (2003) Twenty-year results of the Naples GUN randomized trial: Predictive factors of adjuvant tamoxifen efficacy in early breast cancer. *Clin Cancer Res* 9: 1039–1046

30 Ellis MJ, Coop A, Singh B, Mauriac L, Llombert-Cussac A, Janicke F, Miller WR, Evans DB, Dugan M, Brady C et al. (2001) Letrozole is more effective neoadjuvant endocrine therapy than tamoxifen for ErbB-1- and/or ErbB-2-positive, estrogen receptor-positive primary breast cancer: evidence from a phase III randomized trial. *J Clin Oncol* 19: 3808–3816

31 Polychronis A, Sinnet HD, Hadjiminas D, Singhal H, Mansi JL, Shivapatham G, Shousha S, Jang J, Peston D, Barret N et al. (2005) Preoperative gefitinib *versus* gefitinib and anastrozole in postmenopausal patients with oestrogen-receptor positive and epidermal-growth-factor-receptor-positive primary breast cancer: a double-blind placebo-controlled phase II randomised trial. *Lancet Oncol* 6: 383–391

32 Osipo C, Gajdos C, Cheng D, Jordan VC (2005) Reversal of tamoxifen resistant breast cancer by low dose estrogen therapy. *J Steroid Biochem Mol Biol* 93: 249–256

33 Early Breast Cancer Trialists' Collaborative Group (1998) Tamoxifen for early breast cancer: an overview of the randomized trials. *Lancet* 351: 1451–1467

34 Cuzick J, Powles T, Veronesi U, Forbes J, Edwards R, Ashley S, Boyle P (2003) Overview of the main outcomes in breast-cancer prevention trials. *Lancet* 361: 296–300

35 Early Breast Cancer Trialists' Collaborative Group (2005) Effects of chemotherapy and hormonal therapy for early breast cancer on recurrence and 15-year survival: an overview of the randomized trials. *Lancet* 365: 1687–1717

36 Ferno M, Stal O, Baldetorp B, Hatschek T, Kallstrom AC, Malmstrom P, Nordenskjold B, Ryden S (2000) Results of two or five years of adjuvant tamoxifen correlated to steroid receptor and S-phase levels: South Sweden Breast Cancer Group and South-east Sweden Breast Cancer Group. *Breast Cancer Res Treat* 59: 69–76

37 Bardou VJ, Arpino G, Elledge RM, Osborne CK, Clark GM (2003) Progesterone receptor status significantly improves outcome prediction over estrogen receptor status alone for adjuvant endocrine therapy in two large breast cancer databases. *J Clin Oncol* 21: 1973–1979

38 Konecny G, Pauletti G, Pegram M, Untch M, Dandekar S, Aguilar Z, Wilson C, Rong HM, Bauerfeind I, Felber M et al. (2003) Quantitative association between HER-2/neu and steroid hormone receptors in hormone receptor-positive primary breast cancer. *J Natl Cancer Inst* 95: 142–153

39 Arpino G, Weiss H, Lee AV, Schiff R, De Placido S, Osborne CK, Elledge RM (2005) Estrogen receptor-positive, progesterone receptor-negative breast cancer: association with growth factor receptor expression and tamoxifen resistance. *J Natl Cancer Inst* 97: 1254–1261

40 Saphner T, Tormey DC, Gray R (1996) Annual hazard rates of recurrence for breast cancer after primary therapy. *J Clin Oncol* 14: 2738–2746

41 Stewart HJ, Prescott RJ, Forrest AP (2001) Scottish adjuvant tamoxifen trial: a randomized study updated to 15 years. *J Natl Cancer Inst* 93: 456–462

42 Fisher B, Jeong JH, Bryant J, Anderson S, Dignam J, Fisher ER, Wolmark N; National Surgical Adjuvant Breast and Bowel Project (2004) Treatment of lymph-node-negative, oestrogen-receptor-positive breast cancer: long-term findings from National Surgical Adjuvant Breast and Bowel Project randomised clinical trials. *Lancet* 364: 858–868

43 Fisher B, Jeong JH, Dignam J, Anderson S, Mamounas E, Wickerham DL, Wolmark N (2001) Findings from recent National Surgical Adjuvant Breast and Bowel Project adjuvant studies in stage I breast cancer. *J Natl Cancer Inst Monogr* 30: 62–66

44 Braithwaite RS, Chebowski RT, Lau J, George S, Hess R, Col N (2003) Meta-analysis of vascular and neoplastic events associated with tamoxifen. *J Gen Intern Med* 18: 937–947

45 Reis SE, Costantino JP, Wickerham DL, Tan-Chiu E, Wang J, Kavanah M (2001) Cardiovascular effects of tamoxifen in women with and without heart disease: breast cancer prevention trial. National Surgical Adjuvant Breast and Bowel Project Breast Cancer Prevention Trial Investigators. *J Natl Cancer Inst* 93: 16–21

46 Robertson JF (2004) Selective oestrogen receptor modulators/new antiestrogens: a clinical perspective. *Cancer Treat Rev* 30: 695–706

47 Stenbygaard LE, Herrstedt J, Thomsen JF, Svendsen KR, Engelholm SA, Dombernowsky P (1993) Toremifene and tamoxifen in advanced breast cancer – a double-blind cross-over trial. *Breast Cancer Res Treat* 25: 57–63

48 Pyrhonen S, Ellmen J, Vuorinen J, Gersanovich M, Tominaga T, Kaufmann M, Hayes DF (1999) Meta-analysis of trials comparing toremifene with tamoxifen and factors predicting outcome of antiestrogen therapy in postmenopausal women with breast cancer. *Breast Cancer Res Treat* 56: 133–143

49 Holli K, Valavaara R, Blanco G, Kataja V, Hietanen P, Flander M, Pukkala E, Joensuu H (2000) Safety and efficacy results of a randomized trial comparing adjuvant toremifene and tamoxifen in postmenopausal patients with node-positive breast cancer. Finnish Breast Cancer Group. *J Clin Oncol* 18: 3487–3494

50 Pagani O, Gelber S, Price K, Zahrieh D, Gelber R, Simoncini E, Castiglione-Gertsch M, Coates AS, Goldhirsch A; International Breast Cancer Study Group (2004) Toremifene and tamoxifen are equally effective for early-stage breast cancer: first results of International Breast Cancer Study Group Trials 12-93 and 14-93. *Ann Oncol* 15: 1749–1759

51 Kim SY, Suzuki N, Laxmi YR, Shibutani S (2004) Genotoxic mechanism of tamoxifen in developing endometrial cancer. *Drug Metab Rev* 36: 199–218

52 Buzdar AU, Marcus C, Holmes F (1988) Phase II evaluation of LY156758 in metastatic breast cancer. *Oncology* 45: 344–345

53 Cummings SR, Eckert S, Krueger KA, Grady D, Powles TJ, Cauley JA, Norton L, Nickelsen T, Bjarnason NH, Morrow M et al. (1999) The effect of raloxifene on risk of breast cancer in postmenopausal women: results from the MORE randomized trial. Multiple Outcomes of Raloxifene Evaluation. *JAMA* 281: 2189–2197

54 Vogel VG, Costantino JP, Wickerham WM, Cronin N, Wolmark N (2002) The study of tamoxifen and raloxifene: preliminary enrollment data from a randomized breast cancer risk reduction trial. *Clin Breast Cancer* 3: 153–162

55 Sporn M (2004) Arzoxifene: a promising new selective estrogen receptor modulator for clinical chemoprevention of breast cancer. *Clin Cancer Res* 10: 5313–5315

56 Early Breast Cancer Trialists' Collaborative Group (1996) Ovarian ablation in early breast cancer: overview of the randomized trials. *Lancet* 348: 1189–1196

57 Davidson NE (2001) Ovarian ablation as adjuvant therapy for breast cancer. *J Natl Cancer Inst Mon* 30: 67–71

58 Leung SF, Tsao SY, Teo PM, Choi PH, Shiu WC (1991) Ovarian ablation failures by radiation: a comparison of two dose schedules. *Br J Radiol* 64: 537–538

59 Coy DH, Schally AV (1978) Gonadotropin releasing hormone analogues. *Ann Clin Res* 10: 139–144

60 Millar RP, Lu Z, Pawson AJ, Flanagan CA, Morgan K, Maudsley SR (2004) Gonadotropin-releasing Hormone Receptors. *Endocrine Rev* 25: 235–275

61 Kiesel RA, Rody A, Greb RR, Szilagy A (2002) Clinical use of GnRH analogues. *Clin Endocrinol* 56: 677–687

62 Emons G, Grundker C, Gunthert AR, Westphalen S, Kavanagh J, Verschraegen C (2003) GnRH antagonists in the treatment of gynecological and breast cancers. *Endocr Relat Cancer* 10: 291–299

63 Prowell TM, Davidson NE (2004) What is the role of ovarian ablation in the management of primary and metastatic breast cancer. *The Oncologist* 9: 507–517

64 Klijn JG, Blamey RW, Boccardo F, Tominaga T, Duchateau L, Sylvester R (2001) Combined tamoxifen and luteinizing hormone-releasing hormone (LHRH) agonist *versus* LHRH agonist alone in premenopausal advanced breast cancer: a meta-analysis of four randomized trials. *J Clin Oncol* 19: 343–353

65 Early Breast Cancer Trialists' Collaborative Group (1998) Polychemotherapy for early breast cancer: an overview of the randomized trials. *Lancet* 352: 930–942

66 Aebi S, Gelber S, Castiglione-Gertsch M, Gelber RD, Collins J, Thurlimann B, Rudenstam CM, Lindtner D, Crivellari D, Cortes-Funes H et al. (2000) Is chemotherapy alone adequate for young women with estrogen-receptor positive breast cancer? *Lancet* 355: 1869–1874

67 Dellapasqua S, Colleoni M, Gelber RD, Goldhirsch A (2005) Adjuvant endocrine therapy for premenopausal women with early breast cancer. *J Clin Oncol* 23: 1736–1750

68 Kaufmann M, Jonat W, Blamey R, Cuzick J, Namer M, Fogelman I, deHaes JC, Schumacher M, Sauerbrei W (2003) Survival analyses from the ZEBRA study. Goserelin (Zoladex) *versus* CMF in premenopausal women with node-positive breast cancer. *Eur J Cancer* 39: 1711–1717

69 Castiglione-Gertsch M, O'Neill A, Price KN, Goldhirsch A, Coates AS, Colleoni M, Nasi ML, Bonetti M, Gelber RD (2003) Adjuvant chemotherapy followed by goserelin *versus* either modality alone for premenopausal lymph node-negative breast cancer: a randomized trial. *J Natl Cancer Inst* 95: 1833–1846

70 Schmid P, Untch M, Wallwiener D, Kosse V, Bondar G, Vassiliev L, Tarutinov V, Kienle E, Luftner D, Possinger; K TABLE-study (Takeda Adjuvant Breast cancer study with Leuprorelin Acetate) (2002) Cyclophosphamide, methotrexate and fluorouracil (CMF) *versus* hormonal ablation with leuprorelin acetate as adjuvant treatment of node-positive, premenopausal breast cancer patients: preliminary results of the TABLE-study (Takeda Adjuvant Breast cancer study with Leuprorelin Acetate). *Anticancer Res* 22: 2325–2332

71 Jakesz R, Hausmaninger H, Kubista E, Gnant M, Menzel C, Bauernhofer T, Seifert M, Haider K, Mlineritsch B, Steindorfer P (2002) Randomized adjuvant trial of tamoxifen and goserelin *versus* cyclophosphamide, methotrexate and fluorouacil: evidence for the superiority of treatment with endocrine blockade in premenopausal patients with hormone-responsive breast cancer – Austrian Breast and Colorectal Cancer Study Group Trial 5. *J Clin Oncol* 20: 4621–4627

72 Boccardo F, Rubagotti A, Amoroso D, Mesiti M, Romeo D, Sismondi P, Giai M, Genta F, Pacini P, Distante V et al. (2000) Cyclophosphamide, methotrexate, and fluorouracil *versus* tamoxifen plus ovarian suppression as adjuvant treatment of estrogen receptor-positive pre-/perimenopausal breast cancer patients: results of the Italian Breast Cancer Adjuvant Study Group 02 randomized trial. *J Clin Oncol* 18: 2718–2727

73 Rutqvist L (1999) Zoladex and tamoxifen as adjuvant therapy in premenopausal breast cancer: a randomized trial by the Cancer Research Campaign (CRC), Breast Cancer Trials Group, the Stockholm Breast Cancer Study Group, the South-East Sweden Breast Cancer Group and the Gruppo Interdisciplinare Valutazione Interventi in Oncologia (GIVIO). *Proc Am Soc Clin Oncol* 18: 67

74 Roche H, Mihura J, de Lafontan B, Reme-Saumon M, Martel P, Dubois J, Naja A (1996) Castration and tamoxifen *versus* chemotherapy (FAC) for premenopausal, node and receptor positive breast cancer patients: a randomized trial with a 7 yrs median follow-up. *Proc Am Soc Clin Oncol* 15: 117

75 Roche H, Kerbrat P, Bonneterre J, Fargeot P, Fumoleau P, Monnier A, Chapelle-Marcillac I, Bardonnet M (2000) Complete hormonal blockade *versus* chemotherapy in premenopausal early stage breast cancer patoents with positive hormone-receptor (HR+) and 1–3 node-positive (N+) tumor, results of the FASG 06 trial. *Proc Am Soc Clin Oncol* 19: 72a

76 International Breast Cancer Study Group (2001) Randomized controlled trial of ovarian function suppression plus tamoxifen *versus* the same endocrine therapy plus chemotherapy: is chemotherapy necessary for premenopausal women with node-positive, endocrine responsive breast cancer? *The Breast* 10 (S3): 130–138

77 Davidson NE, O'Neill AM, Vukov AM, Osborne CK, Martino S, White DR, Abeloff M (2005) Chemoendocrine therapy for premenopausal women with axillary lymph node-positive, steroid hormone receptor-positive breast cancer: results from INT 0101 (E5188). *J Clin Oncol* 23: 5973–5982

78 Szymczak J, Milewicz A, Thijssen HA, Blankenstein MA, Daroszewsli J (1998) Concentration of

sex steroids in adipose tissue after menopause. *Steroids* 63: 319–321

79 Miller WR, Mullen P, Sourdaine P, Watson C, Dixon JM, Telford J (1997) Regulation of aromatase activity within the breast. *J Steroid Biochem Mol Biol* 61: 193–202

80 Lipton A, Santner SJ, Santen RJ, Harvey HA, Feil PD, White-Hershey D, Bartholomew MJ, Antle CE (1987) Aromatase activity in primary and metastatic human breast cancer. *Cancer* 59: 779–782

81 Smith IE, Dowsett M (2003) Aromatase inhibitors in breast cancer. *N Engl J Med* 348: 2431–2442

82 Campos SM (2004) Aromatase inhibitors for breast cancer in postmenopausal women. *The Oncologist* 9: 126–136

83 Sainsbury R (2004) Aromatase inhibition in the treatment of advanced breast cancer: is there a relationship between potency and clinical efficacy? *B J Cancer* 90: 1733–1739

84 Lonning PE, Pfister C, Martoni A, Zamagni C (2003) Pharmacokinetics of third-generation aromatase inhibitors. *Sem Oncol* 30 (S14): 23–32

85 Geisler J, King N, Dowsett M, Ottestad L, Lundgren S, Walton P, Kormeset PO, Lonning PE (1996) Influence of anastrozole (Arimidex), a selective, non-steroidal aromatase inhibitor, on *in vivo* aromatisation and plasma oestrogen levels in postmenopausal women with breast cancer. *Br J Cancer* 74: 1286–1291

86 Johannessen DC, Engan T, Di Salle E, Zurlo MG, Paolini J, Ornati G, Piscitelli G, Kvinnsland S, Lonning P (1997) Endocrine and clinical effects of exemestane (PNU 155971), a novel steroidal aromatase inhibitor, in postmenopausal breast cancer patients: a phase I study. *Clin Cancer Res* 3: 1101–1108

87 Geisler J, Haynes B, Anker G, Dowsett M, Lonning P (2002) Influence of letrozole and anastrozole on total body aromatization and plasma estrogen levels in postmenopausal breast cancer patients evaluated in a randomized, cross-over study. *J Clin Oncol* 20: 751–757

88 Dowsett M, Jones A, Johnston SR, Jacobs S, Trunet P, Smith IE (1995) *In vivo* measurements of aromatase inhibition by letrozole (CGS 20267) in postmenopausal patients with breast cancer. *Clin Cancer Res* 1: 1511–1515

89 Bertelli GF (2005) Sequencing of aromatase inhibitors. *Br J Cancer* 93 S1: S6–9

90 Mitwally MF, Casper RF (2002) Aromatase inhibition for ovarian stimulation: future avenues for infertility management. *Curr Opin Obstet Gynecol* 14: 255–263

91 Buzdar AU, Jonat W, Howell A, Jones SE, Blomqvist CP, Vogel CL, Eiermann W, Wolter JM, Steinberg M, Webster A et al. (1998) Anastrazole *versus* megtrol acetate in the treatment of post-menopausal women with advanced breast carcinoma: results of a survival update based on a survival analysis of data from two mature phase III trials. Arimidex Study Group. *Cancer* 83: 1142–1152

92 Kauffman M, Bajetta E, Dirix LY, Fein LE, Jones SE, Zilembo N, Dugardyn JL, Nasurdi C, Mennel RG, Cervek J et al. (2000) Exemestane is superior to megestrole acetate after tamoxifen failure in postmenopausal women with advanced breast cancer: results of a phase III randomized double-blind trial. The Exemestane Study Group. *J Clin Oncol* 18: 1399–1411

93 Mouridsen H, Gersanovich M, Sun Y, Perez-Carrion R, Boni C, Monnier A, Appelfstaedt J, Smith R, Sleeboom HP, Janicke F et al. (2003) Phase III study of letrozole *versus* tamoxifen as first-line therapy of advanced breast cancer in postmenopausal women: analysis of survival and update of efficacy from the International Letrozole Breast Cancer Group. *J Clin Oncol* 21: 2101–2109

94 Bonneterre J, Thurlimann B, Robertson JF, Krzakowski M, Mauriac L, Koralewski P, Vergote I, Webster A, Steinberg M, von Euler M (2000) Anastrozole *versus* tamoxifen as first-line therapy for advanced breast cancer in 668 postmenopausal women: results of the Tamoxifen Arimidex Randomized Group Efficacy and Tolerability Study. *J Clin Oncol* 18: 3748–3757

95 Paridaens R, Therasse P, Dirix L, Beex L, Piccart M, Cameron D, Cufer T, Roozendaal K, Nooij M, Mattiacci MR (2004) First line hormonal treatment for metastatic breast cancer with exemestane or tamoxifen in postmenopausal patients – a randomized phase III trial of the EORTC Breast Group. *Proc Am Soc Clin Oncol* 23:6

96 Howell A, Cuzick J, Baum M, Dowsett M, Forbes JF, Hoctin-Boes G, Houghton J, Locker GY, Tobias JS for the ATAC Trialists' Group (2005) Results of the ATAC (Arimidex, Tamoxifen, Alone or in Combination) trial after completion of 5 years' adjuvant treatment for breast cancer. *Lancet* 365: 60–62

97 Dowsett M, Cuzick J, Wale C, Howell A, Houghton J, Baum M (2005) Retrospective analysis of time to recurrence in the ATAC trial according to hormone receptor status: an hypothesis-generating study. *J Clin Oncol* 23: 7512–7517

98 The Breast International Group (BIG) 1-98 Collaborative Group (2005) A comparison of letrozole and tamoxifen in postmenopausal women with early breast cancer. *N Engl J Med* 353: 2747–2757

99 Coombes RC, Hall E, Gibson LJ, Paridaens R, Jassem J, Delozier T, Jones S, Alvarez I, Bertelli G, Ortmann O et al. on behalf of the Intergroup Exenestane Study (2004) Exemestane improves disease-free survival in postmenopausal patients with early breast cancer after two to three years of tamoxifen: a double blind randomized trial. *N Engl J Med* 350: 1081–1092

100 Coombes RC, Hall E, Snowdon CF, Bliss JM (2004) Intergroup Exemestane Study: a randomized trial in postmenopausal patients with early breast cancer who remain disease-free after two to three years of tamoxifen-updated survival analysis. *Breast Cancer Res Treat* 88: S7

101 Jakesz R, Jonat W, Gnant M, Mittlboeck M, Greil R, Tausch C, Hilfrich J, Kzasny, Menzel C, Samonigg H et al. on behalf of the ABCSG and the GABG (2005) Switching of postmenopausal women with endocrine responsive early breast to anastrozole after 2 years' adjuvant tamoxifen: combined results of ABSCG trial 8 and ARNO 95 trial. *Lancet* 366: 455–462

102 Boccardo F, Rubagotti A, Puntoni M, Guglielmini P, Amoroso D, Fini A, Paladini G, Mesiti M, Romeo D, Rinaldini M et al. (2005) Switching to anastrozole *versus* continued tamoxifen treatment of early breast cancer: preliminary results of the Italian Tamoxifen Anastrozole Trial. *J Clin Oncol* 23: 5138–5147

103 Goss PE, Ingle JN, Martino S, Robert NJ, Muss HB, Piccart MJ, Castiglione M, Dongsheng T, Shepeherd LE, Pritchard KI et al. (2003) A randomized trial of letrozole in postmenopausal women after five years of tamoxifen therapy for early-stage breast cancer. *N Engl J Med* 349: 1792–1802

104 Goss PE, Ingle JN, Martino S, Robert N, Muss HJ, Piccart MJ, Castiglione M, Dongsheng T, Shepherd LE, Pritchard KI et al. (2005) Randomized trial of letrozole following tamoxifen as extended adjuvant therapy in receptor-positive breast cancer: updated findings from NCI CTG MA-17. *J Natl Cancer Inst* 97: 1262–1271

105 Rose C, Vtoraya O, Pluzanska A, Davidson N, Gersanovich M, Thomas R, Johnson S, Caicedo JJ, Gervasio H, Manikhas G et al. (2003) An open randomised trial of second-line endocrine therapy in advanced breast cancer: comparison of the aromatase inhibitors letrozole and anastrozole. *Eur J Cancer* 39: 2318–2327

106 Winer EP, Hudis C, Burstein HJ, Wolff AC, Pritchard KI, Ingle JN, Chlebowski T, Gelber R, Edge SB, Gralow J et al. (2005) American Society of Clinical Oncology Technology Assessment on the use of aromatase inhibitors as adjuvant therapy for postmenopausal women with hormone receptor positive breast cancer: Status Report 2004. *J Clin Oncol* 23: 619–629

107 Wakeling AE, Dukes M, Bowler J (1991) A potent specific pure antiestrogen with clinical potential. *Cancer Res 51:* 3867–3873

108 Nicholson RI, Gee JM, Manning DL, Wakeling AE, Montano MM, Katzenellenbogen BS (1995) Responses to pure antiestrogens (ICI 164384, ICI 182780) in estrogen-sensitive and-resistant experimental and clinical breast cancer. *Ann NY Acad Sci* 761: 148–163

109 Robertson JFR, Harrison M (2004) Fulvestrant: pharmacokinetics and pharmacology. *B J Cancer* 90 (S1): S7–S10

110 Hu XF, Veroni M, De Luise M, Wakeling A, Sutherland R, Watts CK, Zalcberg JR (1993) Circumvention of tamoxifen-resistance by the pure antiestrogen ICI 182780. *Int J Cancer* 55: 873–876

111 Robertson JF, Nicholson RI, Bundred NJ, Anderson E, Rayter Z, Dowsett M, Fox JN, Gee JM, Webster A, Wakeling AE et al. (2001) Comparison of the short-term biological effects of 7alpha-[9-(4,4,5,5,5-pentafluoropentylsulfinyl)-nonyl]estra-1,3,5, (10)-triene-3,17beta-diol (Faslodex) *versus* tamoxifen in postmenopausal women with primary breast cancer. *Cancer Res* 61: 6739–6746

112 Howell A, Abram P (2005) Clinical development of fulvestrant (Faslodex). *Cancer Treat Rev* 31: S3–S9

113 Robertson JF, Osborne CK, Howell A, Jones SE, Mauriac L, Ellis M, Kleeberg UR, Come SE, Vergote I, Gertler S et al. (2003) Fulvestrant *versus* anastrozole for the treatment of advanced breast carcinoma in postmenopausal women: a prospective combined analysis of two multicenter trials. *Cancer* 98: 229–238

114 Howell A, Pippen J, Elledge R, Mauriac L, Vergote I, Jones SE, Come SE, Osborne CK, Robertson JF (2005) Fulvestrant *versus* anastrozole for the treatment of advanced breast carcinoma: a prospectively planned combined survival analysis of two multicenter trials. *Cancer* 104: 236–239

115 Howell A, Robertson JF, Abram P, Lichinitser MR, Elledge R, Bajetta E, Watanabe T, Morris C, Webster A, Dimery I et al. (2004) Comparison of fulvestrant *versus* tamoxifen for the treatment of advanced breast cancer in postmenopausal women previously untreated with endocrine thera- py: a multinational, double-blind, randomized trial. *J Clin Oncol* 22: 1605–1613

116 Sørlie T, Perou CM, Tibshirani R, Aas T, Geisler S, Johnsen H, Hastie T, Eisen MB, Rijn M, Jeffrey SS et al. (2001) Gene expression patterns of breast carcinomas distinguish tumor sub- classes with clinical implications *Proc Natl Acad Sci* 98: 10869–10874

117 Paik S, Shak S, Tang G, Kim C, Baker J, Cronin M, Baehner FL, Walker MG, Watson D, Park T et al. (2004) A multigene assay to predict recurrence of tamoxifen-treated, node-negative breast cancer. *N Engl J Med* 351: 2817–2826

118 Paik S, Shak S, Tang G, Kim C, Joo H, Baker J, Cronin M, Watson D, Bryant J, Costantino J et al.; NSABP, Pittsburgh, PA; Genomic Health Inc, Redwood City, CA (2004) Expression of the 21 genes in the Recurrence Score assay and prediction of clinical benefit from tamoxifen in NSABP study B-14 and chemotherapy in NSABP study B-20. *Breast Cancer Res Treat* 88: 24

119 Goldhirsch A, Wood WC, Gelber RD, Coates AS, Thurlimann B, Senn HJ (2003) Meeting high- lights: updated international expert consensus on the primary therapy of early breast cancer. *J Clin Oncol* 21: 3357–3365

120 Goldhirsch A, Glick JH, Gelber RD, Coates AS, Thurlimann B, Senn HJ (2005) Meeting high- lights: international expert consensus on the primary therapy of early breast cancer 2005. *Ann Oncol* 16: 1569–1583

Index

The MDT-Series
Milestones in Drug Therapy

The discovery of drugs is still an unpredictable process. Breakthroughs are often the result of a combination of factors, including serendipidity, rational strategies and a few individuals with novel ideas. *Milestones in Drug Therapy* highlights new therapeutic developments that have provided significant steps forward in the fight against disease. Each book deals with an individual drug or drug class that has altered the approach to therapy. Emphasis is placed on the scientific background to the discoveries and the development of the therapy, with an overview of the current state of knowledge provided by experts in the field, revealing also the personal stories behind these milestone developments. The series is aimed at a broad readership, covering biotechnology, biochemistry, pharmacology and clinical therapy.

Forthcoming titles:

Tamoxifen and Beyond, V.C. Jordan (Author), 2006
Pharmacotherapy of Obesity, J.P.H. Wilding (Editor), 2006

Published volumes:

TNF-alpha Inhibitors, J.M. Weinberg, R. Buchholz (Editor), 2006
Aromatase Inhibitors, B.J.A. Furr (Editor), 2006
Cannabinoids as Therapeutics, R. Mechoulam (Editor), 2005
St. John's Wort and its Active Principles in Anxiety and Depression, W.E. Müller (Editor), 2005
Drugs for Relapse Prevention of Alcoholism, R. Spanagel, K. Mann (Editors), 2005
COX-2 Inhibitors, M. Pairet, J. Van Ryn (Editors), 2004
Calcium Channel Blockers, T. Godfraind (Author), 2004
Sildenafil, U. Dunzendorfer (Editor), 2004
Hepatitis Prevention and Treatment, J. Colacino, B.A. Heinz (Editors), 2004
Combination Therapy of AIDS, E. De Clercq, A.M. Vandamme (Editors), 2004
Cognitive Enhancing Drugs, J. Buccafusco (Editor), 2004
Fluoroquinolone Antibiotics, A.R. Ronald, D. Low (Editors), 2003
Erythropoietins and Erythropoiesis, G. Molineux, M. Foote, S. Elliott (Editors), 2003
Macrolide Antibiotics, W. Schönfeld, H. Kirst (Editors), 2002
HMG CoA Reduktase Inhibitors, G. Schmitz, M. Torzewski (Editors), 2002
Antidepressants, B.E. Leonard (Editor), 2001
Recombinant Protein Drugs, P. Buckel (Editor), 2001

Glucocorticoids, N. Goulding, R.J. Flower (Editors), 2001
Modern Immunosuppressives, H.-J. Schuurman (Editor), 2001
ACE Inhibitors, P. D'Orleans-Juste, G. Plante (Editors), 2001
Atypical Antipsychotics, A.R. Cools, B.A. Ellenbroek (Editors), 2000
Methotrexate, B.N. Cronstein, J.R. Bertino (Editors), 2000
Anxiolytics, M. Briley, D. Nutt (Editors), 2000
Proton Pump Inhibitors, L. Olbe (Editor), 1999
Valproate, W. Löscher (Editor), 1999

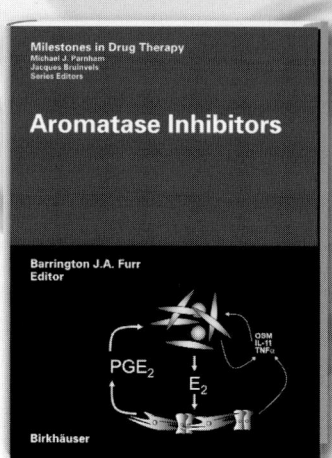